# Mountain Environments

We work with leading authors to develop the
strongest educational materials in geography,
bringing cutting-edge thinking and best learning
practice to a global market.

Under a range of well-known imprints, including
Prentice Hall, we craft high quality
print and electronic publications which help
readers to understand and apply their content,
whether studying or at work.

To find out more about the complete range of our
publishing please visit us on the World Wide Web at:
www.pearsoneduc.com

# Mountain Environments

## Romola Parish
Department of Geography, University of St Andrews

*An imprint of* **Pearson Education**

Harlow, England · London · New York · Reading, Massachusetts · San Francisco · Toronto · Don Mills, Ontario · Sydney
Tokyo · Singapore · Hong Kong · Seoul · Taipei · Cape Town · Madrid · Mexico City · Amsterdam · Munich · Paris · Milan

**Pearson Education Limited**
Edinburgh Gate
Harlow
Essex CM20 2JE
England
and Associated Companies throughout the world

*Visit us on the World Wide Web at:*
www.pearsoneduc.com

First published 2002
Second impression 2002

ISBN 0 582 41911 5

**British Library Cataloguing-in-Publication Data**
A catalogue record for this book is available from the British Library

**Library of Congress Cataloging-in-Publication Data**
Parish, Romola.
  Mountain environments / Romola Parish.
    p. cm. — (Ecogeography series)
  Includes bibliographical references (p. ).
  ISBN 0–582–41911–5 (pbk.)
  1. Human ecology.  2. Mountains.  3. Mountain people.  I. Title.  II. Series.
  GF57 .P37  2002
  910′.9143—dc21                                    2001032160

Typeset in 11/12pt Adobe Garamond by 35
Transferred to digital print on demand, 2007
Printed and bound in Great Britain by
CPI Antony Rowe, Chippenham and Eastbourne

Logging, pulping and manufacturing processes are
expected to conform to the environmental regulations
of the country of origin.

# Contents

# List of plates

# List of figures

# List of tables

# Series preface

Ecogeography marries ecology with geography. It places ecosystems in a spatial context. Its practitioners – ecogeographers – investigate the distribution and structure of ecosystems at various scales, with particular attention given to the spatial and dynamical relations amongst biotic and abiotic components of ecosystems. They usually focus on the 'landscape scale', defined as areas from tens of square metres to about 10 000 square kilometres, or the size of Cheshire, England. However, ecogeographical studies cover all spatial scales, from a few square metres to the entire globe. Regional-scale investigations view the 'big picture', probing the distribution, structure and dynamics of ecosystems in Cheshire-sized regions to continents and the entire planetary surface. The aim of the Ecogeography Series is to provide discursive accounts of the ecogeography of the world's major ecological regions – tropical, temperate, desert, Mediterranean, polar and subpolar, mountain, riverine, lacustrine, subterranean, urban, coastal and marine. Zonal climates circumscribe some of these 'regions', but the rest are distinctive in other ways.

Each book in the series examines the nature of an ecological 'region'. Individual authors are free to bring their own slant to content, their own predilections to emphasis, but they all prosecute the ecogeographical theme. In doing so, they cover such topics as climate, topography, soils, plants and animals, communities, ecosystems, land use and environmental problems. Note that the human response to the 'regions' is an important element of the books. The Ecogeography Series is about people and environment, and is not a set of regional physical geographies.

Books in the Ecogeography Series at once provide good depth and breadth of coverage. Breadth comes from an outline of where the ecological 'regions' are found and what they are like. Depth comes from selected examples and case studies, and from carefully chosen suggestions for further reading. The books are suitable for geographical and environmental courses worldwide, and fill a gap in the market – there is, for the most part, a lack of undergraduate books with a regional focus on the people and environment theme. They are designed for students' needs and pockets, providing second-year and third-year undergraduates with weighty but wide-ranging volumes on specific ecological 'regions' at an affordable price.

Richard Huggett
June 2000

# Preface

The year 2002 has been designated 'International Year of the Mountains' by the United Nations in an effort to raise awareness of the richness and diversity of the mountains of the world, to celebrate their uniqueness and to bring to light the true nature of some of the current issues surrounding their management and development. One might ask, "Why mountains?" This book is a contribution to the answer to such a question.

Mountains are distinctive, upstanding features of the land surface, occurring on every continent. They are defined primarily on the basis of altitude, which influences all aspects of environment and livelihood. Few countries are without mountains of some sort, and a large proportion of the world's population is dependent directly or indirectly on mountains for all or part of their livelihoods. Major rivers originate in mountains, and they often have higher precipitation than adjacent areas, making them critical sources of water. They are extremely rich in biodiversity, partly as a result of the diversity of environmental conditions on mountains. This rich resource requires conservation through sustainable use for the benefit of local, national and global users. Mountains are sensitive indicators of climatic change. Zones of vegetation and land uses have marched up and down mountainsides in response to changing climatic conditions in the past. Now in this era of concern over climatic change and environmental response, biodiversity, agricultural activity, tourism and physical stability are all threatened in different ways and to uncertain degrees.

Mountains are home to substantial populations who have adapted continually through time to changing resource availability, population pressures and internal and external social, economic and political change. The resilience of many of these populations, and their profound wisdom and adaptability complement their cultural diversity and is itself a rich resource requiring acknowledgement and protection. For those who visit mountains – and for many who do not – they hold a fascination and awe, a sense of danger, isolation, hazard and extreme climate. The cultures of the indigenous populations hold as much appeal as the natural beauty and ruggedness of the landscape. The role of mountains in the world's tourist industry is growing, and needs to be integrated with other economic activities and development and conservation needs.

Mountains and their populations are both vulnerable to change and yet have a degree of resistance to it. They are faced with ever greater pressures and stresses – population growth; integration into the global capitalist economy;

climatic change and the effects of national and international development initiatives. However, this increasing exposure of mountains and their communities to the wider world also reveals to external communities the complexity and diversity of the environment and its people, their resilience and flexibility and the coping strategies as well as the areas of crisis. The interaction of many different elements – history, economics, anthropology, geomorphology, geology and climate – is revealed as intricate, complex and dynamic. It is not individual forms and processes of the environment and institutions, but the combination of these processes which makes mountains distinctive.

This book cannot tackle all aspects of mountains. It does give a quite comprehensive overview of the nature of the physical and social environments and the way in which these elements interact in the context of the resources available in the mountains. Only selected areas of current concern are tackled – development, tourism and conservation. These are other issues, other mountain regions and other aspects of mountain environments and livelihoods that are not covered here: to do so would have required the compilation of several volumes. Rather, what this book does do is to give the reader a taste of the intricacies of mountains and an overview of the main issues. It does not seek to perpetuate the romantic vision of idyllic lifestyles close to nature, nor to promote a dichotomy between Western science and ignorant hill farmer, nor to prescribe the right way to manage these areas (there is no one right way) but to open the mind to alternatives, to the interweaving of many narratives – physical, social, cultural, economic, anthropological, historical, meteorological, religious, technological – into a tapestry in which all the insignificant as well as the obvious colours play a part. The case studies add flesh to these bones, and it is hoped that the reader will be encouraged to explore some aspects further, having gained an introduction to mountains in a holistic sense, and an understanding of how the aspects in which they are particularly interested relate to mountains as a whole.

Mountains mean different things to different people: a playground for the rich; a homeland and source of life; a source of trouble and strife, drugs and political resistance to governments; a source of floods and sediments to lowlanders; a source of medicine, environmental information and new crops to scientists; a source of hazardous events to seismologists and aid agencies; a source of literary, spiritual or physical inspiration to visitors. This book is a reflection of what mountains mean to me, and whilst it is impossible to convey in their entirety the significance and importance of mountains and their people, this book is a beginning, as I trust it will be for its readers.

I take this opportunity to thank the many colleagues at Sussex, St Andrews and elsewhere in the UK and overseas, with whom I have worked. Particular thanks are due to Don Funnell for his sustained support, encouragement and collaboration – I could not have asked for a better colleague. I also thank my father for his practical help, support and unfailing belief in me.

Romola Parish
August 2001

# Publisher's acknowledgements

We are grateful to the following for permission to reproduce copyright material:

Table 1 from UNEP-WCMC Mountains and forests: global statistical survey (**12 December 2000**) http://www.orghabitats/mountains/statistics/htm, reprinted by permission of UNEP-WCMC; Figure 1 after map from *Mountain Forum Bulletin*, Vol. 1, Issue 2, September 1998, reprinted by permission of The Mountain Institute; Figure 1.2 after Figure 5.8 from *Environmental Systems: An Introduction, Second Edition*, reprinted with kind permission of Kluwer Academic Publishers (White, I.D., Mottershead, D.N., Harrison, S.J. 1992); Figure 1.8 after Figure 8.4 from *Environmental Hazards: Assessing Risk and Disaster*, published by Routledge reprinted by permission of Taylor and Francis Ltd (Smith, K. 1992); Table 3 from High mountains as human habitat from *Human Impact on Mountain Environments* edited by N.R.J. Allan, G.W. Knapp and C. Stadel, Rowman and Littlefield Publishers, Inc. (Grötzbach, E. 1988); Figure 3.7 after Figure from The Pleistocene change of vegetation and climate in Tropical South America from *Journal of Biogeography*, 1, Blackwell Science Ltd (van der Hammen, T. 1974); Figure 7.1 after Figure 4 from *Conservation Farming on Steep Lands*, reprinted with permission from the Soil and Water Conservation Society (Moldenhauer, W.C. and Hudson, N.W. eds 1988); Figure 7.3 after Figure 1 from Oukaimedene, Morocco: A high mountain *agdal* from *Proceedings of a Conference on Common Property Resource Management*, published by National Academy Press reprinted by permission of National Academy of Sciences (Gilles, J.L., Hammoudi, A. and Mahdi, M. 1986); Figure 9.1 after Figure 2.4 from Die Organisation von Raum und Zeit – Bevölkerungswachstum, Ressourcemanagement und angsßte Landnutzung im Bagrot/Karakoram from *Petermanns Geographische Mitteilungen* 139(2). Klett-Perthes, Justus Perthes Verlag (Ehlers, E. 1995); Figure 9.2 after Figure 8.1 from The importance, status and structure of *Almwirtschaft* in the Alps from *Human Impact on Mountain Environments* edited by N.R.J. Allan, G.W. Knapp, and C. Stadel, Rowman and Littlefield Publishers, Inc. (Penz, H. 1988); Figure 9.3 after Figure on p. 99 from Human adaptation to environment in Zangskar from *Himalayan Buddhist Village: Environment, Resources, Society and Religious Life in Zangskar, Ladakh*, published by the University of Bristol Press reprinted by permission of H. Osmaston (Osmaston, H., Frazer, J. and Crook, S. 1994); Table 10.3 from *Investing in Mountains: Innovative Mechanisms and*

*Promising Examples for Financing Conservation and Sustainable Development*, Synthesis of a Mountain Forum Electronic Conference in Support of the Mountain Agenda, Mountain Forum, The Mountain Institute and the Food and Agriculture Organization of the United States, reprinted by permission of The Mountain Institute (Preston, L. ed. 1997); Table 11.2 from http:// www.unesco.org/whc/heritage.html, UNESCO World Heritage Centre, reproduced by permission of UNESCO; Table 11.3 from *Parks on the Borderline: Experience in Transfrontier Conservation*, IUCN – The World Conservation Union (Thorsell, J.W. ed. 1990); Table 11.4 from *Mountain People, Forests, and Trees: Strategies for Balancing Local Management and Outside Interests*, Synthesis of an Electronic Conference, April 12–May 14, 1999, Mountain Forum and The Mountain Institute, reprinted by permission of The Mountain Institute (Butt, N. and Price, M.F. eds 2000); Table 12.1 from *Community-Based Mountain Tourism: Practices for Linking Conservation with Enterprise*, Synthesis of an Electronic Conference, April 13–May 18, 1998, Mountain Forum and The Mountain Institute, reprinted by permission of The Mountain Institute (Godde, P. ed. 1999); Figure 12.3 after Figure 2 from Highways to the sky: The impact of tourism on South Asian mountain culture from *Tourism Recreation Research*, Vol. 13(1), reproduced by permission of Professor Tej Vir Singh, Editor-in-Chief of Tourism Recreation Research.

Figure 1, Table 10.3, Table 11.4 and Table 12.1: Promising examples of initiatives that support sustainable investments, tourism, and forestry were provided by the Mountain Forum network through reports published by The Mountain Institute. The Mountain Forum is a global network of people and organizations concerned with mountain communities, environments, conservation, and sustainable development. The Mountain Institute is a global field-based organization that works with local partners to advance mountain cultures and preserve mountain environments.

While every effort has been made to trace the owners of copyright material, in a few cases this has proved impossible and we take this opportunity to offer our apologies to any copyright holders whose rights we have unwittingly infringed.

# Introduction

## Why mountains?

Mountains are essentially identified by altitude; they are upstanding elements of the landscape, imbued with their own mythology and embedded in our psyche as places of both danger and refuge. The common perception of inhospitable terrain and climatic extremes, together with cultural designation as sanctuaries of strange and fearsome deities and rare creatures, is countered by a perception of mountains' purity – unspoiled and untainted by the modern world, they become places to be sought out for rest and recreation. Neither is totally true. Mountain people are considered variously as backward, irresponsible peasants; the quaint, photographic capital of tourism; or innovative, resilient entrepreneurs. Mountains are landscapes of contrast, both sought after and avoided.

Mountains and hills over 1000 m in height cover some 27 per cent of the world's surface (Mountain Agenda, 1997: Figure 1, Table 1) and support about 10 per cent of its population directly, whilst a further 50 per cent rely on them for water (Denniston, 1995). Some 33 million people live in the Himalayas, 26 million in the Andes and 11 million in the Alps – which comprise 2 per cent of the world's total population (Bätzing et al., 1996). Very few countries have no mountains; even the UK, where the Scottish Highlands are relatively low compared with the major ranges of the world, recognizes the different character of its hill lands and peoples from its lowland environments. This difference comprises a number of 'specificities' – i.e. those characteristics which make mountains 'special' and thus worthy of specific study, although it is only recently that they have been so considered.

Mountains have a growing significance in the modern world. The Mountain Agenda (1997) identifies three areas in which they have particular importance in the twenty-first century: as water towers, as bastions of biodiversity and for recreation. There have been considerable efforts to raise awareness of mountain constraints and opportunities through the setting up of mountain-specific research foci and policy initiatives. The Mountain Agenda campaigned successfully for the incorporation of a specific chapter on mountains in Agenda 21 established at the Rio Conference in 1992; the resultant Chapter 13 outlines the specific needs and possibilities of mountain development

1

1. Hindu Kush
2. Himalayas
3. Kunlun Mountains
4. Tien Shan, Zeravshanskiy Khrebet
5. Altai Mountains
6. Qilian Shan
7. Ural Mountains
8. Pyrenees
9. Alps
10. Apennines
11. Atlas Mountains
12. Dinaric Alps
13. Carpathians
14. Balkan Mountains
15. Caucasus Mountains
16. Ethiopian Highlands
17. Mitumbar Mountains
18. Mutchinga Mountains
19. Kjollen Range
20. Brooks Range
21. Alaska Range, St Elias Mountains
22. Coast Mountains
23. Cascade Range
24. Sierra Nevada Range
25. Rocky Mountains
26. Sierra Madre Occidental
27. Vancouver Island
28. Sierra Madra del Sur
29. Appalachian Mountains
30. Ozarks
31. Andes
32. Jura Mountains
33. Guiana Highlands
34. Brazilian Highlands
35. Great Dividing Range
36. Dawna Range, Bilauktaung Range
37. Truong Son Range
38. Pegunungan Barisan
39. Pegunungan Iran
40. Central Cordillera
41. Kyushu-sanchi
42. Aravalli Range
43. Sierra de Gredos, Sierra de Guadarrama
44. Sudetes
45. Rhodope
46. Wicklow Mountains
47. Cambrian Mountains
48. Northwest Highlands
49. Grampian Mountains, Cairngorm
50. Karakoram
51. Macdonnell Ranges
52. Southern Alps
53. Baffin Island
54. Pegunungan Maoke
55. Japan Alps
56. Madagascar
57. Massif Central
58. Laurentides
59. Drakensberg
60. Sayan
61. Chang-pai Shan
62. Novaya Zemlya
63. Vatnajökull, Hofsjökull, Langjökull, Myrdals Jökull
64. San Bernandino Mountains
65. Kamchatka
66. Chersky Range, Verhoyansk Range
67. Koryak Range
68. Jabal Akhdar
69. Jamaica
70. Eastern African Highlands
71. Taurus Mountains
72. Asir
73. San Gabriel Mountains
74. Pontic Mountains
76. Adamawa Mountains
77. Sardinia
78. Western Ghats
79. Eastern Ghats
80. Wuyi Shan
81. Tasmania
82. Cordillera Central
83. Pennine Chain, Cheviot Hills
84. Khrebet Dussye Alin, Khrebet Turana
85. Yablonovyy Khrebet
86. Great Khingan Range
87. Sierra Maestra
88. Pinar del Rio
89. Arakan Yoma Range
90. Khasi Naga Hills
91. Sicily
92. Cordillera de Talmanca
93. Taiwan
94. Pamirs
95. Corsica
96. Mindanao
97. Maya Mountains
98. Jebel Marra
100. Otway Ranges
101. Flinders Ranges
102. Gory Putorana
103. Sikote-Alin Range
104. Cameron Highlands
106. Pindhos
107. Cordillera Cantabrica
108. Zagros Mountains
109. Kimberley Plateau
110. Tibesti
111. Ahaggar
112. Matmata Mountains
113. Sri Lanka
114. Min Shan, Tapa Shan
115. Tsinling Shan
116. Ningling Snah, Daxue Shan, Bayan Har Shan, Dalou Shan, Hengduan Shan, Qionglai Shan, Xiqing Shan
117. Coastal Range
118. Stirling Range
119. Sierra Madre Oriental
120. Tibetan Plateau
121. Heard Island
122. Antarctic Peninsula
123. Transantarctic Mountains
124. Queen Maud Land
125. Enderby Land
126. Greenland

**Figure 1** Mountains of the world. *Source:* After map in *Mountain Forum Bulletin*, Vol. 1, Issue 2, September 1998.

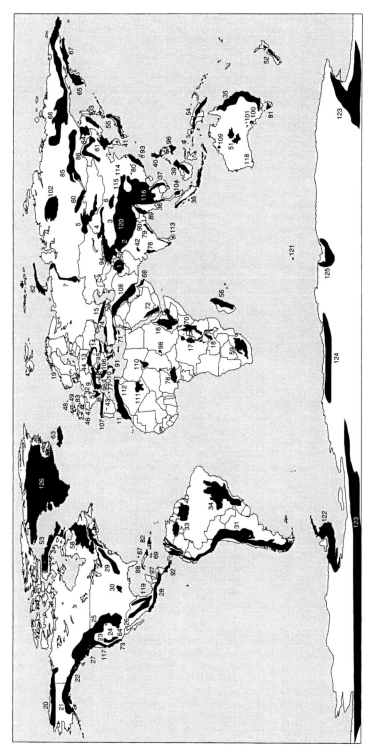

**Figure 1** Map of the world's mountains. *Source*: After map in *Mountain Forum Bulletin*, Vol. 1, Issue 2, September 1998.

Table 1 Distribution of the world's mountains (mountain area statistics in km²).

| Region | >4500 m | 3500–4499 m | 2500–3499 m | 1500–2499 m slope >2° | 1000–1499 m slope >5° or relative relief >300 m | 300–900 m + relative relief >300 m | Total |
|---|---|---|---|---|---|---|---|
| North America | 197 | 11 417 | 200 830 | 1 092 881 | 1 104 529 | 1 840 140 | 4 249 994 |
| Central America | 38 | 968 | 67 127 | 353 586 | 259 367 | 412 215 | 1 093 301 |
| Caribbean | | | 32 | 2 809 | 5 528 | 38 322 | 46 691 |
| South America | 154 542 | 583 848 | 374 380 | 454 417 | 465 061 | 970 707 | 3 002 955 |
| Europe | 73 | 225 | 497 886 | 145 838 | 345 255 | 1 222 104 | 2 211 308 |
| Africa | | 4 859 | 101 058 | 559 559 | 947 066 | 1 348 382 | 2 960 997 |
| Middle East | 40 363 | 40 363 | 128 790 | 339 954 | 906 461 | 733 836 | 2 870 539 |
| Russian Federation | 31 | 1 122 | 31 360 | 360 503 | 947 368 | 2 961 976 | 4 302 360 |
| Far East | 1 409 259 | 741 876 | 627 342 | 895 837 | 683 221 | 1 329 942 | 5 687 477 |
| Continental Southeast Asia | 170 445 | 107 974 | 97 754 | 211 425 | 330 574 | 931 217 | 1 849 389 |
| Insular Southeast Asia | 22 | 4366 | 34 376 | 120 405 | 157 970 | 599 756 | 916 895 |
| Australia | | | | 385 | 18 718 | 158 645 | 177 748 |
| Oceania | | | 41 | 7 745 | 29 842 | 118 010 | 155 638 |
| Antarctica | 17 | 1 119 112 | 4 530 978 | 165 674 | 144 524 | 327 840 | 6 288 145 |
| TOTAL | 1 774 987 | 2 704 557 | 6 877 376 | 4 600 234 | 2 808 000 | 3 135 007 | 12 604 031 |

*Source:* WCMC, 2000.

**Table 2**   UNCED 1992 Agenda 21, Priority Actions: Chapter 13, Mountain Development.

---

A. *Generating and strengthening knowledge about the ecology and sustainable development of mountain ecosystems.*

This programme area aims to survey, record, understand processes and conservation techniques and to build networks and coordinate efforts to protect ecosystems.

B. *Promoting integrated watershed development and alternative livelihood opportunities.*

This area aims to develop appropriate land-use planning, promotion of income-generating activities and development of technology and institutions to mitigate natural disasters.

---

(Table 2). The Mountain Forum, an electronic forum established in 1997, ICIMOD (The International Centre for Integrated Mountain Development, based in Kathmandu) and other bodies have sprung up to coordinate and disseminate research initiatives on mountain issues. This has culminated in the designation by the UN of 2002 as the 'International Year of Mountains', which is concerned with the implementation of Chapter 13, not only raising awareness of the problems and constraints of mountain environments, but also celebrating their diversity and rich cultural heritages. It is interesting to note that in 1997 (five years after Rio + 5) there were still calls to put the words of Chapter 13 into action, when several of the objectives in that Chapter were timetabled for completion by 2000!

Mountains have long held a fascination for many cultures, and are associated with images of power, sanctity, fear and tranquillity, with beauty and with disasters and extreme weather conditions – all of which are valid. Perceptions of mountains change with culture, gender, race, familiarity and technology, both in time and space. Ancient maps portray mountains as the homes of gods and evil powers, and as the source of abundant water, which is particularly important in those areas of the world where mountains represent oases in dry deserts (see inside back cover). These ancient maps represented mountains as barriers between lands – for example, the Atlas bordering the Mediterranean region, full of untamed savages and barbarous hordes. Certainly, for some peoples they have been areas of refuge, enabling them to retreat from invading forces and to avoid being conquered.

The initiatives stimulated by both mountain-specific fora, and by wider development and environmental management agendas in developing countries have concentrated disproportionately upon agricultural development, land degradation and the management of land, water and forests. This reflects the main concerns of these areas, where mountains are densely populated and subsistence agriculture the traditional mainstay of the economy. Agriculture is also the starting point of many policies in Europe aiming to counteract emigration and abandonment and also to encourage economic alternatives to the burgeoning tourist industry. The geographic and economic marginal status of many mountains in Europe has earned them 'Less Favoured Region' status, entitling them to subsidies and development funding. Tourism brings its own opportunities and problems, which occur in all mountainous regions to an increasing extent as the rich urban elite turn to mountains as a playground and resting place from the rigours of the commercial world. These are two extremes of a great diversity of issues surrounding mountains, but as the most prominent, and associated with critical questions of water and biodiversity, they are perhaps the best-known areas of concern, not just to the local population, but to all people in the world. It is these issues that have been behind the impetus to bring mountains increasingly into a global focus by identifying them within global policy structures such as Agenda 21.

## What are mountains?

Mountains are, at their simplest, upstanding areas of relief. This simple definition, however, requires qualification – how upstanding? Do high altitude areas of flat land count? What other characteristics denote a mountainous environment? A commonly accepted policy definition of a mountain is an area of higher *relative relief* than its surrounding lowland, i.e. by an arbitrary altitude category. If an altitude of 1000 m is accepted as this cut off point, it includes the main ranges – Rockies, Alps, Atlas, Himalayas, Andes, etc., but excludes apparently mountainous areas such as the Scottish Highlands which share many other characteristics – such as an arctic climate at the summit – with many other mountain regions. These areas are also beset with similar economic issues such as declining agriculture and tourist development to, for example, the European Alps. The concept of relative relief is perhaps more important than absolute relief, though. That is, a high, level plateau is not necessarily mountainous in the same way as a lower-altitude area with steep slopes and sharp relief common to mountain-valley landscapes, and which we more commonly associate with the term 'mountain'. However, the high plateau may share similar high-altitude climate conditions to those which prevail in more dissected landscapes.

One of the key features of mountain geography is that of *verticality* – i.e. the changes of environment with altitude; altitude is really the essence of mountains. Verticality is expressed in many different contexts – vegetation zones from foot to summit, changes in climatic parameters, zones of predominant land-use types and management strategies, and patterns of geomorphological processes. It is a concept which is revisited many times within this book. Verticality was a central part of the study of mountains by early geographers such as Troll, and later by Uhlig (1978), Brush (1976a), etc., who sought to classify mountains by their altitudinal characteristics. The most obvious was the change in environmental conditions with altitude, and thus mountains could be classified geomorphologically, climatically or ecologically by the presence of certain indicators – permanent snow, permafrost, air pressure and composition or floristic elements similar to arctic flora. Thus, the German literature differentiated between *Hochgebirge* (high mountains such as the Andes and Himalayas) and *Mittelgebirge* (middle-altitude mountains and highlands) (Grötzbach and Stadel, 1997). However, individually these classifications tended to reflect certain ranges – they were spatially explicit, despite

**Table 3** Cultural/historical classification of mountains.

---

1. Old world mountains
   - Relatively densely populated, traditional subsistence agriculture base. Intact traditional subsistence, overpopulation. Agriculture either settled (Himalaya–Karakoram–Hindu Kush) or interacts with nomads (Middle East, High Atlas).
   - Declining traditional systems with expansion into new areas, especially tourism (Alps, Pyrenees).
   - Large collective agricultural systems with recent development in some areas of new activities (Russia, China). Collective systems may be replaced with cooperatives or privatization (Tadjikistan, Kyrghizstan).
2. New world mountains
   - Areas of European colonization overseas; relatively sparsely populated, market-orientated agriculture, forestry and tourism. Some areas preserved as wilderness (North America, New Zealand).

---

*Source*: After Grötzbach, 1988.

parallels being drawn between them to establish a commonality amongst mountain characteristics. Others classified mountains culturally or demographically, recognizing the occurrence of populations and their histories (Grötzbach, 1988). One such example is given in Table 3.

Defining mountains by reference to their relative physical altitude above adjacent lowlands is also reflected in highland–lowland linkages of trade and culture, as well as environmental processes. Thus a different character could be identified in cultural, social or economic environments which could be considered specifically mountain (see Table 4). In reality, though, these factors do not suddenly change at the point of interface, but rather form a continuously changing 'catena' across this divide. Therefore, whilst it is possible to consider mountain environments independently of the adjacent landscape, their setting or context provides a wider meaning to them.

Mountain environments are both complex with regard to the interactions between different physical and human elements, and diverse, not only at a global scale, with variations from tropics to poles, but also at a local scale, with, for example, the effect of aspect on soil conditions. This applies both to the physical environment and the cultural context of mountain environments. Whilst classifications may seek to generalize within identified groups of mountain environments or cultures, this simplifies a great diversity of real conditions and contexts. Some characteristics of traditional mountain economies and cultures are ubiquitous, as identified in Table 4, but each category encompasses a great diversity of variations, and many have been substantially modified in recent decades.

There are other more general characteristics which are common to mountains. *Complexity* is manifest in the interactions between physical processes and human activities at all scales in shaping the adaptability and resilience of environment or people, and thus, like *diversity,* pervades all aspects of the study of mountains. The traditional economy is one of pluri-activity, making

**Table 4**   Ubiquitous adaptations and common characteristics.

1. Agro-economy:
   - Mixed, integrated, diverse agro-pastoral economy
   - Mainly subsistence
   - Exploitation of a range of ecological niches and zones
   - Vertical and horizontal exchange networks
   - Migration as a permanent or temporary survival strategy
   - Some form of *Alpwirtschaft* – labour organization
2. Indigenous management systems:
   - Communal management of critical or low-productivity resources
   - Individual ownership of terraced plots
   - Citizenship determining rights of access to resources
   - Tribal/village council with elected representatives
   - Terraces, manuring and soil conservation practices
   - Irrigation systems and water allocation

*Source*: After Brush, 1976b.

different uses of different altitudinal and ecological niches and combining subsistence agriculture with a variety of other activities. This complexity increases the difficulty of tackling so-called crises in mountain degradation and economy, but also gives rise to innovation and risk avoidance.

The concept of *Alpwirtschaft* emerging from early studies of livelihoods in the European Alps (Netting, 1981; Penz, 1988; Viazzo, 1989) refers to the management of labour and land resources to integrate the various activities into a harmonious whole, making the best use of all available resources. This traditional pattern of activity involves the division of labour by sex and age, in order to manage irrigation and cropping of widely dispersed plots and altitudinal zones – high pasture, forests, agricultural terraces, lowland grazing and household management. This requires considerable organization at household and community level, and the division of responsibility for different production activities. This is a particularly important characteristic of mountain economies as it is the basis of human survival and of the interaction between human and physical environments.

Whilst it is impossible to do justice to all the diversity of mountain environments, this book seeks to represent it, and as the reality of complexity is difficult to demonstrate within the scope of this book, a glimpse of the interactions between different aspects of mountains must suffice. What this book does attempt to do, however, is to represent diversity and complexity in the context of that key mountain specificity, verticality, which is the essence of what makes mountains different.

## What crises?

One of the reasons why mountains have become more visible to the wider world is the perception of imminent environmental and economic collapse due to outmigration, deforestation and poor land management, leading to degradation and collapse of societies. Another key reason is the growing importance of mountains for recreation and the conflict between this and conservation. Much more recent work has ascertained that whilst the landscape is fragile and some traditional economies stressed, many are very resilient and innovative and there is in reality no such impending disaster.

The problems and issues facing different regions of the world are different. Grötzbach's proposed classification of mountain regions distinguished between Old World densely populated mountains and New World sparsely populated ones (Table 3). The categories are associated with different aspects of conservation, sustainable agriculture, etc. arising from their different conditions. Western Europe, for example, is dominated by tourism, rural emigration, abandonment and issues of agricultural subsidies and employment, whereas the Himalayas and Andes are more concerned with managing resources such as land and water to support large and growing populations, and transitions of traditional into capitalist economies. New World sparsely populated mountains are more concerned with managing 'pristine wilderness' environments, limiting tourism and access to certain areas, managing underdevelopment and controlling mining activities. Some areas face major problems of mining and pollution (Chile) whilst others are confronted by those of political stability and ethnic marginalization (Myanmar, Thailand, the Basque region). Some are a focus for debates over global intervention in forest protection, which conflicts with national needs for lumber and local traditional forest use (Nepal, Thailand). All these issues are variations on similar themes, with different aspects predominating and so case studies are used in this book to illustrate particular perspectives where appropriate.

The penetration of capitalism into mountain regions has been one of the most rapid and prevalent agents of economic change. The effects of this are twofold: first, the economy is no longer predominantly self-sufficient and subsistence exchange economics have been replaced by those based on money. This has meant either a change in orientation of agricultural activities towards market-based production rather than subsistence, or their replacement by alternative activities, particularly tourism. In Hunza, northern Pakistan, for example, cropping has been dominated by production of seed potatoes for the Punjab, whilst flour and rice are now bought and transported up the Karakoram Highway. In many parts of the Alps, such as in Switzerland and Austria, either tourism has taken over, or specialist production (organic, *appellation controllée*) to fill specialist niches is re-established, or is maintained by EU subsidies at artificial levels. The second aspect of this economic change is the fact that, as mountain regions are more fully integrated into the wider world, they can rely on outside assistance in the event of calamity.

A second aspect is demographic change. The outmigration of young people leaves the old to manage, or the absence of males leaves women and children to continue traditional activities. The rural exodus in many areas has caused abandonment of land, as in Greece and Portugal. Where alternative economic activity is possible, this is less acute. But change in use of high pastures from grazing to skiing and construction in areas prone to floods and avalanches leads to problems of hazard management. Elsewhere again, political moves such as the permanent settlement of nomads (the Druze in Syria and Berbers in Morocco) and the enclosure of traditionally open, seasonally occupied grazing lands have caused changes in the livelihood strategies of the population, and often cultural dislocation. Intervention by development agencies has often failed, due to a lack of appropriate strategies, lack of extension work, poor communication and misunderstanding of the locals' needs. The exclusion of people from protected areas is being replaced by development where local populations are integrated into plans for conservation of mountain resources and landscapes. Externally driven changes in environmental conditions, such as climatic change and pollution, have changed the sensitive balance between survival and collapse, or even enhanced the potential productivity of mountain areas. Civil conflicts have caused degradation both directly, and through abandonment.

All these changes, however, have been occurring in some form throughout time. Climate changed substantially during the Quaternary ice ages. Invasion, subjugation and interaction between ethnic groups have resulted in opening and closing of mountain passes and routes; in the exchange of ideas and technology and periods of stagnation and retrenchment; of the alternate rise and fall of different aspects of the rural economy – pastoralism, then settled agriculture, specialist production and abandonment – cycles known in many areas, such as the Mediterranean, even since classical times. Mountain regions, therefore, are not a collection of static, traditional, introverted, encapsulated ethnic museums, but living, dynamic, responsive organisms, which have always been able to adapt and change, to take on new ideas, to survive and thrive. Even today, the theories of Himalayan degradation have been muted and the myths debunked, to reveal that environmental degradation is more a case of constant change incorporating periods of degradation and recovery; that mountain populations are not ignorant peasants but wise, innovative survivors from whom Western science and urban elites have much to learn.

In terms of policy, some countries such as Nepal, Bhutan, Chile and Switzerland are almost entirely mountainous, and therefore orientated politically and economically towards a mountain-determined (albeit topographically defined) theatre of operation. Other countries such as France, Britain, Kenya and Indonesia are only partly mountainous and thus national development, infrastructure and economic initiatives are primarily sectorally determined and may be less appropriate for mountain contexts. Such countries do not always consider their mountain regions as first and foremost mountainous, but areas of marginality, poverty and underdevelopment relative to the rest of the country. This perpetuates the conception of mountains as backward areas, which culturally is extended to their inhabitants. In such countries, it may not be appropriate to put forward specifically mountain approaches to national development in the same way as in the former group, but it may be difficult to incorporate such regions into national plans without recognizing the specific conditions and needs of such areas. This development dilemma in terms of a *politique de montagne* is a critical issue facing some European countries today. It also reflects the problems of the post-modern, post-structuralist attention to detail which, whilst privileging the local, fails to connect the locales together into some coherent whole.

## What is a 'Mountain Environment'?

Perhaps the new vision of mountain – and, indeed, any marginalized – environment and people is that in the future they may hold the advantage over lowland or urban areas. They are repositories of local wisdom regarding environmental conditions and sustainable management. Mountains may in the future possess physical conditions which are more favourable to live in, and for agricultural productivity, if the surrounding lowlands are increasingly affected by droughts under warmer climatic scenarios. We should recognize and respect this, and also the right and need for the local populations to determine their own way forward. We need to appreciate too how their environments and cultures can enrich our own.

Within this is a recognition of the dichotomy between 'us' (the scientifically informed West/North) and 'them' (the disadvantaged and backward South/East). This dichotomy lies at the heart of unsuccessful development and ideally a new vision of the future embraces and celebrates difference and different types of wisdom and knowledge, different ways of doing things, and seeks to build bridges across the divide, and to exchange ideas and technology. The complexity and diversity of the detail of mountain environments, and the existence of certain common specificities and characteristics is an excellent context in which to consider these differences, and how each has its own value. A study of mountain environments needs to incorporate both difference and commonality in many different spheres – physical, economic, cultural, etc. But it needs to be more than the traditional regional geography and be global in its scope in order to achieve this, reaching out both spatially and temporally, and into other disciplines.

The study of mountains is essentially one which integrates and assimilates a number of disciplines; Forsyth (1998), for example, explains the need to integrate approaches. Development methodologies such as integrated watershed management are based on a consideration of both human and physical perspectives. Whilst this does not devalue the importance of detailed quantitative climatic or anthropological studies, for instance, it means that in the process of integration, some of the specific detail may be lost, with a consequent risk of simplification and reduction of diversity and complexity. However, in its place is a greater understanding of the interactions of different parts of the physical and human environment, and the creation of a dynamic image of all the facets which contribute to the concept of a mountain. Thus, whilst it is important to study the cogs that make up a machine, it is the application of the cogs, and their interaction which make the machine work. Although mountains may occupy their own chapter in Agenda 21, other chapters, such as those dealing with sustainable agriculture (Chapter 14), biodiversity (Chapter 15), the role of farmers (Chapter 32), women (Chapter 24), etc. (Ives *et al.* 1997) also relate to mountain environments and peoples.

Whilst integration is accepted as a critical factor in, for example, development approaches to environmental and economic problems, the resultant loss of detail in local studies has been criticized. Mountain environments are a particular case in point here; they have substantial variations in population and environmental characteristics, not only between but also within valleys, This highlights the issue of scale, and the importance of the local, both in the understanding of physical processes, and in human behaviour. It is the local community, household or individual and often sub-valley level which is the most significant unit of operation. Thus, in understanding the diversity of mountains, it is often necessary to hold in one's mind the images of many contrasting or similar local vignettes of mountain life and environment, rather than to overgeneralize distinctive characteristics. However, there are substantial similarities between mountain regions and the adaptations of human endeavour which enable populations to be sustained on apparently diverse and scarce resource bases. Thus it is possible to describe some general aspects, such as terracing and water management, and the altitudinal influence of temperature and precipitation, but at the same time to illustrate the diversity of difference, and local complexities which make mountains both fascinating areas of study, but also almost bafflingly complicated in terms of presenting solutions to problems perceived differently by different actors concerned.

This is the approach of this book: it attempts to present a coherent overview of physical and human environments (Parts 1 and 2) and their interaction, the resources which form the basis of the constraints and opportunities offered by such environments for human use (Part 3), and an introduction to some of the critical issues of debate which focus on specifically 'mountain' problems – conservation, tourism, degradation (Part 4). Whilst these last are not confined to mountains – any more than the physical, economic and cultural processes – certain combinations and characteristics of these processes and their interaction construct inherently and identifiably 'mountain' (specificities) which justify the consideration of mountains as a unit of study.

This book seeks to explore implicitly the ideas of tradition and modernity, ignorance and wisdom, environment and culture, development and change. It seeks to provide an overview of the characteristics of mountains – what makes them unique and precious in their own light, but also how they fit into the wider world. Case studies occur throughout, and offer more detailed vignettes of these issues, adding flesh to these bones. Some areas turn up frequently (Morocco, Pakistan and Yemen, for example), being places in which I have worked, and this provides a glimpse of several facets of each of these areas.

**Part 1**

# Evolution of mountain landscapes

# Introduction: stability and instability

Mountains are well known as regions of extreme climate and hazardous conditions. Recent publicity over deforestation, degradation and the apparent imminent collapse of productive environments needs to be placed in the context of natural processes of formation and modification and the geomorphological, climatic and ecological environment. These three elements of the physical environment (Chapters 1–3) form a framework for considering resource availability and utilization (Chapters 7–9), and to some extent shape the culture and social organization of mountain communities (Chapters 4–6).

Physical processes of mountain regions are characterized by cyclicity and oscillations between stability and instability. Cycles of events may be diurnal, as in the aseasonal tropics where the greatest variation in environmental conditions occurs between day and night; seasonal variations occur in the temperate and high latitudes, whilst cycles of events such as drought or floods may occur on annual, decadal or much longer timescales of thousands of years, as for the Quaternary glacial–interglacial cycles, and on geological timescales of millions of years, as with the formation of mountains themselves. All these different cycles are superimposed and the complexity of resultant processes gives rise to diverse environments where small differences in slope angle, aspect or altitude can have great impacts on soils, vegetation, erosion and potential for land use. High-magnitude, low-frequency events such as tectonic uplift, earthquakes, major floods and glacial activity may occur at widely spaced intervals, resulting in periods of intense geomorphic activity (instability) and intervening periods of relative quiescence (stability). This creates a sort of 'punctuated dynamism' in landscape evolution, operating at a variety of timescales.

The concept of stability applies to the physical stability of landscapes, and also to their ecology (Gigon, 1983) and to the agricultural systems in operation (Winiger, 1983). The issue of stability and instability of mountain regions and their sensitivity and vulnerability to external forces of change, such as climate, has generated considerable academic interest. The landscape of mountains, with its high relative relief, large areas of bare rocky slopes, and abundant loose material means they are highly susceptible to erosion, especially under extreme climatic conditions – such as snow melt and heavy rainfall. Add to this the frequency of earthquakes and volcanic activity arising from their location in tectonically active areas, and it is easy to recognize the inherent instability of mountain environments. Whilst human activity has been blamed for enhancing natural rates of degradation, it is often the case that the management strategies of indigenous populations are able to cope admirably and effectively with the hazards of mountain living; this is explored further in Part 3.

The stability of a landscape is determined by its resistance and response to external forces of change (Allison and Thomas, 1993; Brunsden, 1993). The resistance comes from the geological character of the landscape – the rigidity, complexity and morphology of different rock types and structures, relief and elevation. The environment is able to absorb the effects of change by dissipating energy; for example, the deposition of sediment by a river as it enters a lake, or the dissipation of wave energy on a beach. In mountains, however, where steep slopes predominate, processes such as streamflow have an exceptionally high potential energy which is able to erode rapidly and to carry considerable quantities of material. Thus, mountain environments tend to be highly dynamic landscapes. The environmental history of mountains is also important – for instance, the legacy of past earth movements and the Quaternary glaciations mean that the geomorphology is always out of equilibrium, which contributes to their instability and propensity for change.

Our perception of how stable or unstable a mountain region is, is complicated by the fact that several different forms of stability exist. They describe different types of response according to the magnitude of the force of change and the response of the environment and its ability to absorb the change. Stability may be *cyclic*, such as the seasonal or diurnal temperature changes or variations in streamflow, or it may be *elastic*, reflecting a temporary change in response to a one-off event (an earthquake or a pollution event), with an eventual return to the original state (Gigon, 1983). Alternatively, the conditions may be *constant*, reflecting little change over time where the environment is able to absorb the effects of change without having to respond. All these different states and conditions may be superimposed at any one time, and the same factors may generate different responses in different areas (for example, the different responses of vegetation to global warming). The scale of observation in time or space also affects our perception of stability. What is important about mountains is that, although the actual processes operating are no different from those in any other environment, it is the influence of altitude in particular which gives them an image of enhanced activity and hazard.

Chapter 1

# The making and shaping of mountains

Mountain building commonly occurs in three stages. First, a tectonic event such as uplift or volcanic activity creates elevated relief. Second, erosion processes reduce this elevation, whilst continued uplift may both counteract and enhance this. In the third phase erosion dominates and the net result is a gradual lowering of elevations. Thus, early geomorphologists' perception of mountains was as ephemeral landforms (Powell, 1895) with a 'lifecycle'. The tectonic and erosion forces interact. Recent work (Pinter and Brandon, 1997) emphasizes the importance of erosion operating both in destructive modes by removing material, and in constructive modes by instigating uplift. Crustal adjustment to the removal of material by denudation or by deglaciation causes rebound of the crust (isostasy) and thus rejuvenation of the denudation processes. This recognition of the creative role of erosion views mountains as active, continually rejuvenated landforms. The uplift of land masses to form mountains interferes with atmospheric circulation and thus modifies climate on continental scales (Chapter 2). The interaction of tectonic, geomorphologic and climatic forces in creating and shaping mountains creates a series of positive and negative feedback effects, giving a degree of internal control. For example, the uplift of the Himalayas limits the effects of the monsoon to the southern parts of the range, whilst the rainshadow effect causes much drier conditions on the Tibetan Plateau, with the result that it is considerably less dissected than the southern Himalayan slopes.

## Formation

Mountains are formed primarily as a result of tectonic plate movements: the collision of plates causes compression and uplift of intervening sediments together with the subduction of one plate and associated volcanic activity (Figure 1.1). They may also form at divergent plate boundaries due to the upwelling of magma through the weakened crust – as in the case of the mid-Atlantic ridge which extends some 15 000 km and rises up to 4000 m above the ocean floor. Mountain building results in complex geological formations, with intense folding and faulting, intrusion of igneous material, metamorphic activity and consequently a varied lithology. The association of mountain building with plate boundaries, many of which are actively moving today, accounts for the frequent earthquake and volcanic activity and the sharp relief with which they are associated (Figure 1.2).

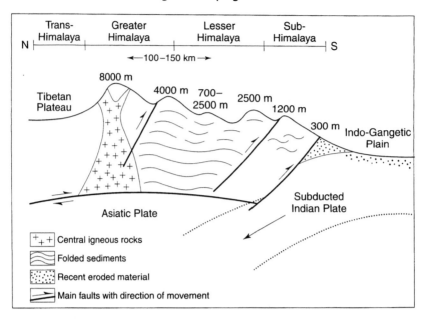

**Figure 1.1**   Simplified N–S section of the Himalayas showing the subducted Indian plate, and folding and faulting of the intervening sediments to form the mountains.

There is a perception that there has been enhanced tectonic activity over the last 40 million years, together with a phase of major climatic change during which the cycles of glacial and interglacial conditions prevailed (Isaacs, 1992). The link between the uplift of mountains and climatic change resulting from altered patterns of atmospheric circulation is well established but it is difficult to determine which 'came first'. Continental drift can induce climate change by a change in the distribution of land and sea, and climate change could induce mountain building by enhancing erosion rates and thus uplift.

Fold mountains are formed by tectonic forces bending and shaping the earth's crust (Figure 1.3). They technically occur where the axes of the folds are generally parallel to the range, but faulting and cross-folds introduce complexity into the system and result in a diversity of outcrops and thus variable erodibility. Different groups of rocks occur: *flysch* and *molasse* sediments are uplifted marine and continental sediments respectively. They are often soft sandstones and shales and may be rapidly eroded. Intensely folded strata form *nappes* – bent-over folds, which introduce crest weaknesses exploited by denudation processes. Block faulting creates uplifted (horst) and lowered (graben) blocks such as those of the uplifted Black Forest and Vosges and down-faulted Rhine valley of Europe. These areas are characterized by broad flat valleys and dissected upland plateaux.

Volcanic mountains can be classified by the type of material and the nature of eruptive activity (Figure 1.4). Basalt cones produce runny lava and less explosive activity, forming gentle shield cones such as those of Hawaii. Vesuvian, Pelean and similar types are explosive and produce gas, ash and thick andesitic

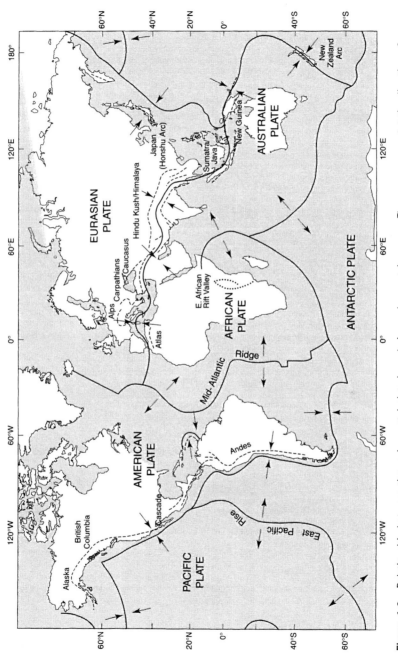

**Figure 1.2** Relationship between the main crustal plate boundaries and mountain ranges. The arrows represent the direction of plate movement (modified after White *et al.*, 1992 and Strahler and Strahler, 1992).

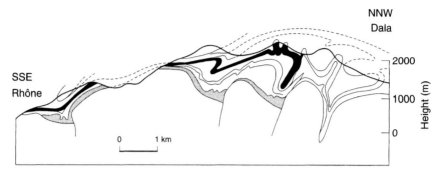

**Figure 1.3** Simplified SE-NW section of the Aar Massif in the European Alps, showing the complex folding and faulting of sediments which occurs in mountain building. Surface erosion exposes a variety of rock types at the surface, making these formations difficult to interpret. The dotted line marks the continuation of selected strata (after Baer, 1959).

a) Hawaiian (e.g. Mauna Loa & Mauna Kea)

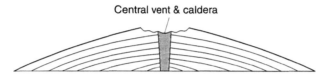

b) Composite cone (e.g. Mount St Helens & Fujiyama)

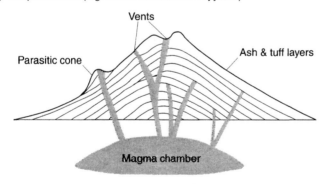

c) Caldera (e.g. Napak, Uganda)

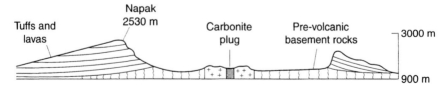

**Figure 1.4** Types of volcanic cone: a) Hawaiian cones have gentler slopes and form from runny lava with gentle eruptions. b) Composite cones with alternating ash and lava layers tend to be steeper and are associated with more explosive activity. c) Old volcano vents may collapse to form a caldera, leaving the rim upstanding as higher ground – this example is Napak, southern Uganda (adapted from King, 1949).

## Box 1.1   The Himalayas

The Himalayas are complex and young, extending 2500 km northwest–southeast between north India and south Tibet with associated ranges of the Pamir, Karakoram, Hindu Kush complex extending north and west into Pakistan and Afghanistan, and the Zagros mountains of Iran. They join the mountains of Turkey and the Eastern Mediterranean and also extend into Southeast Asia. They contain the world's highest mountain, Everest (8848 m), and many over 7000 m. The Himalayas were formed by the convergence of the Indian and Asian plates, beginning 40–50 million years ago. The intervening sediments were squeezed upwards as some 2 km of the Indian plate was subducted at rates of 150–200 mm/yr; since then the rate has been 50 mm/yr. Current rates of uplift vary: in Papua New Guinea it is estimated at 1000–3000 mm/1000 years and in the Karakoram 4000 mm/1000 years. The highest Himalaya forms a central range with successively lower parallel ranges – the Middle Mountains and foothills to the south, each of which is separated from the others by a major thrust fault, and to the north the Tibetan Plateau, which is a high altitude desert plain dipping gently northwards (Figure 1.1). Climatic change from 8 million years ago strengthened the Asian monsoon on the southern side, which has enhanced the rates of erosion in the south and east, and increased the flux of sediment by a factor of 13, resulting in the formation of the great river plains of India. This continual stripping of sediment from the mountains has contributed to their continued uplift at the highest rates in the world.

*Sources*: Gerrard, 1990; Goudie, 1995; Pinter and Brandon, 1997.

lava, forming steep, complex cones. Welded materials such as tuffs take longer to erode than uncemented ash and scoria. Most volcanoes have complex histories – Fujiyama in Japan actually comprises two more recent cones superimposed on a Pleistocene one. A buried soil between the oldest and the two younger cones indicates a period of quiescence lasting between 5000 and 10 000 years. Despite the activity of volcanoes, the rich, free-draining and fertile soils make them attractive places for human settlement.

The geological history of each mountain range is unique. In Boxes 1.1, 1.2 and 1.3 an outline of three major ranges is given.

## Box 1.2   The Andes–Rockies

The Andes–Rockies ranges of the Americas involved a collision between the Pacific and North American plates. The North American Cordillera (Figure

*(continued)*

*(continued)*

1.5) system is 600–900 km wide and comprises two parallel ranges with intermontane plateaux and basins. To the east lie the Rockies and to the west is the Pacific coast Cordillera stretching from Alaska to California, where it includes the Sierra Nevada, with a general altitude of around 4400 m. The western side of North America is much more tectonically active than the east. The Pacific range continues into Central America, incorporating the volcanic Caribbean islands, and then into South America and the Andean range.

The Andes rise steeply from a narrow coastal plain, at maximum 250 km wide. They were formed by the collision between the Nazca oceanic plate and the South American plate 50–60 million years ago. In northern South America the Andes reach 4500 m and comprise parallel dissected valleys. In Colombia

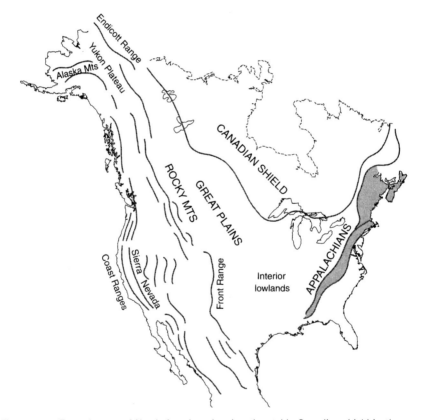

**Figure 1.5** Tectonic map of North America, showing the stable Canadian shield in the north and east, the Appalachians in the southeast and the west coast mountain ranges (adapted from Holmes and Holmes, 1978). (See also Figure 1.2 for the directions of plate movements forming these ranges.)

*(continued)*

*(continued)*

the range comprises the Eastern and Western Cordillera with a valley in between. In Ecuador the range is at its narrowest: 150–200 km. In Peru and Bolivia the range widens again into two ranges bordering the altiplano, a large, dry, expanse 3600 m high with interior drainage. Lakes Titicaca and Poopo were uplifted as part of the altiplano block. Moving south, the range turns due south from the Atacama desert, the driest place on earth, into Argentina and Chile. The range becomes single-crested and altitude and width decline southwards as climate becomes more temperate and then arctic. The whole Andean range is still very active. The Bolivian Andes were estimated to have risen at rates of 100–200 mm/1000 years from 20 to 40 million years ago, with rates in the last three million years of as much as 700 mm/1000 years. In Peru the rate of uplift is currently estimated at 4000–5000 mm/million years. Frequent earthquakes and volcanic activity occur.

*Sources*: Robinson, 1967; Troll, 1968; Gerrard, 1990; Summerfield, 1991.

## Box 1.3   The European Alps

The European Alps extend across Europe and merge with other chains in the Alpine forelands to north (the Carpathians, Balkans and Caucasus) and south (the Pyrenees, Catalonian, Cantabrian and Iberian mountains of Spain, the Jura of France, the Apennines of Italy and the Dinaric Alps) (Figure 1.6). The complexity of the Alpine arc arises from the rotation of micro-plates of the Mediterranean and the overthrusts and nappes formed from compression. Much of the uplift of the Alps occurred in the Pliocene and Quaternary at rates of 1–2 mm/yr. Four main glaciations have successively modified the landscape, eroding the fractured nappes and exploiting the weaker sedimentary rocks, forming a complex landscape without clear boundaries.

The Alps can be simplified into four main areas: the Central High Alps composed of hard crystalline rocks forming the highest peaks, such as Mont Blanc in Switzerland (4807 m) and the Matterhorn (4478 m). These areas have sharp relief. To the north and south and west are relatively soft calcareous schists which have eroded into smooth longitudinal valleys including those of the Rhone and Rhine. Further out from this are the Alpine limestones, with arid surfaces, steep relief and slow weathering. Such limestones occur in Greece and the Dinaric Alps (reaching 2917 m and 2522 m, respectively). Finally, surrounding this is a narrow belt of soft flysch sediments forming rounded low hills such as the Carpathians in Bulgaria and Slovenia.

*Sources*: Gerrard, 1990; Selby, 1985.

*(continued)*

*(continued)*

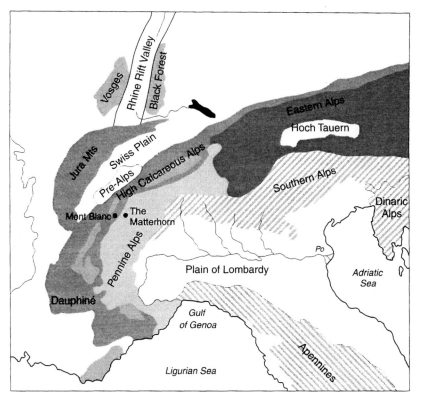

**Figure 1.6** Simplified sketch of the main tectonic units of the European Alps. To the north is the Rhine rift valley which is a graben between the upthrust horsts forming the Vosges and Black Forest mountains. Mont Blanc lies in the calcareous high Alps (adapted from Holmes and Holmes, 1978).

## Modification: processes of denudation

Geological structure, climate, altitude and latitude are the main controlling factors of denudation processes. Steep slopes and high relief produce high potential energy and thus erosive power of fluvial and gravity processes. Large temperature ranges give rise to frost weathering and insolation weathering processes. In the literature there is often an overemphasis on catastrophic events as the main agents of erosion, but the seasonal cycles of mid and high latitudes and continual denudation in the tropics are just as important over longer timescales. There is great variability in time and space due to diurnal and seasonal fluctuations in climate and thus in streamflow and sediment production. Aspect and altitude also control rates; for example, the dry, arid Pamir is much less denuded than the wetter southern Himalaya, and in the European Alps different rates of weathering and soil formation occur between south- and north-facing slopes.

There is considerable uncertainty about true rates of denudation due to the paucity of data, the high spatial and temporal variability and the impact of events such as major floods. The dry Karakoram range has an estimated mean rate of 1 mm/yr compared with Darjeeling, which is much wetter and has an estimated range of 0.5–5 mm/yr. However, one flood event in Darjeeling in 1968 caused by 1266 mm rain in three days gave rise to local rates of 10–20 mm in that year. Further events occurred in 1980 and 1984 and frequently since then (Chattopadhyay, 1981). The influence of earthquakes is important as it creates loose sediment and contributes to its redistribution, but Fookes *et al.* (1985) found that frequent earthquakes did not correlate with the highest erosion rates, as climate and vegetation were also important factors.

## Agents/processes of denudation

The dissection of mountain landscapes occurs by various agents of denudation picking out faults and lines of weakness in the rocks, and reworking and transporting sediments down valley and eventually out of the mountain system.

### Weathering

Thermal weathering processes include 'cold' frost weathering and 'hot' granular disintegration and spalling. Frost shattering of bare rock faces is closely associated with periglacial and glacial activity and produces angular scree. It operates by the freezing of water in crevices, which breaks off fragments of rock. Insolation weathering, resulting from the heating of rock surfaces by the sun and cooling at night, is common in granitic rocks which are subjected to high insolation intensities giving rise to high rock surface temperatures and rapid temperature changes. Unloading, arising from the release of pressure, may also occur. Chemical weathering processes are limited by low temperatures at high altitudes, but cause deep weathering in the tropics. Rapp (1960) demonstrated in his study of the Karkevegge mountains in Sweden that chemical weathering and solute transport were more important than often thought in alpine environments.

The formation of desert varnish in the High Atlas has been attributed to hotter conditions in the past. It is an important protection for the otherwise weak sandstone. The varnish is an accumulation of ferric and manganese oxides on the surface and has several stages of formation but requires water, which draws the salts out by evaporation (Robinson and Williams, 1992). Such features are thought to take between 3000 and 10 000 years to develop (Elvidge, 1979; Oberlander, 1994). Frost weathering is common, even in the tropics. It may occur all year round at high altitudes or be limited seasonally to a few months per year at higher latitudes. Likewise, chemical weathering may be limited to the summer months.

### Wind erosion

Wind erosion is facilitated by higher windspeeds and reduced drag and turbulence at high altitudes. The degree of abrasion damage to rocks depends on their type: sandstones in the High Atlas, for example, are particularly susceptible as they are poorly cemented. In the tropics and at lower altitudes, wind speeds are inhibited by vegetation, which increases turbulence and also protects the ground surface. Wind has a greater impact on vegetation and soils than on the rocks themselves. In many parts of the Himalayas, for example in Ladakh and Tibet (Osmaston, 1994), loess (windblown silt) forms an important component of soils at high altitude, where it is deposited, but elsewhere wind action strips soils of their fine component, leaving them stony and infertile. Vegetation is also subject to wind abrasion, desiccation and exposure (see Chapter 3).

## Water

Water is perhaps the most important agent in the erosion and transport of sediments. Streamflow is very powerful in mountains due to the high hydraulic gradients resulting from elevation. In the upper reaches, the flow is turbulent, with frequent hydraulic jumps (waterfalls). Lower down the valley, more sediment is entrained and carried. Where there is little vegetation or soil to impede runoff at higher altitudes and on steep slopes, streamflow responds rapidly to rainfall and melting. Thus, hydrographs peak strongly after rainfall events and with diurnal and seasonal melt regimes, making flow erratic. Sediment may be carried as bed load, suspended load or as solute. The sediment includes loose slope material from rockfalls, landslides, frost and other processes, glacial debris, and alluvium. Sediment tends to have periods of storage in the valley system, gradually being reworked and moving down valley in a series of jumps in response to periods of higher flow and transport energy. The coarse debris system in mountains is considered by some to be 'closed' – i.e. limited to reworking of materials within the valley, with only occasional major floods being able to transport material outside the confines of the valley (Barsch and Caine, 1984). As streamflow can quickly subside, alluvial banks, terraces, braided channels and other depositional features are both common and dynamic valley-floor features. The fine debris load, by contrast, is 'open' in that silts and clays are carried beyond the valley, forming lowland flood plains such as those of the Indus, Ganges and Brahmaputra rivers flowing from the Himalayas.

## Glaciers

Glaciers operate on longer timescales, move more slowly than rivers, but transform the landscape completely. The world distribution of mountain glaciers depends on climate and altitude. The Karakorams have the largest concentration of ice cover of any mountain range, despite being located in the very arid region of the Himalayan chain. Glaciers can be classified by size and form into cirque, valley or sheet and by temperature into warm-based (which have liquid at the base from pressure melting and are more liable to slide) and cold-based glaciers (which are frozen to the underlying rock). Glaciers may be active, passive or 'dead' (stagnating). Active glaciers are growing and moving, passive glaciers have a balance in accumulation and ablation and may not be actively moving. Glaciers erode by plucking rock from rock faces, and by grinding rock surfaces to create the U-shaped and hanging valleys, pyramidal peaks, sharp ridges and other features so familiar in glaciated landscapes. The average erosion rate of glaciers varies between about 100 and 30 000 mm/1000 years, based on sediment load of meltwaters flowing from selected glaciers (Drewry, 1986; Warburton and Beecroft, 1993). In the Punjab Himalaya, denudation rates for the Raikot glacier were estimated as 4600–6900 mm/1000 years compared with 1400–2100 mm/1000 years for the whole basin, which seems a fairly typical ratio (Gardner and Jones, 1993). The rate depends on the bedrock lithology and structure and the rate of movement of ice. In the Appalachians, it is estimated that during the last glacial period, rates of erosion were 1210–3040 mm/1000 years, compared with rates of 20–30 mm/1000 years for present-day fluvial erosion (Braun, 1989).

Deposition from glaciers occurs by meltwater depositing fans, kames and eskers, or by stagnating ice depositing moraines. Moraines may block valleys, creating lakes, or may run along valley sides, forming terraces on the slopes. The sediments deposited comprise mainly glacial till, which is unsorted and may have a high clay component derived from rock flour and fluvio-glacial debris consisting of sorted sands and gravels. Depositional landscapes often have an 'untidy' appearance of uneven hummocks and lumps. These features are gradually reworked by fluvial erosion, but many survive from the last glacial period, giving a source of evidence for the extent of glaciation and patterns of activity.

## Snow

Snow is an important characteristic of high mountains. Permanent snow lies on high summits above the 0°C isotherm, whilst seasonal snow accumulates and melts each year. Snow patches persist on shady slopes and nivation weathering may occur, although this tends to be slow. The main denudation effect comes when the snow melts. Thick snow on steep slopes is prone to avalanche, which can carry sediment ('dirty' avalanches). The amount of material moved varies with the nature of the avalanche and the bedrock or sediment available. Annual accretion rates of material from avalanches varies from 0.04 to 40 mm/yr in both Spitzbergen (André, 1990) and the Rockies (Luckman, 1978; Gardner, 1983). Rapp (1960) rates avalanches as the third most powerful process in Karkevegge in arctic Sweden, after solute transport and landslides. They represent a hazard to populations (see below). Snow cover can insulate the soil and protect plants, but late-lying snow retards growth.

## Mass wasting

Mass wasting, in the form of soil creep, landslides, rockfalls, mudslides and other slope processes, is a ubiquitous feature of mountain landscapes, and is a focus of study and concern regarding safety, damage to construction and degradation of agricultural land (Plate 1.1). There is a wide range of slope processes. They may be primary, occurring as a result of the accumulation of

**Plate 1.1**  The landscape from the Karakoram Highway. The cultivated land surrounding the village is being eroded and undercut by the river below. The sediments are weak sands and gravels. Debris slides are common – for example, crossing the road at lower left and an irrigation channel at upper right. Scattered remnants of forests occur on the higher slopes.

debris on a slope exceeding its angle or mass of stability. Alternatively, they may be secondary effects triggered by earthquakes, volcanic activity or as a result of meteorological events such as typhoons and monsoon rainfall. The rate of movement of material varies from a slow creep which is common on all slopes with a regolith and vegetation cover, in the order of 10–100 mm/yr (for example, the Drynock slide in British Colombia moved less than 0.3 m in five years (Brunsden and Allison, 1986)) to rapid and massive mudslides which wreak destruction on settlement (for example, the Huascaran mudslide in 1972 which buried Yungay village, Peru, killing 17 000 people, moved at 480 km/hr in 1970). Rates of creep in temperate latitudes are higher in continental climates (15 mm/yr) than in maritime climates (0.5–2 mm/yr) due to the freezing of the ground in the continental environment (Young and Saunders, 1986). Creep tends to affect only the upper 500 mm of soil. Slides and slumps may be shallow-seated, affecting the loose regolith layers, or deep-seated, arising from failures in the geological strata. Shallow slides may be anchored by trees but deep-seated slides cannot be controlled. Rockfalls occur from free faces, caused by failure at joints and bedding planes. This may be in response to unloading following deglaciation, by undercutting at the foot of cliffs by rivers or by frost weathering of exposed cliffs. The material forms large fans and cones of scree, which may become stabilized and vegetated and eventually occupied and cultivated. Rockfall debris and the movement of scree as a debris avalanche is rapid and destructive, with rates of movement of 10–1000 m/s (Brunsden and Alison, 1986).

Other forms of slides and slumps may be triggered by rainfall, which causes saturation and collapse. This may be enhanced or reduced by terracing practices (see Chapter 7). The susceptibility of mountain landscapes to mass wasting depends on the nature of the lithology, dip of strata, sediment type and cementation, vegetation type and cover as well as the nature of the trigger event. In tropical areas, deep weathering produces a deep regolith, but the wetter conditions favour vegetation growth, which stabilizes the soil. Finally, in periglacial areas, solifluction and gelifluction processes occur, whether continuously or seasonally. Solifluction lobes may move at rates of 10–20 mm/yr (for example in Spitzbergen; Gerrard, 1990) but vary between 10 000 and 100 000 mm/1000 yrs, which is faster than soil creep (Young and Saunders, 1986). The rate is dependent on moisture content, so silty soils move fastest whilst sandy soils drain too freely and clays are too cohesive. These processes form characteristic lobes and terraces on slopes.

## Zonation of geomorphological processes

The distribution of the different processes of denudation is complicated, but a general pattern occurs (Brunsden and Allison, 1986). Climate and aspect as well as elevation are strong controlling factors in determining the pattern of processes. Human activity may both enhance and retard these processes.

The cryo-nival zone (zone 1 on Figure 1.7) lies at the high summits where ice and snow predominate for much of the year in seasonal latitudes, and all year on some tropical mountains (Barsch and Caine, 1984). Glacial erosion, periglacial processes and nivation predominate, together with wind abrasion and frost shattering of exposed rock faces. Abundant rocky material accumulates as cones and talus slopes, or is incorporated into glaciers. The low temperatures and aridity limit chemical weathering and fluvial processes may be intermittent and limited to summer months.

The upper slopes (zone 2) below the cryo-nival zone are dominated by frost weathering and rockfalls forming scree. There tends to be minimal soil formation on steep slopes as the material is not anchored and is prone to periodic instability. It is the slopes of 30–40° which are the most dangerous, as these are shallow enough to allow material (soil, rocks, snow) to accumulate, but too steep for the the accumulations to maintain stability, thus failures are common. In addition, the steepness allows the rate of movement to be rapid. This zone is characterized by seasonal snow cover in temperate latitudes and thus tends to be used for forests, summer grazing and, increasingly, for winter sports. Avalanches of snow are common, and meltwater and rainfall events trigger torrential streamflow, undercutting talus slopes and scree cones, causing mass movement.

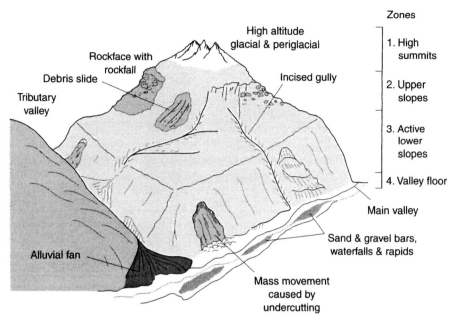

**Figure 1.7** Zonation of geomorphological processes (adapted from Fookes *et al.*, 1985).

The lower slopes (zone 3) are zones where deposition becomes more important. Alluvium and colluvium accumulate, soils form and this zone is often permanently occupied and modified by terracing and irrigation activities. Deep weathering occurs in warmer climates and more sheltered conditions, and soil creep, landslides and mudflows are common.

## Box 1.4    Geomorphological processes in the Hunza valley, Pakistan

The high altitude and extremes of climate in this region mean that a variety of processes operate, illustrating the interaction between geology and climate. Frost weathering occurs daily in mid-winter at 1200–1500 m at Gilgit. Skardu at 2100–2500 m marks the limit of frequent daily winter frost. At 3000 m, there are between one and three months of frost a year, whilst above 5000 m, thaw is rare. In the summer, however, rock surface temperatures at 4000 m may reach 30°C and, with the presence of moisture, chemical weathering can occur. Rock surface temperatures can reach 40°C (20°C under cloudy conditions) and may be 24°C warmer than air temperatures. This, together with the large temperature range, causes physical insolation weathering. Annual precipitation is low and dominated by high-altitude winter snow so streamflow is determined primarily by snow and glacial melt. Snowmelt forms a peak in the spring, whilst glacial melt from the abundant glaciers creates a larger peak in later summer. Diurnal

*(continued)*

*(continued)*

fluctuations also occur, with peaks in later afternoon after the heat of the day. The predominance of bare rock allows rapid runoff, so streams show the characteristic flashiness of mountain hydrographs in response to melt rates and rainfall events. In this region, the monsoon has no effect, so 40–70 per cent of the flow occurs in July and August, determined by glacial melt, compared with 60–80 per cent of the flow which comes from monsoon rainfall in the Eastern Himalayas. The augmentation of river flow by glaciers means that there is a high suspended sediment load – estimated at 4800 t/km/yr from the Gilgit and Hunza rivers, whereas the Indus, which has a catchment four times larger, carries only 30 per cent more suspended sediment. The Hunza river yields 90 t/km/yr of solute load, but although this is three times the world average, it comprises only 2 per cent of the river's load. Glaciers also present a substantial hazard – the high ice coverage of the area and active movement have resulted in frequent damming of the rivers, collapse and flood peaks, and mass movement processes represent a continual threat to human activity (Plate 1.2).

*Sources*: Hewitt, 1968; Whalley *et al.*, 1984; Ferguson, 1984; Kreutzmann, 1994.

**Plate 1.2** The Hunza valley, northern Pakistan. In the centre of the frame the dark line of vegetation marking an irrigation channel has been disrupted by an active debris slide, showing the vulnerability of these structures. In some cases the channels are constructed as tunnels beneath the surface to minimize damage (see Plate 8.2).

On the valley floors (zone 4), reworking of sediment deposited from the slopes and upper valley occurs. Streams are often prone to flooding, limiting the potential use of these areas. Undercutting of slopes may contribute to their collapse.

## Hazards

Mountains are inherently hazardous environments and have a disproportionate number of hazard events. The active geomorphological and tectonic processes, and extreme and variable climate mean that catastrophic events are common. It is also true, however, that such events only become hazards once they affect human populations and activities. Thus, whilst avalanches, floods, earthquakes and so on are fundamental characteristics of mountain environments, it is only when they interfere with human activity and result in deaths, disruption and economic loss that they become recognized as hazards (Plate 1.3).

Hazards may arise from glacier movements, floods, landslides, earthquakes and volcanic activity. The physical and economic characteristics of mountains mean that in many areas access to stricken regions is difficult, and the costs of recovery disproportionately high. Hazards have both human and natural causes, and there is not necessarily anyone to blame – a difficult conception in today's increasingly litigious society. The interaction between human and natural

**Plate 1.3** The closure of the Karakoram Highway is a frequent occurrence. Here a series of substantial landslides was triggered by heavy rain in September 1995. Heavy earth-moving equipment is clearing the road despite continuing activity of the slopes, as these roads are a lifeline to many villages up valley.

processes and activities is difficult to separate (Blaikie and Brookfield, 1987), and in many cases whilst both are present, one may dominate at a given period: for example, in the European Alps, fourteenth-century soil erosion was primarily caused by natural processes, but in the eighteenth century it was attributed to changes in the agricultural economy which enhanced natural rates of erosion. Hazards may be classified into those which are mountain-specific (e.g. avalanches) and those which are mountain-generated, affecting areas outside the mountain regions (downstream flooding and sedimentation). Many hazards have multiple causes and effects. Hewitt (1997) also identifies war, refugees and disease as hazards, but these are considered elsewhere.

A survey of hazards showed that between 1950 and 1990, 97 per cent affected mountain lands, 75 per cent affected the foothills and peripheral regions where concentrations of populations are located, and 71 per cent were associated with dryland regions (Mediterranean, Western Himalaya, Middle East) (Hewitt, 1992). An inventory of hazards in Hunza (Kreutzmann, 1994) shows that between 1830 and 1993 there were 125 damming events, 53 of which were caused by glaciers, 28 avalanches and 18 mudslides. There were 102 earthquakes which affected Afghanistan but not Hunza, even though their epicentres were in Pakistan. Thus the effect was relatively localized.

According to a survey of hazards (OFDA, 1988), earthquakes are the most frequent, followed by floods. Secondary landslides, outbursts, fires, etc. often cause more damage than the original earthquake, but tend to be under-represented in surveys as they are not the primary cause of the disaster. In Baluchistan an earthquake in 1930 killed some 30 000 people, destroying settlements which were located on weak alluvial soils. At Spitak in Armenia, an earthquake in 1988 killed 3600, but in Northridge, California, 1994 only 50 died. In Kansu, China in 1920 an earthquake triggered a landslide killing some 180 000 people, engulfing ancient and new towns and villages. The recent earthquakes in Colombia (January 1999) and El Salvador (January 2001) both triggered landslides which devastated city suburbs and agricultural land. Both events were estimated to have killed around 1000 people and in the latter case some 100 000 people were made homeless by the event. These examples also reflect the fact that warning and coping systems are much more effective, and mountain populations much lower in the developed world than in the developing world. In addition, physical accessibility and the availability of financial and technological assistance also tend to be more limited in developing countries.

Floods are the second most frequent and devastating hazard in all mountain regions. These include annual floods from monsoon rainfall and floods caused by extreme rainfall events either in season or out, as in Darjeeling in 1968, for example (see above). In the High Atlas, the Ourika valley experienced a devastating flood in August 1995 as a result of heavy rain; houses, agricultural terraces and roads were destroyed and deep sediment fans deposited across the valley floor. Within four days the road access had been re-established with the aid of government and EU funding, but compensation payments are far short of the real value of land and livestock lost to farmers, as well as tourist income (Johnstone, 1997).

Glacial outbursts are frequent hazards in regions where glaciers are numerous and active. In the Hunza valley in the Karakoram, some 35 outbursts have been recorded in the last 200 years (Hewitt, 1982), seven of which have occurred in Shimshal valley since 1884 (Gerrard, 1990). Outbursts occur when ponded water is released due to collapse of the blockage, which might be a landslide or terminal moraine or even the glacier itself. Such a blockage occurred for one hour behind the Batura glacier in Hunza in 1974, and caused a clear floodwave downstream. Some 30 glacial advances across headwaters have been recorded in this region during the last two centuries. Glacial fluctuations due to climate change affect the pattern of glacial hazards. Advancing ice during the Little Ice Age blocked rivers, destroyed irrigation and communication systems, and changed the local ecological conditions. When the ice retreated, different hazards prevailed, with deposition of material and stagnating ice and flooding.

Volcanoes are recognized as hazardous environments and yet, despite their activity, they continue to be attractive sites for settlement, either due to pressure of population and shortage of land, or because of their rich, fertile, easy to work soils. Vesuvius, Etna and other Mediterranean volcanoes are well known, as are those around the Pacific 'ring of fire', such as Mount Pinatubo in the

---

## Box 1.5    The Alps: a hazardous landscape

The Austrian Alps have suffered various significant geomorphological events, which in more recent times have increasingly been classified as hazards, due to changes in human activity. Kofels in the Austrian Alps suffered a major landslide 8700 years ago, flowing into the Otzal. During the thirteenth century and from the seventeenth to the nineteenth century the expansion of glaciers during the Little Ice Age cut off communications between valleys and destroyed irrigation and agricultural systems and drainage. In 1969 a debris flow destroyed 42 of 44 check dams and caused $1.8 million damage. In 1987, widespread floods resulting from late snow melt, unusually heavy and late falls and sudden and rapid melting from a rise in temperature of 15°C as arctic conditions ceased, resulted in flood damage to most major valleys. In Otzal all post-1915 constructions were destroyed. In 1999, the Austrian Alps, along with the French Alps and other regions, suffered the worst avalanches since records began; 2 m snow fell in four days, stranding 25 000 tourists in ski resorts. However, the predicted trend towards later and thinner snow cover in the Alps brings other problems as this region is economically highly dependent on its tourist industry. Finally, the tragic fires in the road and rail tunnels in Mont Blanc in the Swiss Alps and elsewhere in 2000 reinforce the problems of access which not only impede rescue operations but which necessitate the construction of tunnels in the first place.

*Sources*: Aulitsky *et al.*, 1994; Lamb, 1988; Schwarzl, 1990.

Philippines and Mount Soufrière in Montserrat, which erupted in 1995. The different types of volcano will determine the nature of the hazard. Basaltic volcanoes have more frequent, gentle eruptions; for example Mount Ruapelu on North Island, New Zealand has had 55 small eruptions in a period of 37 years. This is compared with Vesuvian, Pelean and Krakatoan types which erupt violently after long periods of quiescence. Secondary hazards including mass movements are also a problem; a mudslide triggered by a volcanic eruption in Nevada del Ruiz, Colombia in 1985 killed 20 000; the volcano had been dormant since 1845.

## Hazard perception and management

The impact of hazards has increased due to economic and demographic pressures to expand onto more marginal land or to put areas of land to new uses which interfere with the natural absorption process of mountain landscapes. For example, the development of winter sports has placed human activity right in the path of avalanches, and whilst they may not be more frequent, they now have greater impact. Likewise, floods have more devastating effects, partly because of the modifications to channels, which may reduce the capacity of the channel to absorb and dissipate flood waters, but also because of permanent developments in the valley floor area, which are, again, directly in the path of flood waters. There is some debate over the effects of deforestation and afforestation activities in controlling or enhancing flood frequencies (see Chapter 10). In historical times societies had considerable knowledge of frequent avalanche corridors so they could avoid these areas; only using them in the summer. However, in many areas where winter tourism has been established, the rapidity and frequency of avalanches has required substantial mitigation and management strategies. In the same way, the location of historical settlements above flood level minimized the risk of inundation, but expansion onto valley floors now increases the effect of this hazard.

Rockfalls and landslides are a constant problem. Small-scale slumps in terraces are rapidly mended to prevent them affecting lower terraces. Economic pressures in Nepal are encouraging farmers to irrigate previously rainfed terraces. Traditionally, rainfed terraces were located on steeper slopes as they were more prone to failure and thus less suitable for paddy rice. The conversion of these to irrigated terraces increases the numbers of, albeit small, failures placing further stress on labour supplies which are already under pressure (Gerrard and Gardner, 2000). Larger slumps and slides are abandoned until stabilized.

Settlements have been abandoned due to major landslides or glacial outbursts; four such events are recorded in the Hunza valley by Kreutzmann (1994), although some of these sites were later reoccupied. More recent developments such as road building have been associated with increased landslide hazards due to the steep cuttings in weak rocks. The Karakoram Highway is frequently blocked by such collapses after rain. There have been a number of studies which illustrate the increase in landslides in association with road building or logging activities – for example, that of Swanston and Swanson (1976) in Canada (see Chapter 10).

Hazard planning requires community organization, and, increasingly, state intervention takes over from the less formally structured traditional responses of individuals and family units. External aid for disasters has developed a culture of dependency amongst some societies, which replaces their previous internal coping strategies and resilience (Tobin, 1999). This is an inevitable consequence of integration into wider capitalist economies and has both positive and negative effects upon societies. This is partly the result of the increase in stakeholders – those who have undertaken or invested in the development and require insurance payouts, for example. It is no longer a case of just a few locally based livelihoods, but an international newsworthy and fund-raising arena. The growing importance of liability – either in insurance or in mitigation strategies – has made the solution of these problems more difficult in some areas. In 1999, avalanches in the European Alps resulted not only in loss and damage to property in the Alps, but complaints from tourists that their insurance cover only covered too little snow, not too much!

The most frequent strategy of hazard management is a mapping and zoning exercise. In Nepal this was undertaken in order to ascertain the connections between certain landslide events with land use, vegetation and geology (Kienholz *et al.*, 1984). Various calculations of density per unit area and the use of aerial surveys and base mapping were employed to analyse the distributions. This has also been applied with regard to typhoon hazards in Taiwan (Petley, 1998) and to zones on volcanoes such as Mount Soufrière, which has a 'forbidden zone' in which residence is technically prohibited. However, the need for land is such that many have defied authority and remained in these areas in order to survive.

In the European Alps the classification of land into zones which determine what developments may take place is critical to the safe development of tourism and other facilities. An Austrian Forest Law of 1975 defines both the degree of risk and what constitutes avalanche and torrent hazards. All areas are zoned into Red (no developments), Yellow (no developments which encourage groups of people) and White ('safe'). These are based on a 150-year return period of an event of defined magnitude (Aulitsky, 1994). Similar zoning systems occur elsewhere in the Alps (Table 1.1). Expert and public opinions are incorporated into planning decisions, although political and economic pressure may lead to a creative interpretation of a hazard, or be based purely on the law and not on the environmental situation, thus permitting developments where they should not occur (Kienholz, 1984).

These zoning systems are useful and effective guidelines, but they are not infallible. The 1999 avalanches in the Alps caused destruction and loss of 80 lives and destroyed buildings in the white zone. In France the event was the worst since 1908, and in Austria since records began in 1954. Thus, although the zoning system gives the impression that an area is known to be safe, in reality there is no area of complete safety. Although Austrian development is controlled by law, there is no such legal control in Germany, Italy or the former

Table 1.1  The Swiss avalanche zoning system.

1. RED ZONE (High hazard)
   - Avalanches with return intervals of <30 years
   - Avalanches with impact pressures >3 t/m$^2$ and return intervals ≤300 years
   No buildings or winter parking facilities
   Special bunkers needed for equipment

2. BLUE ZONE (Potential hazard)
   - Avalanches with impact pressures <3 t/m$^2$ and return intervals of 30–300 years
   No public buildings that encourage gatherings of people
   Private houses must be strengthened to withstand the impact
   The area may be closed during periods of hazard activity

3. WHITE ZONE (No hazard)
   - Very rarely affected by small pressures of up to 0.1 t/m$^2$
   No building restrictions

*Source*: After Perla and Martinelli, 1976; Smith, 1992.

Yugoslavia, and consequently developments in these areas carry a potentially greater, although unquantified, risk. This reflects differences and changes in the perception of hazards and of acceptable risk in mountain regions, and the fact that economic forces may override higher degrees of development control and make higher levels of risk more acceptable (Price and Thompson, 1997).

Europe has instigated the EUR-OPA Major Hazard Agreement in 1993 (Council of Europe, 1993) which seeks to coordinate response and warning strategies in the event of major disasters, particularly for the benefit of the southern Mediterranean countries where there are proportionately more hazards and less finance. On an international scale, the UN, governments and charities such as Oxfam and the Red Cross are mobilized to provide relief and technological expertise in the event of such disasters.

With respect to earthquakes and volcanoes, complex monitoring systems are in place, but mainly in countries such as the USA – and again where centres of populations are located (e.g. San Francisco) rather than in distant mountains. The USA has a sophisticated on-line earthquake monitoring programme and effective warning systems. Mount St Helens was effectively predicted (1980) by measurements of ground temperature, earth tremors and topographic bulging, although the direction of the blast could not be predicted. However, the low population of the region made this less of a hazard than on a densely populated active volcano such as Mount Soufrière in Montserrat. This case is renowned for the legalities which ensued over the obligations of the UK to provide aid for rehabilitation on the basis of Montserrat's colonial connection with the UK dating from 1632.

The management strategies of floods and avalanches range from passive avoidance and warning systems to active physical constructions designed to divert and channel flow (Figure 1.8), to slow the rate of flow or to induce small avalanches by dynamiting to minimize build-up and thus avalanche risk

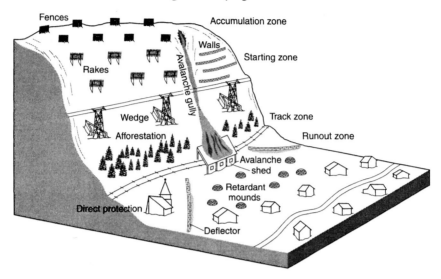

**Figure 1.8** Various physical methods of controlling avalanches. In the upper accumulation zone the structures are designed to retain snow; in the lower track and runout zones they are designed to deflect the path of the avalanche away from buildings etc. (from Smith, 1992).

(Smith, 1992). Floods are not subject to the same legal definition as avalanches and torrents in Austria, and have only begun to be placed in a legal context in Europe since 1983. However, the constraints of applying legal solutions are demonstrated by the fact that, as floods have not been officially defined, it is impossible to apply the law effectively.

## Key points

- Mountain building occurs by the uplifting of the earth's crust and of intervening sediments between converging plates. Igneous activity and earthquakes continue in areas of active plate movement.
- Erosion plays an important part in rejuvenating mountain landscapes, contributing to their continued uplift. Erosion is also destructive, lowering the mountains and depositing sediment on lowland flood plains.
- The rates and processes of weathering and erosion differ with altitude and latitude, being controlled by climate and elevation as well as rock type and structure. Water and ice are the most powerful agents of erosion. Variations in time and space are substantial.
- The geological history, including the landscape form and geology and its history of glaciation, means that many mountain landscapes are in disequilibrium with their current environment, making them extremely dynamic.
- Mountain landscapes are commonly perceived as hazardous due to earthquakes, volcanoes, mass movements, floods, etc. Mountains have a dis-

proportionately high frequency of natural disasters, and problems of access make it difficult to cope with the effects.

- The increasing development of areas of mountains such as valley floors and upper slopes, particularly for the growing tourist economy, has increased the hazardous effect of floods and avalanches, which can give the impression of more frequent and larger events.

- The economics of many mountain areas has meant that a higher degree of risk may be deemed acceptable in the interests of short-term financial gain. Whilst more effective warning and mitigation systems can be applied, they do not guarantee safety; there is no area of zero risk in mountain landscapes.

# Chapter 2

# Mountain climate and climate change

Climate is perhaps the most important single factor operating in mountain environments. It determines the nature and rate of geomorphological processes and the human activities which may take place – the crops which can be grown, tourism and hydro-power potential, etc. The factors controlling climate operate on a variety of temporal and spatial scales, which interact to produce a mosaic of environmental conditions on the ground. This chapter outlines the global and valley-scale factors that determine the pattern of climate, before examining climatic change in the past and future.

## Global factors

### Latitude

Latitude controls seasonality, day length and angle of incidence of the sun's rays (which is also determined by topography). Thus, it is a major factor (with altitude) controlling solar radiation receipts, and as most of the energy on mountain systems is derived directly from the sun's heat, this is a critical influence. In mid June, days are 12 hours 7 mins at Mount Kenya in the tropics, 13 hours 53 mins at Mount Everest (subtropics), 15 hours 45 mins at the Matterhorn in the Swiss Alps and 20 hours 19 mins at Mount McKinley in Alaska; mid-December days are correspondingly shorter at higher latitudes. Thus the total receipts of solar radiation in high-latitude summers may exceed those of the tropics, particularly if cloud prevails in the latter. During winter, however, considerably less radiation is received, which corresponds with much lower temperatures, a higher proportion of precipitation falling and accumulating as snow and periglacial activity occurring for part of the year. All of these restrict the growing season to only a few weeks in the highest latitudes.

As a consequence of latitude, the seasonal temperature range becomes much more important in higher latitudes, whereas daily range is more important in the tropics; Ecuador has a daily range of 11°C and a seasonal range of 1°C, but in Siberia the values are 5 and 60°C, respectively. Latitude also determines the global-scale atmospheric circulation systems which affect mountain regions, although large mountain masses have the capacity to modify atmospheric circulation. Thus the Himalayas are influenced by the subtropical high pressure cells between 20° and 40° N, contrasting with Equatorial low pressure, and

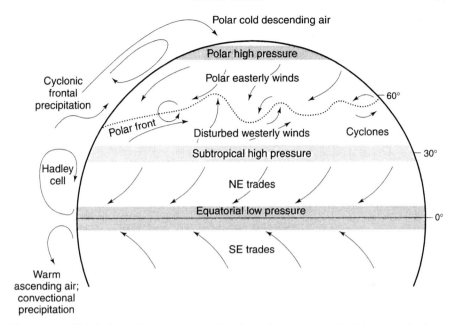

**Figure 2.1** Global circulation patterns showing the main pressure and wind systems in the northern hemisphere. These are reflected in southern hemisphere circulations. The Himalayas straddle the subtropical high pressure belt, whereas the Andes cut across these circulation patterns (see Figure I.1).

subpolar lows at higher latitudes. The locations of the trade winds, jet stream, and the intertropical convergence zone where the trade winds converge are determined by latitude, and affect the distribution and amount of precipitation and snowline altitudes. These are well illustrated by the Andes and Himalayas ranges (Figure 2.1).

## Altitude

Altitude is a fundamental factor influencing most aspects of climate. Generally, with increasing altitude, the air gets thinner (less dense) and there is a reduction in oxygen, carbon dioxide, water vapour and particles. As a consequence, vapour pressure is reduced, and so too the capacity of the air to absorb and retain heat, so that the air temperature responds rapidly to direct solar heating – hence the rapid fall in temperatures if the sun clouds over. There are changes in precipitation with altitude (see *Precipitation* on page 47) and increases in wind speed, as there is less friction between wind and land surface.

## Continentality

Continentality – i.e. the location of mountains relative to the sea – is important. The ocean has a moderating influence on climate and serves as a source of water and condensation nuclei, thus rainfall is often higher on coastal mountains where the prevailing winds blow off the sea – such is the case on the western side of the Rockies and Andes. Another example is Hawaii where the average rainfall is 650 mm/yr but on Mount Waialeale on Kauai it is 12 344 mm on the windward side. The maritime influence reduces the diurnal and seasonal temperature ranges, so continental interior climates tend to have greater temperature ranges and are drier, giving rise to higher tree and snowlines.

## Topography

The topographic influence on regional climates by interfering with upper air circulation, is important. These barrier effects are responsible for the pattern of the Indian monsoon – the Himalayas prevent the southward penetration of cold air from the north, whilst deflecting the northward-moving monsoon winds westwards. As these winds are limited to a maximum of 5000 m altitude they cannot cross the Himalayan ranges. The monsoon has its greatest effect on the eastern Himalayas, and diminishes westwards. The northern flanks of the mountain range and the Tibetan Plateau are very arid as they lie in the rain shadow and are dominated by dry continental air from the north. Likewise, the Andes effectively separate the Pacific coastal air from the South American interior, and the Atlas Mountains act as a boundary between the dry African continental and the relatively wet Mediterranean climates.

Barrier effects are also responsible for generating and deflecting regional winds. Pressure differences on each side of a mountain range mean that air will pile up against it, be pushed up, usually with precipitation, and then move down the other side, often rapidly, as a dry wind. Examples of these regional winds include the bora, a cold dry wind blowing from southwest Russia to the Adriatic; the oroshi of Japan; the chinook, which is a warm wind blowing eastwards across the Great Plains of America in winter, and the foehn of Europe, which can blow both north or south and is a dry wind gusting up to 160 km/hr, raising temperatures in its path. These winds are associated with steep pressure gradients across the mountains, rainfall on the windward side and clear, transparent, dry air on the lee side. Lenticular clouds can also form in the lee of mountain ranges as a result of disturbance of the air creating a wave effect. Deflection of regional winds is also important as windspeeds are greatly increased if relatively shallow air masses are forced through mountain passes and around high land. Such is the case of the mistral, a hot dry wind that blows southwards across France and is deflected round the east side of the Massif Central and then southwest along the Alpes Maritimes, and the bise, where cold polar air is blocked in spring and channelled by the Jura and Alps (Wanner and Fürger, 1990). This wind may reach gale force and forces uplift, causes orographic precipitation as the air crosses the Alps, and then gives rise to the southern foehn on the lee side (Figure 2.2).

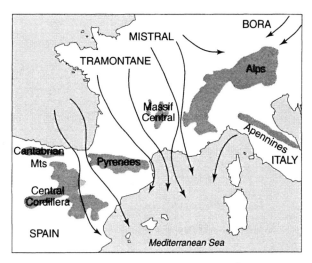

**Figure 2.2**  Regional winds in western Europe: the tramontane (deflected east of the Pyrenees) and mistral (east of the Massif Central). The bora is also shown. The foehn blows perpendicularly off the Alps to north or south.

## Mountain mass

Finally, the mountain-mass effect enables mountains to generate their own climates. Large mountain masses such as the Himalayas and Tibetan Plateau are sufficiently massive for their absorption of heat in the summer to create a high pressure cell that weakens and reverses the westerly jet stream and contributes to the generation of the monsoon. The Tibetan Plateau may be 5–6°C warmer at 5000 m than at the equivalent altitude in the tropics. The aridity of the air means that skies are clear, insolation is intense and much hotter conditions prevail than in the tropics, where the humidity means that cloud cover reduces radiation receipts. By contrast, isolated mountains such as the East African volcanic peaks tend to have climatic conditions much closer to those of free air as their mass is insufficient to generate their own climate, although they do modify local climatic conditions.

## Main climatic elements

The actual weather experienced in mountain areas is determined by patterns of temperature and precipitation, which are influenced by the global factors considered above. Temperature is controlled primarily by solar radiation, and precipitation by elevation and topography. Topography, in particular slope angle and aspect, is a very important control over climate and determines conditions on a valley (meso) scale. Each valley can generate its own climate, and local storms and differences in growing conditions can vary widely between and within a single valley. This creates a diverse mosaic of environmental conditions which is reflected in the soils and vegetation and exploited

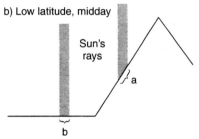

a) High latitude, midday

Area a is smaller and receives higher intensities
of solar radiation

b) Low latitude, midday

Area b is smaller, receiving higher intensities
of solar radiation

**Figure 2.3** The effect of topography and latitude on solar radiation. In the high latitudes at midday the overhead sun is more perpendicular to slopes, which thus receive higher intensities of radiation than does flat land. This is reversed in the tropics.

by farmers. Within a valley, climatic conditions can form identifiable layers; a 'topoclimate' some 250 m deep is influenced by aspect, altitude and slope geometry and a 'microclimate' 15 m deep is modified by vegetation and slope materials (Barry, 1992).

## Solar radiation

Two aspects of radiation are important: intensity and quality. The intensity of radiation depends on the angle of incidence of rays as they hit the ground; perpendicular rays will be more intense than those hitting sloping ground (Figure 2.3). Thus slope angle is crucial but latitude also has an influence, as at higher latitudes the angle of incidence is more perpendicular to slopes than to level ground, so mountains may receive more energy than lowlands. However, this may be countered by higher reflectivity (albedo) of snow-covered land. The albedo of clean snow on glaciers is 95 per cent, but that of dirty snow and ice may be only 20–40 per cent and the snow is thus more prone to melting (Gerrard, 1990). The absorptive capacity of the air reduces with altitude (see altitude effects, page 41); for example, the concentration of particles may be 25 000/cm$^3$ at 0–500 m, but only 80/cm$^3$ at 5000–6000 m. However, the thinner air means that in clear skies, the surface may receive 21 per cent more

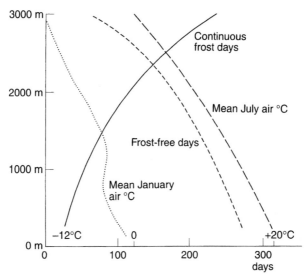

**Figure 2.4** The relationship between elevation, temperature and frost for the Western Alps (data from Geiger, 1965).

radiation at 3000 m than at 200 m. Aspect is also important in determining intensity; in the northern hemisphere south-facing slopes are warmer and drier as they receive more radiation than north-facing slopes; at 50° latitude, the difference may be a factor of four (Frank and Lee, 1966). Consequently, snow and treelines are higher, soils drier, and settlement and agriculture more abundant on south-facing slopes. East- and west-facing slopes also differ, although this is linked more closely to clouds (page 46).

The quality of radiation also changes: there is an increase in the ultraviolet (UV) component with altitude. UV light retards the growth of plants, which must adapt by developing thick cuticles and growing and reproducing in the spring when intensities are less. In addition, UV is linked with skin cancer in humans; in alpine environments there is 50 per cent more UV in summer and 120 per cent more in winter than at sea level, hence the need for protective action to be taken by people undertaking winter sports. Clouds act as a filter by reducing both intensity and UV.

## Temperature

Temperature declines with increasing altitude at a lapse rate of approximately 1–2°C/300 m (Figure 2.4). If temperature decreased linearly with altitude it would be 30–35°C cooler at 5000 m, but this simple relationship is complicated by a range of factors such as vegetation, cloudiness, aspect and winds. Temperature ranges may be seasonal or diurnal (see *latitude*, page 40). Inversions often occur in still conditions where cold air drains down slope and collects in the bottom of a valley, pushing warmer air up slope. This presents ecological difficulties for plants and so frost-sensitive crops and vegetation tend to be found on slopes above the inversion layer. A temperature inversion may also give rise to the concentration of pollutants in the base of the valley, which may be dissipated by thermal winds (Hanna and Strimantis, 1990).

## Clouds

Clouds reduce radiation and temperatures so if they occur on a regular basis they can affect growing conditions in mountains. As many plants are at their physiological limits, the impact of clouds can severely affect their chances of survival and influence patterns of agricultural activity (see Chapter 7). On some coastal mountains, clouds are an important source of moisture as wet maritime air moves up slope and condenses at dew point. At dew point (which lies at 1300–2400 m in the Sierra Madre Oriental on the east coast of Mexico, for example) dense and floristically rich cloud forest occurs which is totally dependent on this source of moisture.

Clouds may form and disperse on a daily cycle in many mountains, for instance in the Tyrol (Tucker, 1954), where small cumulus clouds develop on south-facing slopes and rise to form much larger masses in the afternoon over the main valley ridges. In Papua New Guinea stratoform clouds tend to form at 1800–2000 m in the afternoon from convective evaporation of the morning in a humid climate. These dissipate within an hour. Cumulus forms above 3500 m, bearing rain, but skies clear within 2–3 hours of the rainfall (Barry, 1992). Thus, there is an intravalley recycling of moisture. Fog stratus may form in still conditions in valley bottoms. In the European Alps, for example, this must be burnt off before the sun can warm the soil on east-facing slopes in the morning, thus delaying growth. In the summer, storm cumulus may shade the west-facing slopes, effectively curtailing optimum growing conditions there. In tropical climates, where humid conditions prevail, the formation of cumulus clouds each afternoon means that agricultural crops such as rice are limited in altitude to about 1500 m, whereas in the subtropical Himalayas, which do not have these daily cloud formations, rice may grow at 2500 m (Uhlig, 1969, 1978).

Orographic cap clouds are considered responsible for acid deposition in the Western Sudety Mountains on the Polish–Czech Republic border as they coincide altitudinally with the most severe defoliation of the forests at 800–1200 m (Dore et al., 1999). Defoliation has an uneven distribution, being at higher elevation on the windward side where cap clouds form 240 days per year (Sobik and Migala, 1993). Similar incidences occur in the Catskill Mountains of North America (Weathers et al., 1995).

## Relative humidity and evapotranspiration

Both relative humidity and evapotranspiration decrease with altitude; at 2000 m relative humidity is 50 per cent of the value at sea-level, less than 25 per cent at 5000 m, and less than 1 per cent at 8000 m. Thus, despite increases in precipitation with altitude at least up to dew point, and the presence of glaciers,

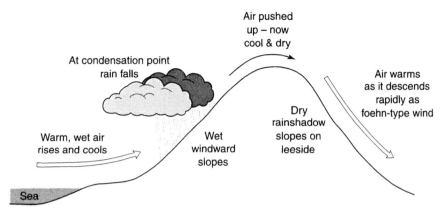

**Figure 2.5** The formation of orographic rainfall and lee-slope winds such as the foehn.

mountain summits tend to be arid. Winds add to the dehydrating effect, requiring special adaptations in mountain fauna and flora to survive.

## Precipitation

Precipitation, either as rain or snow, tends to increase with altitude – at least up to the level of condensation of clouds, which may be marked by a zone of maximum precipitation, above which it decreases rapidly. The main causes of precipitation are orographic, convectional and cyclonic. Orographic precipitation is formed from the forced ascent of air (Figure 2.5). Where the prevailing winds come from an ocean, the difference between windward and lee is dramatic; for example, westerly winds from the Pacific form rain over the coastal ranges of North America but on the lee side the winds are very dry, having an extra desiccating effect and contributing to the aridity of the Nevada Desert and Death Valley. The zone of maximum precipitation lies at 1800 m in Cameroon, West Africa, but 1500 m in East Africa and between 900 and 1600 m in the tropical Andes. Convectional rainfall is mostly associated with the tropics, where intense heating during the day in a humid climate causes thunderstorms in the afternoon. Cyclonic rainfall is associated with the convergence of air masses and the passage of fronts, and is most common in the mid latitudes.

The zone of maximum precipitation is very clearly delineated in some areas, as in the case of cloud forests, but diffuse in other areas. It occurs between 500 and 1000 m in the tropics, but up to 2000–3000 m where clouds are less extensive. Precipitation generally increases up to 3000 or 3500 m in the mid latitudes, where trade winds increase wind speed and uplift compared with the higher and lower latitudes. It may be highest at the crest which is partly a function of the lower altitudes of mid-latitude mountain ranges.

The snowline marks the limit of permanent snow, above which periglacial and frost conditions prevail. A seasonal snowline marks the level of winter snow in mid- and high-latitude mountains. Snowlines generally correspond with the 0°C isotherm. It may be as low as sea level in polar regions, 1700 m in

Scandinavia, 3300 m in Central Europe, 4500 m in moist Equatorial mountains and 6500 m in the dry Tibetan Plateau and southern Andes. The Tibetan Plateau has a significant mountain mass effect which creates warmer and more arid conditions, which is also the case for parts of the central Andes in the Atacama desert. The aridity of this area arises from the convergence of anti-cyclonic air masses and the cold Humboldt Current, which reduces evaporation and is enhanced by the barrier/rain shadow effect. No glaciation or persistent periglacial activity occurs above 6500 m, and vegetation is minimal as soils are too sparse (Messerli et al., 1993). There is an increase in the proportion of pre-cipitation falling as snow with both altitude and latitude. Its persistence is dependent on accumulation (depth) and rate of freezing and melting, which is determined by aspect in particular, but also by vegetation, and turbulence (Clark et al., 1985). It also varies, like precipitation, with the direction of the prevailing winds, which determine the amount of available moisture; in west British Colombia, Canada, snowlines are around 1600 m, but 2900 m inland and 3100 m in the eastern Canadian Rockies. Snow has a high albedo which slows melting and retards spring growth. However, it also has an insulating effect, protecting vegetation, and even animals from extremely cold air tem-peratures; the trans-mission of heat is 120 mm in dry snow, and only 40 mm in wet snow (Gerrard, 1990). Other forms of precipitation include fog (see page 71 Box 3.2 and page 249 Box 11.2), rime and other frozen forms.

## Winds

Finally, winds play an important part in defining mountain valley climates. Apart from the large-scale atmospheric wind circulation patterns considered above, and the regional winds created by mountain barrier effects, there are local valley-scale wind circulation patterns. These are mostly thermal winds generated by differential heating of the adjacent plains or sea relative to the mountains.

There are three main types. First, mountain–valley winds which comprise anabatic and katabatic winds. Anabatic winds blow up valley during the day due to thermal heating of adjacent lowlands, particularly in the afternoon, and may be moist, creating clouds later in the day. At night a down valley drainage of cold air – a katabatic wind – occurs, which may create a temperature inver-sion in calmer conditions, as the cold air collects in the valley bottom. Second, there are slope winds which rise up slope during the day from heating of the valley floor and return as a cold air drainage down slope at night. Both the up valley and up slope winds have an anti-wind circulation, creating circulation cells. Such patterns determine the formation and dissipation of clouds (Figure 2.6). Pollutants may be carried up valley by anabatic winds from urban and industrial centres in the lower valley or adjacent plains. They can accumulate in areas with less free circulation, such as tributary valleys or behind topographic obstacles, which causes problems of air quality (Warner-Merl, 1999). Finally, in glaciated valleys there is a constant very chill glacier wind which drains down the valley. It is most noticeable in still, calm conditions and can have severe ecological effects, stunting growth and perpetuating frost.

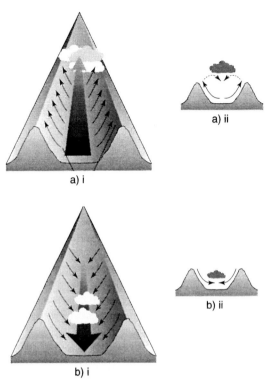

a) ii

a) i

b) ii

b) i

**Figure 2.6**   Mountain–valley and slope winds: a)i shows daytime up-valley and up-slope winds generated by thermal heating of plains and valley floor. Cumulus clouds are forming at the head of the valley where the uplifted air condenses. This may produce local afternoon rain and has a shading effect. a)ii shows a cross section of the slope wind circulation cell. b)i shows night time cool air drainage. Basal temperature inversions and low-level stratus may occur in still conditions (b)ii. The stratus must burn off in the morning before the sun can warm the ground.

## Climatic change

Mountains are important indicators of climate change, due to their sensitivity; small changes can have significant effects on geomorphology and ecology. Mountain environments are also very vulnerable to change, given the rapid response of geomorphological processes to climatic events. Many species exist at their ecological limits and relatively small changes in conditions can cause extinction. Change is perhaps the norm for mountain climates, but there have been periods of greater latitudinal and altitudinal shifts in climatic conditions in the past. The main period of change was the Quaternary ice age during which there was a global expansion and stagnation of glaciers in cycles of about 100 000 years, followed by the Holocene warming phase of the last 10 000 years. On shorter, century timescales, the Little Ice Age of the fourteenth to nineteenth centuries was also a global phenomenon, as is the current concern with the enhanced carbon dioxide levels and its effect on climate and ecosystems. This section considers the process and impacts of past changes, some of the problems of prediction of future climate in mountains, and an outline of the impacts.

## Box 2.1   East African mountain climates

The climate of East Africa is dominated by three monsoon winds – a dry northeasterly monsoon in the northern hemisphere winter, a shallow south-easterly summer monsoon and a wet monsoon from the Equatorial Atlantic. Seasonal rainfall correlates with these seasonal wind patterns and with shifts in the ITCZ (Inter-Tropical Convergence Zone), which is a low-pressure zone created from the convergence of the northeasterly and southeasterly trades, creating convectional rainfall.

Local relief controls are important in controlling cloud and hence precipitation. The lowlands around Kilimanjaro receive 500 mm/yr, northeast of the mountain 750 mm, but 1500–2000 mm to the south and southwest. The precipitation of the Kenyan Highlands, including Mounts Kenya and Elgon, varies between 700 and 1300 mm. The altitude of maximum rainfall occurs at 2500–3000 m on southeast Kenya with 2500 mm/yr. Most precipitation falling above 4500 m (700–800 mm) is snow and hail. Precipitation on Mount Kenya is highest on the western and southern aspects, reflecting the origins of the monsoon-bearing winds. The foothills to the west of the Ruwenzori lie in the rain shadow with less than 750 mm precipitation; but east of the mountain where the prevailing wind occurs, this increases to 2500 mm. The zone of maximum precipitation is lower in the west at 1000–1400 m, but higher in the upper western slopes where snow falls equivalent to 2500 mm water. The height of the 0°C isotherm is 4750 m in East Africa.

The annual temperature range decreases from the subtropics to the Equator. On Mount Kenya the annual range is 2°C and daily range 10–20°C. The annual range in the Ruwenzori is also modified by greater cloud cover. There is radiation asymmetry from north to south and also from east to west from diurnal circulations, convection clouds and rainfall maxima in the afternoon. The west-facing slopes thus receive less sun than the east-facing. Clouds are important in maintaining tropical glaciers as ablation is primarily radiation-controlled.

*Sources*: Hastenrath, 1981; Osmaston, 1989a, b; Rosqvist, 1990; Winiger, 1981.

## Quaternary mountain climates

The reconstruction of past environments is based on the mapping of geomorphological features such as moraines, which enable the extent of glaciers and position of equilibrium line altitudes (ELAs) to be estimated. The ELA is the altitude at which accumulation of snow on a glacier is equal to ablation (melting) and can be used to estimate the extent and volume of former and present ice masses. In addition, lake and ocean sediment and ice cores provide information on vegetation changes, atmospheric composition and sedimentary environments. These data are used as proxies to aid the simulation of future climatic changes. In reconstructing the Quaternary, it is often the last, or most extensive glacial features which survive; each passage of an advancing glacier destroys the evidence of previous advances in its path.

Messerli and Winiger (1992) reviewed the patterns of change in African mountains 18 000 years ago during the last glacial, and 8000 years ago around the Holocene thermal optimum. Some 18 000 years ago, altitudinal belts were depressed due to the expansion of ice and snow on summits, whereas in the Holocene, conditions were warmer and wetter than at present. In the Atlas Mountains, there are no glaciers today, but there is evidence of terminal moraines at 2600–2700 m and of periglacial activity at 2000 m. Dresch (1941) estimated that the snowline was 800 m lower than today, corresponding to a reduction in temperatures of 4°C. In the Tibesti Mountains, frost weathering occurred below 1800 m, with an estimated reduction in winter temperatures of 10°C; the interior of the continent thus experienced greater fluxes in temperature regimes. Summers were drier and 6–8°C cooler. In the Ethiopian Highlands, there was an expansion of ice caps down to 3750 m in the Simen Mountains and 3100–3200 m in Bale (Hurni, 1989). In East Africa, Mount Kenya has terminal moraines at 3100 m and Kilimanjaro at 3300–3600 m (Messerli and Winiger, 1980). During this time the southern monsoon was much weaker, creating arid conditions during the coldest period. The southward shift of the polar front resulted in drier conditions in North Africa (Rognon, 1987) and an increase in seasonal and altitudinal differences in the Saharan mountains (Messerli and Winiger, 1980). Around 8000 years ago, the climate was much wetter, with the mountains receiving more precipitation. In the Middle Atlas of Morocco, evergreen oaks were supported (Lamb *et al.*, 1989). During this time the ITCZ was much further north, bringing humid tropical air as far as the Atlas Mountains and the Mediterranean. During the winter, the polar front brought precipitation to the northern Sahara region. There were weaker differences between seasons, and also between highland and lowland climates.

Similar patterns of change occur in other mountain regions. In the Atacaman Andes, which are presently extremely arid, conditions in the last cold maximum around 18 000 years ago were 7°C cooler than at present. Between 17 000 and 11 000 years ago, lake levels were 5–10 m higher, indicating higher precipitation, which continued into the early Holocene (11 000 – 7000 years ago) with summer temperatures 3.5°C warmer than at present. However, from 3000 years ago the climate has become increasingly arid, and at present the groundwater recharge is minimal.

## The Little Ice Age

The Little Ice Age (LIA) was a period of global re-advance of glaciers and cooler conditions. It occurred broadly between the thirteenth and nineteenth centuries and was preceded by the warm period known as the Medieval maximum (AD 900–c.1200) (Grove, 1996). The cooling at the beginning of the LIA moved westwards between 1200 and 1500, and warming moved eastwards before the LIA between 800 and 1000 and afterwards between 1700 and 1900. There is considerable detailed information from the European Alps, as the palaeoecological and geomorphological sources of data are augmented by documentary evidence of changing environmental conditions. Detailed studies of this period in this region have been made (e.g. Ladurie, 1972; Lamb, 1988; Grove, 1988). During the Medieval maximum period there is no mention of sea ice around Iceland and Greenland and the distribution of Norse settlements in Scandinavia lay in areas later overrun by ice and which, in many cases, have not been inhabited since. On the Great Plains of North America conditions were moister and more favourable to human use from 700 until 1200, when the climate became increasingly arid. This is attributed to the increasing prevalence of westerly winds, which brought moist air to the Rocky Mountains, leaving the area inland in the rain shadow.

In the European Alps, documentary evidence of glacier movements comes from patterns of trade between valleys – the Wallis/Valais canton was settled by German immigrants who maintained contact with their mother region over the mountain passes until they became inaccessible. Some passes which had been open during Roman times fell into disuse. Lakes were blocked by advancing glaciers and outbursts became frequent; twelve outburst affected the Saas valley alone between 1589 and 1772. There are episodes of retreat during this period, for example, just before and after 1500 (Patzelt, 1974). In the Upper Valais, water supply ducts which had been constructed to carry glacier melt fell into disrepair around 1300 as the climate became wetter and in some areas they were overrun by ice. However, on the southern side of the Alps, in Italy, they were still under construction and in use up until 1450. Gold mines in the Hohe Tauern of Austria were abandoned in AD 1300. In central Europe, about one-third of cultivated land was lost to avalanches and ice, which is reflected in records of tax changes between 1630 and 1640 (Lamb, 1988).

In the twentieth century there was a period of warming from 1900 to 1950 of 2°C in summer temperatures, marked by a general retreat of alpine glaciers from 1900 to 1960. An advance in the 1980s is attributed to higher precipitation. In Sonnblick in the Austrian Alps, snow cover lasted an average of 82 days in the period 1910–1925 but only 53 days from 1958 to 1970, with a 0.5°C rise in summer temperatures. Such changes reflect the importance of the few long climatic records which are available. The paucity of such records is a major obstacle to the accurate understanding and prediction of change processes and patterns.

## Predicting future change

There is an enormous literature dedicated to the analysis of models simulating physical processes of climatic change based on a doubling of $CO_2$ and changes to other parameters (see for example the IPCC studies, Houghton et al., 1996). The main instrument is the General Circulation Models, but these have been criticized because of their inaccuracy (i.e. the great variation in predictions between different models), and their inability to cope with and account for complex and often unknown feedback mechanisms within the climate system (Beniston, 1994). This arises in part from the paucity of data on which they are based, hence the preponderance of studies in the Alps and North America, where records are the longest and most abundant. Finally, the models have relatively poor resolution and the effect of topographic smoothing reduces the simulated impact of high mountain systems on global circulation, leading to underestimates or no estimates at all on suitable spatial and temporal scales for prediction of complex mountain conditions. It has been questioned whether these models are the best tool and represent 'good science', particularly in view of the fact that they are not actually very helpful for policy making (Shackley *et al.*, 1998).

Improvements in resolution, in the coupling of atmospheric and biospheric models, in downscaling of GCMs to regional level and upscaling and nesting of local and regional models into GCMs have all enabled more accurate scenarios to be estimated. However, these are scenarios, not predictions, and the complexity of global and local climates, particularly in varied topographic regions such as mountains, means that only a general idea can be proposed of the potential changes, and their potential impacts.

Predictions are also hampered by the fact that present-day analogues of past conditions do not always exist, so past conditions cannot always be a good guide to the future. Any given process of change can produce different environmental responses in time and space, and the interdependence between climate, geomorphology, soils and vegetation introduces lag times in response, which control or modify that response. For example, vegetation change is dependent on soils as well as climate, and in reconstructing the past or predicting the future it is not always possible to determine which may be the key controlling factor. For this reason, the use of forest ecotones (the transition zones between major vegetation types) as an indicator of climate change has been criticized as the lagged response of forest due to soil conditions does not realistically reflect the true rates of climate change (Kupfer and Cairns, 1996).

The traditional scientific modelling approach is based on physical processes and there is great difficulty in assessing the socio-economic impacts of changes on human activity. This arises partly from the fact that human behaviour is constantly adapting to change in various forms so there is no clear starting point (Berkhout and Hertin, 2000). Second, consideration of the impact of physical environmental changes includes some assessment of risk and strategy, perceptions of which change according to economic, social and political circumstances. Thus, where efforts have been made to integrate social and environmental factors, the trend is towards a less prescriptive and more prospective approach; i.e. a scenarios-based approach as distinct from one based on descriptive modelling (Godet, 1997). These approaches, however, need to be linked in order to achieve a balance between scientific (quantitative) rigour and (qualitative) flexibility (Fontela, 2000).

## Potential impacts of future change

The main response to climatic change in mountains, as indicated by proxy climate indicators of past changes, is a shift in altitudinal zones of geomorphic processes and of vegetation communities. The most recent estimates of predicted global climate change is of a warming of 1.5–5.8°C for the period 1990–2001 (IPCC, 2001). This is a considerable increase in the 1995 estimate of 1–3°C. The increase in prediction is closely linked to the increase in global economic activity. The feedback effects of dust, $SO_2$ and cloudiness in counteracting this are not clearly understood, but it is thought that the cleaning of industrial processes and more active enforcement of pollution regulations have reduced the number of particles emitted and thus the role of this feedback. This effect is enhanced by the lack of significant volcanic eruptions, an important source of dust, in recent years. The current projections indicate greater warming on the land than the sea, with a maximum in high-latitude winter (Kattenberg et al., 1996). In addition, there is expected to be a reduction in diurnal ranges and increases in winter precipitation at higher latitudes and in the Asian monsoon.

On a local scale, however, the complexity of mountain environments means that this general picture may be considerably modified. For example, spatial variations similar to those found in the Canadian Rockies (see Box 2.2) have been determined for the Alps with respect to their effect on forest growth (Fischlin and Gyalistras, 1997). In the subalpine region near Berne, Switzerland (northern Alps), little impact on the forest is expected from increases in winter precipitation and summer warming. At St Gotthard, in the transition zone between the northern and southern Alps, there is currently abundant precipitation and although optimal forest conditions are expected from a doubting of $CO_2$, the limiting factor is soil development which lags behind. At Sion in the more arid central Alps, the climate currently supports forest but under

---

### Box 2.2   Climate change and erosion in western Canada

Five mountain ranges with different hydrological regimes demonstrate different responses to climate change in the past and the present. As many areas have been set aside as national parks, the influence of human activity in masking the evidence of past and present environmental change is much reduced.

Glacial fluctuations occurred throughout the Holocene. Mature forest grew up valley of the present Athabasca and Dome glacier snouts during the time of the Holocene thermal optimum, between 7500–8300 BP and 6000–6500 BP, respectively. Advances are recorded around 4000 BP, 2800–3000 BP and 1500–1700 BP. The LIA advances began around AD 1150, dated by dendrochronology of overridden trees. The maximum advances occurred around AD 1700, with a period of particularly rapid retreat during the last 100 years.

(continued)

*(continued)*

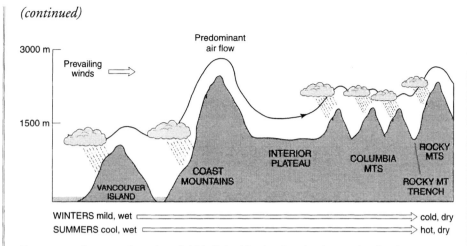

**Figure 2.7** Transect of southern British Columbia showing the changes in climatic conditions with distance from the coast. The main source of water is the Pacific Ocean and, to a lesser extent, the Gulf of Georgia. Orographic precipitation is reduced and occurs at successively higher altitudes with each range westwards (adapted from Chilton, 1981).

In recent times it is even more difficult to assess the effects of climate change due to human activities modifying the landscape. Runoff and avalanches respond rapidly to changes in precipitation but vegetation, glaciers and the sedimentary system show a lagged response. This creates a complicated picture of current conditions which contributes to uncertainties in future predictions of increased precipitation and temperatures by 2050.

Following a transect across the Rockies (Figure 2.7), the west coast and Vancouver Island currently have little periglacial activity and no glaciers. Tundra exists above 4000 m where rainfall is highest, reflecting the maritime influence. Little impact of increased rainfall and temperature is expected in the future. Inland, in the St Elias and coastal mountains, future conditions may increase the already active surging glacier activity and increase the rates of chemical weathering. Warming of the northeast slopes may raise the treelines and permafrost zones due to drying of the soil. In the subarctic dry interior, permafrost occurs near the timberline, which is higher (900 m) than on the coast (300 m). Recent glacial recession has occurred and periglaciation occurs despite the aridity, due to high annual and diurnal ranges. It is here that the greatest impacts are expected, with disruption to pipelines, communications, etc. due to thermokarst development. In the temperate interior alpine region, timberlines are highest and periglacial activity has its lowest limit. Heavy snowfall maintains icecaps in the Colorado mountains and glacial expansion will occur given a rise in precipitation though a reduction in periglaciation may occur due to warmer temperatures.

*Sources*: Slaymaker, 1990; Luckman, 1994; Osborn and Luckman, 1988.

drier conditions will not. In the central/southern Alps at Bever, the predicted response is dramatic, with a completely new forest community; summer warming is greater than winter, with an increase in winter precipitation but no summer decrease.

Any change in climate will have an effect on geomorphological processes and hydrology, causing long-term shifts in the spatial distribution and relative importance of different processes – such as a reduction in frost action and increase in runoff and overland flow. The magnitude and frequency of extreme events, such as major floods and avalanches, are expected to increase and it is this aspect which is likely to have the greatest impact on human activity. Thus, agricultural potential, tourism, energy production and use, water resources, transport and health are all likely to be affected, both in terms of year-on-year changes but also in the increase in hazardous events. The effects vary with altitude and location; the 1/100 year flood in western California arising from winter cyclonic rainfall is three–six times the height of annual floods, but in the alpine Rockies of Colorado, where snow melt dominates the hydrology, the 1/100 year event is only two-three times the magnitude of annual flooding. In the dry foothills, the 1/100 year event from spring or summer thunder storms may be ten times the annual magnitude, which makes it the location of the greatest potential hazard.

Flood frequency generated from climatic change may, however, be tolerated due to economic pressures, making even frequently affected land still viable for use despite increasing losses (see Box 9.7, page 210). The 'Vivian' storm in the European Alps, 1990, or the intense precipitation events of 1987 in the Swiss Alps (Beniston *et al.*, 1995), or even the French and Austrian Alps' avalanches of 1999 represent extreme events which may become more frequent. So too may reduced snowfall occurrences; the European Alps has experienced later first snows and reduced duration, and the lowlands have had between three and four times less snow than previously (Baumgartner and Apfl, 1994; Föhn, 1991).

A warming of mountain climates may cause a reduction in glaciated area, in the extent of permafrost and the depth and duration of snow cover. These will substantially affect the stability of slopes and also agriculture, winter tourism, construction and communications systems. Recent work (presented by Harris at the Institute of British Geographers Conference, Cardiff, 2001) indicates that a melting of the permafrost in the European Alps will have substantial effects on slope stability, which would affect communications networks as well as the winter sports economy. Deep boreholes already indicate warming and even one or two degrees' warming can bring about a thaw and widespread collapse of slopes. In the Victorian Alps of Australia a 1°C increase in temperature would cause a reduction of 50 per cent in snow cover duration, and a 3°C rise would eliminate snow below 1800–2000 m (Whetton *et al.*, 1996). The effect of El Niño Southern Oscillation over the southeast Australian Alps is also likely to reduce snowfall and speed up melting, thus affecting the ski season substantially (Witmer *et al.*, 1986). The movement of glaciers and their relationship to ice-dammed lakes and their catastrophic emptying is well known in mountain regions in the recent past and it is this which causes concern for the future in such areas: the Imja glacier in the Khumbu Himalayas has created a lake 0.7 km² in area and over 100 m deep, which could have potentially catastrophic effects in the event of an outburst. Likewise in the St Elias Mountains of northwest Canada, glacial Lake Melbern formed between 1979 and 1987 and has an area of 12 km². In addition to these changes, climatic change will also affect vegetation, forest fires, nutrient recycling and hence the role of vegetation in slope stability (see Chapter 3).

## Key points

- Mountain climates are characterized by complexity and uncertainty. We have an incomplete understanding of the detailed interactions of different elements within valley systems.
- Mountain climates are shaped by global factors: latitude, altitude, continentality and topographic effects.
- On a more local scale, patterns of temperature and precipitation are determined primarily by insolation, aspect, elevation and geology.
- There is a general reduction with altitude in air density, particles and water vapour, and an increase in precipitation up to condensation level.
- The legacy of past climate change, such as the Quaternary glaciations, has contributed to the dynamism of mountain environments. Past climates also help in the reconstruction of the potential impacts of future climatic change.
- The response of the environment to future warming is poorly defined on a valley scale. Changes in precipitation and temperature are likely to affect slope stability and human activities, although the degree of this effect is highly variable.

# Chapter 3

# Mountain geoecology

The concept of mountain geoecology was first established by Troll in his work on the Himalayas and Andes (Troll, 1971). The term refers to the ecological environment, including soils, topography, vegetation and climate and the way that these interact to produce characteristic ecological patterns (Troll, 1972). Much of this early work emphasized the zonation of vegetation with altitude and although this is simplistic (Figure 3.1) it holds broadly true in many mountain regions. Zones might be discrete, being clearly demarcated, as in the case of the cloud forest which is constrained by available moisture, or diffuse, grading from one vegetation community to another almost imperceptibly.

In many areas, however, the pattern of vegetation assemblages resembles a mosaic as species closely reflect underlying soil conditions and microclimate. The influence of frost hollows, glacier winds and temperature inversions, for example, is reflected in the distribution of cold-sensitive plants which do not thrive in areas affected by these conditions. The zones comprise mainly forests and grasslands of different types, which contain a high species diversity and characteristic forms which have developed in order for plants to survive the conditions on mountain slopes and summits. Many species of mountain flora and fauna are endemic and only occur in restricted locations and associated with particular environmental conditions. Animals have also adapted to survive the extremes of temperature and exposure of mountain environments, as have humans (see Chapter 4).

A large proportion of mountain fauna and flora has originated from the adaptation and mutation of lowland species which have migrated up slope as a result of past climatic changes. For example, the expansion of savanna in central Africa during the Quaternary would have pushed moisture-demanding species up slope, where orographic precipitation could supply the moisture they needed. At the same time, the expansion of arctic conditions in the northern hemisphere enabled arctic flora to reach parts of the Alps, where it has since remained and is now, effectively, cut off from its original source, just as an island is isolated by rising water levels. Elsewhere, species have adapted to cope with drier, wetter or colder conditions; they have remained in their locations and may be the only surviving remnants of ancient assemblages, or unique forms of lowland species. In northern England the 'Teesdale rarities' in the Pennines are relicts of ice age floras, and others exist in the Andes and parts of the Himalayas. The consequence of this adaptation is that mountains

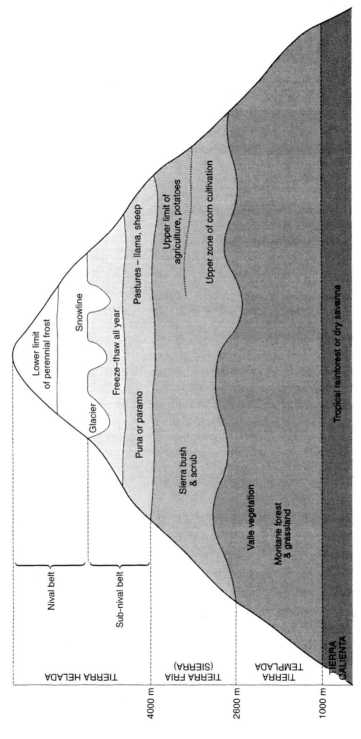

**Figure 3.1** Geoecological zones of the High Andes showing vegetation and land-use belts (adapted from Troll, 1968). See also Figure 3.3.

are characterized by a very high degree of endemism – the occurrence of species or forms unique to those areas. This makes them important repositories of biodiversity and a focus for international concern for their protection.

The simplified natural altitudinal succession from forest to grassland to snow shown in Figure 3.1 has been greatly modified by human activity, especially on mountain slopes which are or have been relatively intensively cultivated by humans. This modification masks the natural response to changing environmental conditions The response of vegetation zones to changing climate is to shift up or down slope or to expand or contract. Such movements in the past have been used for predictions of future ecological conditions. The shift in vegetation zones has implications for conservation of biodiversity under future changing climatic conditions.

This chapter will consider the main characteristics of grasslands and forests and mountain fauna before assessing the potential impacts of climatic change and the effect this might have on conservation of biodiversity.

## Soils

As with so many aspects of the physical environment, soils vary considerably in texture and character with altitude, latitude, parent material and topographic factors such as aspect. Soils are also closely related to vegetation cover, and this interdependence gives rise to a mosaic reflecting micro-environmental conditions. On high summits soils tend to be thin, poorly developed and stony. They are affected by desiccation and aeolian erosion. They accumulate best on gentler and less exposed slopes but on impermeable substrates they tend to become waterlogged. Soils and vegetation form self-perpetuating cycles of development; alkaline soils form on carbonate rocks and support different vegetation – broadleaved evergreen or deciduous species, whereas acid soils developing on igneous rocks will support heathers and conifers, which produce acidic humus which maintains the acidity of the environment.

In high alpine environments the changes in air composition and the low temperatures slow the rates of microbial activity and hence the recycling of nutrients. In forests, nutrients tend to be stored in the plants rather than the soils. Conifer needles and other thick-cuticled leaves accumulate and break down only very slowly. Whilst this limits the nutrient availability to the plant, it also insulates the surface of the soil and protects the bases of tree trunks. Generally, with increasing altitude there is an increase in organic matter, due to slower decomposition rates, and also nitrogen, carbon and phosphorus, indicating an increase in acidity (Buol *et al.*, 1973). At the same time there is a corresponding decrease in alkaline minerals such as calcium, magnesium and potassium. Freeze–thaw processes producing surface polygons and stripes tend to sift the surface materials, bringing the coarser material to the top. This pattern is accentuated by the wind removing smaller particles, giving soils little chance of developing (Mellor, 1979).

In tropical regions where temperatures are warmer and seasons negligible, there is generally very little organic matter accumulation, as recycling of nutrients can occur rapidly. However, where rainfall is high, soils can quickly become leached and again most of the nutrients are stored in the plant material. This is why the fertility of these soils is short-lived and thus shifting cultivation needs to move location after only a few years' use. The Karen people in northern Thailand, for example, use newly cleared swiddens for one year of rice cultivation followed by six to ten years' recovery (Trakarnsuphakorn, 1997). The organic matter component increases with altitude in tropical mountains because as temperatures fall so does the rate of recycling. On granitic rocks, deep weathering can occur, producing abundant, if acidic, 'B' horizons, but such soils can have a high clay content and are thus susceptible to waterlogging.

In middle and higher latitudes, brown forest soils tend to form, especially on more calcareous rocks such as those of the southern Alps, with acidic podsols on granitic rocks. Most mountain soils are lithosols, being stony, well drained, infertile and often acidic (Retzer, 1974). Soils developing on volcanic ash deposits, such as those in the southern Andes and north California can be very fertile but also are rather dry. Soil evolution is linked with time of development and the older soils are influenced more by age than by substrate and other factors. In particular, it has been noted that older – and hence better developed – soils are more common on stable slopes as they are less often disturbed by slope processes (Buol et al., 1973). Aspect is particularly important; in the northern hemisphere soils tend to be moister but cooler on north-facing slopes and hence have less microbial activity and are more susceptible to leaching.

The rates of soil formation vary widely. In the tropics, the rate of chemical weathering is relatively rapid, but in temperate latitudes, much depends on the parent material – sedimentary rocks break down much more rapidly; on rates of weathering and on slope stability. Arid mountain regions have limited soil development and often a residual stony layer on the surface which serves to armour the soil, reducing erosion and preserving moisture. Contemporary land uses have done much to modify natural soils, both by improvement but also by changing the vegetation (e.g. deforestation) and enhancing natural erosion rates (see Chapters 7 and 10).

## Alpine grasslands and meadows

The high summits and upper slopes of mountains, particularly those areas in middle and high latitudes which are seasonally snow-covered, are dominated by grasses and herbs. At lower altitudes, these may grade into open woodland, or the boundary between grassland and forest may be sharply delineated (see *Timberline*, page 71). At their upper limits, plants become smaller and sparser with altitude as soils become thinner and less continuous and the meadows grade into arctic tundra dominated by mosses and lichens on rocky outcrops.

The growth forms of tundra, meadow and grassland species of mountain regions are all adapted to cope with specific mountain conditions (Troll,

a) *Silene acaulis*                                    b) *Sedum stenopetalum*

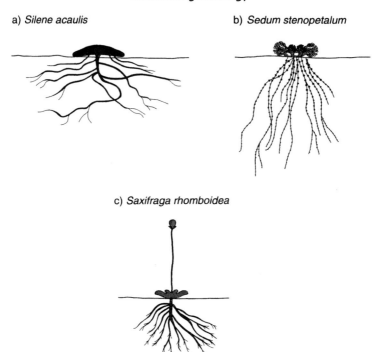

c) *Saxifraga rhomboidea*

**Figure 3.2** Growth forms of alpines. Note the substantial extent of the root area compared with the surface vegetation. The roots in b) have nodules to store nutrients. As the surface plant spreads laterally it puts out new roots (as in a) and b)), a form of vegetative reproduction. The mat in a) traps moisture and insulates the ground surface. The flower stalk in c) rises high above the plant, with pink and white flowers to attract insects.

1972; Rundel, 1994; Körner, 1994) (Figure 3.2). Plants tend to have low growth forms to minimize wind damage and uprooting, and are often shaped by the prevailing winds. Grasses form bunches and tussocks such as Ichu grass (*Festuca* spp.) and herbs may form cushions (e.g. *Azorella*) or have a low rosette form (*Lobelia*) from which a flower stalk appears in summer. These physical forms serve a number of purposes: the compact growth reduces desiccation and retains moisture in the centre of clumps and tussocks on which not only the plant but also insects survive. The clumps also insulate sensitive flowering stalks from extreme cold such as late frosts when growth has begun in the seasonal latitude spring. In Equatorial mountains, rosette forms may close up at night time to protect young buds from night frosts. Leaves often have thick cuticles which protect them from desiccation and from surface damage. Woolly leaves are insulated by holding a layer of warmer air close to the surface of the leaf. Thick cuticles are often associated with fewer stomata (the breathing holes of the leaf), so the undersides of the leaves which are protected from the wind often have more stomata to compensate (Körner and Cochrane, 1983). Foliage tends to be dense, and often leaves are small, which increases the surface area available to photosynthesize.

During the growing season in mid and high latitudes, plants must not only awake from dormancy and reproduce but also lay up stores of food. The thin soils, high winds and aridity of many high summit grasslands mean that plants develop extensive root systems to anchor themselves and store water (Körner and Renhardt, 1987). Sometimes these extend to over six times the surface area of the plant above the ground. Roots may have nodules or thickened storage areas which sustain the plant during periods of short daylight in spring and autumn. With decreasing altitude, as alpine growth forms give way to shrubs and forests, the carbon balance changes from being stored predominantly below ground into storage predominantly above ground in stems and leaves (Körner, 1993).

Reproduction tends to be by offshoots from the parent plant, which can be supported by a shared root system until they are big enough to survive on their own. Reproduction by seeds is rare as dispersal is too risky – the wind may carry seeds far from suitable sites and in high alpines areas there is too little continuous coverage of soil to ensure many seeds reach suitable landing grounds. In addition, seeds take too long to mature and they are not protected and supported by parent plant root systems. Many alpines, however, do produce spectacular flowers in order to attract insects. The flowers may arrive too soon for the right insects to be present; bees especially require warmer conditions than many springtime alpine summits can offer. In addition, to make a seed, male and female flowers need to cross-pollinate. As plants tend to be widely scattered in the alpine regions this becomes less likely.

At lower altitudes where more continuous grasslands occur, there tends to be a more extensive and deeper soil coverage, affording nutrients and moisture as well as anchorage. If vegetation cover is continuous it offers excellent protection from erosion of the soil underneath. These turfs are often able to support grazing, and remain tough and stable unless cut or damaged (Plate 3.1). The productivity of alpine meadows and grasslands is relatively high, given that in mid- and high latitudes all activity is compressed seasonally and in view of the other environmental constraints of low nutrient status, low temperatures and water stress (Körner, 1994). In many areas, grasslands are maintained by grazing and, following abandonment, they may revert to scrubby woodland. Where the timberline is currently artificially maintained at lower altitudes by human activity, forest may regenerate. For example, in the Mediterranean the forest is largely degraded into a maquis, or garrigue, as a result of clearance, but reduction in pastoralism has meant that meadows are no longer maintained and the scrub reverts to forest where feral grazing permits.

In Equatorial regions, growing conditions are maintained all year, and so plants do not have to adapt to seasonal temperature cycles, but need to cope with high diurnal variations in temperatures at high altitude. Photosynthetic cycles may be enhanced by cold nights during which the sugars produced during the day are stored, giving these plants high productivities. This process accounts for the unusual sweetness of fruits in mountain regions, such as the

**Plate 3.1**  High alpine pastures in the Spanish Pyrenees. On the left is an enclosed area of grazing where young cattle are pastured. The faint line of the fence can be traced up the slope. On the right-hand side is much rougher grazing which is part of an area of open access.

apricots of Hunza (Whiteman, 1988). The diversity of species and the form of underground tubers as storage of nutrients has resulted in an abundance of plants useful to humans as staple crops: maize, potatoes, *Chenopodium quinoa*, etc. The high nutrient value of such plants can sustain quite dense human populations, once the art of cultivation or storage is achieved. In the high Andes, tubers are traditionally freeze-dried, using diurnal temperature fluctuations to produce a preserved potato meal which can be stored and eaten later in the year.

The diversity of alpine flora arises from speciation from lowland forms migrating upslope and adapting to new conditions. In the northern hemisphere high latitudes, there is an abundance of arctic species which declines with distance from the poles, representing the limits of migration of these species during previous cold climatic phases; in Scandinavia some 63 per cent of species are of circumpolar origin, but this falls to 35 per cent in the Swiss Alps. In the southern hemisphere, different species predominate (see *Forests*, page 66) as the origins of the flora tended to lie in local speciation rather than long-distance migration. The East African peaks contain up to 80 per cent endemics at high altitude, reflecting their isolation.

## Box 3.1 Andean paramo and puna

In the Andes, two distinct grasslands occur; the dry puna and the moist paramo. In the Ecuador Andes paramo grasslands occur between the permanent snowline at 4500 m and the upper limits of agriculture at 3500 m. The paramo is wet all year round (800–2000 mm/yr) with mean annual temperatures of 1°C as this region has an aseasonal climate, and is used for grazing. It is dominated by tussock grasses such as *Andropogon*, *Calamagrostis* and *Stipa*. Humid grassland also exists above the cloud forest zone but its distribution is limited to areas influenced by the trade winds. As environmental conditions become drier and more seasonal southwards along the Andean chain (Figure 3.3), the wet paramo grades into moist puna (precipitation 500–900 mm/yr) and then dry or salt puna (<400 mm/yr), becoming a thorny desert scrub in the region of the Atacama desert in Chile, comprising thorny shrubs, succulents and sclerophyllous grasses (woody plants with small, leathery, evergreen leaves), including the unique *Polylepsis* trees. Hot dry regions also occur in arid basins inland of the wet mountains in Mexico and Bolivia where strong diurnal winds blow through the gorges, enhancing the aridity, and thus hot dry puna species replace the wetter varieties. The high proportion of endemics in Andean flora makes them important in global biodiversity.

*Sources*: Troll, 1968; Klötzli, 1997.

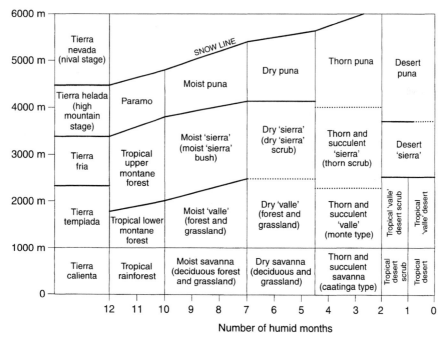

**Figure 3.3** Horizontal and vertical arrangement of geoecological types in the tropical Andes (from Troll, 1968). See also Figure 3.1.

## Mountain forests

Forests are an important and ubiquitous part of mountain landscapes, phys-ically, economically and aesthetically. Physically, they contribute to the stabilization of slopes and reduce the effects of avalanches from grasslands at higher altitudes. They also help to regulate river flow as they store water. How-ever, all these points are hotly debated in the context of discussions over the issue of deforestation (see Chapter 10). Economically, forests provide a wide range of materials for food, fodder, construction, etc., both directly from trees and from other plants which survive in forest ecosystems (see Chapters 7 and 9). Forest use occurs both as a part of traditional agro-economies where the supply of fodder provides a critical link between livestock and crop production, and as a large-scale commercial exercise in the form of logging (e.g. of tropical hardwoods) or plantation agriculture specializing in particular species (e.g. tea, coffee, pinewood), as, for example, in Southeast Asia. In between are a variety of economic levels of exploitation, such as the production of cork from *Quercus suber*, the cork oak, in the Mediterranean, and the production of fruit and nuts from trees grown on individual plots as in Nepal, Pakistan and Morocco.

Finally, forests are an important ecological resource, being, in their natural state, floristically rich and diverse. They contain many medicinal plants, fungi, dyes, honey, etc., which have traditionally been exploited for local use but are increasingly exploited commercially for more global markets. The recognition of the biodiversity contained within forests, and the range of potential uses to which different plants are put makes the issue of deforestation and the owner-ship and control of forest use a highly contentious one at the present time. The actual and potential loss of biodiversity is matched by a loss of indigenous knowledge of the uses of the resource, and of indigenous control over it as international and national institutions try to gain control of forests for uses other than local livelihoods. Conflicts arising over the control of forests are common in many areas, particularly the Himalayan region, where the Chipko movement originated as a protest over external felling of 'their' forests (see Chapters 5 and 10).

There are two main types of mountain forest – coniferous and broadleaved, but they do not occur mutually exclusively. Forest assemblages vary with altitude and latitude, and reflect the different growing conditions in different mountain areas.

Evergreen coniferous trees tend to dominate the higher latitudes and alti-tudes as they are more hardy. The thick cuticles of the leaves reduce desicca-tion and the dark green colour is due to greater concentrations of chlorophyll, which increases the photosynthetic activity of the tree during periods of lim-ited daylight. The down-sweeping branches shed snow before it accumulates too thickly and causes damage. Conifers tend to be relatively fast-growing, with soft wood, which is popular as a source of timber for construction and for paper pulp. Resin is also present in the bark. Their relatively rapid growth makes them attractive as a plantation crop: a stand of pine can mature much faster than tropical hardwoods such as mahogany and teak. Native species of

Table 3.1 Distribution of forests in different altitude classes (areas in km$^2$).

| Forest type | >4500 m | 3500–4499 m | 2500–3499 m | 1500–2499 m; slope >2° | 1000–1499 m; slope >5°; relative relief >300 m | 300–999 m; relative relief >300 m | Total |
|---|---|---|---|---|---|---|---|
| Tropical & subtropical moist forests | 19 359 | 83 597 | 138 808 | 277 008 | 545 741 | 1 172 982 | 2 237 495 |
| Tropical & subtropical dry forests | 183 | 15 054 | 35 113 | 37 167 | 110 257 | 336 204 | 533 978 |
| Temperate & boreal deciduous needleleaf | | | 1 230 | 131 799 | 238 615 | 945 595 | 1 317 239 |
| Temperate & boreal coniferous needleleaf | 2 008 | 22 954 | 150 645 | 455 122 | 799 769 | 1 331 530 | 2 762 118 |
| Temperate & boreal deciduous broadleaf & mixed | 1 713 | 19 832 | 121 098 | 271 155 | 584 131 | 1 249 448 | 2 247 377 |
| Total | 23 263 | 141 437 | 446 894 | 1 172 241 | 2 278 513 | 5 035 759 | 9 098 107 |

*Source:* WCMC, 2000.

**Plate 3.2**   An example of the argan tree (*Argania spinosa*) in southern Morocco. This is a relict from wetter periods in the Holocene. The current aridity reduces regeneration of the trees. It is a common sight to see goats climbing amongst the branches, seeking fodder in this region. The argan produces nuts not unlike almonds in appearance. The oil pressed from the kernels is an orange-brown colour and used for cooking. In the past it was also used for lighting fuel. The crushed kernels produce a nut paste which tastes like peanut-butter with overtones of diesel!

conifer in Europe include pine (*Pinus*), fir (*Abies*), spruce (*Picea*) and larch (*Larix*). Juniper (*Juniperus*) and heathers (*Ericaceae*) tend to occur as a shrub zone in drier areas and in the higher altitude parts of the forests.

Broadleaved deciduous trees predominate in temperate latitudes where losing leaves in winter is an important survival strategy and appropriate for a seasonal climate. Species include ash (*Fraxinus*), elm (*Ulmus*), oak (*Quercus*), hazel (*Corylus*) and beech (*Fagus*) in warmer areas; chestnut (*Castanea*), hickory (*Carya*) and maple (*Acer*) in North America. Birch (*Betula*) often occurs in high latitude mountains and is an early colonizing species. An understorey of holly (*Ilex*) and climbers such as ivy (*Hedera helix*), together with smaller ground plants such as bulbs and fungi, flourishes in forests as they offer sheltered conditions. However, ground cover plants tend to flower early before they are shaded out by the developing leaf canopy each year.

Some evergreen broadleaved trees are found at high altitudes, such as the holm oak (*Quercus ilex*) in the Middle Atlas Morocco, where it occurs together with *Cedrus atlantica*, and many grow elsewhere in the Mediterranean basin. Evergreen broadleaved species are able to photosynthesize all year round where other conditions permit, rather than going into winter dormancy. Endemic species such as the argan tree (*Argania spinosa*) of the Anti Atlas are relics of previous more humid conditions (Plate 3.2). In Southeast Asia, bamboo (*Bambusa spp.*) is an important element of the forest in drier areas (Table 3.2).

In humid tropical latitudes, broadleaved trees may be deciduous despite the continuous growing season. However, they lose their leaves at some time in the year but not necessarily confined to any one season, thus forest stands may comprise trees in bud, in full leaf, and with falling leaves. In the southern hemisphere the southern beech (*Nothofagus*) and the monkey-puzzle tree (*Araucaria*) dominate. Other species such as *Rhododendron*, *Vaccinium* and *Ericaceae* have southern forms of global species. *Senecio* and *Lobelia* have distinctively different forms in the south. Other species include the wild ancestors of cocoa, coffee, kapok and vanilla.

Forests follow a succession, changing species and form with altitude. The density of forests tends to decrease with altitude, with trees becoming more widely spaced as competition increases for nutrients and water. There is usually a reduction in the number of species, and a simplification of forest form from complex layered structures with trees of different heights, to single-layered forests. Species also change with altitude according to their physiological limits. In many cases there is a succession from lowland montane forests containing broadleaved species less tolerant of cold, to forests increasingly dominated by one or two species and greater dominance of conifers with altitude (Figure 3.4).

**Table 3.2** Thailand hill tribe uses of bamboo.

| Latin name | Common name | Occurrence | Use |
|---|---|---|---|
| *Bambusa arundinacea* | Giant thorny bamboo | Lower elevations throughout Southeast Asia and into Pakistan | Musical instruments, arrows, spears, water pipes (i.e. 'hubble-bubble'), pig pens, chicken coops. Edible shoots |
| *Bambusa pallida* | Graceful bamboo | Higher elevations northern Thailand and its border with Myanmar | Strips used to tie bundles together |
| *Bambusa polymorpha* | Greyish-green bamboo | Lower mixed forests with teak and moderate to higher elevations in the north | Thick stems are used for heavy walls, bridges, houses, fences |
| *Bambusa tuldoides* | Verdant bamboo | Mid to high altitude in north and west, extends into Myanmar and China | Long-lasting and resistant to weevils, so used for handles of agricultural tools, ties for house construction, rafters and shoulder poles for loads |
| *Bambusa ventricosa* | Buddha bamboo | Native to China but imported to Thailand | Tobacco pipes |

*(continued)*

Table 3.2 (cont'd)

| Latin name | Common name | Occurrence | Use |
|---|---|---|---|
| Cephalostachyum pergracile | | In mixed forests throughout northern Thailand with teak | Mats, hats, ties, roof poles. Used for cooking rice. Edible shoots |
| Cephalostachyum virgatum | | Mixed forests with teak | Contains an irritant 'poison', so used for hunting. Not edible. Used for water pipes, musical instruments, weaving of mats |
| Dendrocalamus brandisii | | Only above 1000 m in far north Chiang Rai Province | Mien tribe use it for construction; Akha and Lahu avoid it; Karen use it only for matting |
| Dendrocalamus giganteus | Giant bamboo | Very large variety in mixed forests at lower altitudes in northwest Thailand | House posts; lasts well if kept damp by water poured into a hole in the side. Flooring, rice steamers, water troughs |
| Dendrocalamus hamiltonii | | Common in mixed forests throughout Southeast Asia | Food, houses, water carriers, household items, cages, drainpipes, for the ancestral altar, weaving, etc. |
| Dendrocalamus latiflorus | Sweet bamboo | Native to China and Myanmar, cultivated in north Thailand | Tallest species used; floors, walls (when split), rice cookers, containers and measuring scoops |
| Dendrocalamus strictus | Male bamboo | Dry mixed forests of western and northern Thailand | Water supply pipes, construction, planting sticks, crossbows, utensils, looms, fire-hardened weapons, fibres twisted into rope. Edible shoots |
| Gigantochloa spp. | | Far northern areas | Edible shoots |
| Oxytenanthera albo-ciliata | Field bamboo | Mixed forests in lowlands and highlands | Implements, handles, bird traps, baskets, mats, harness fittings, weirs for fishing and irrigation. Edible |
| Thyrostachys siamensis | | Pure stands occur with hardwoods in west and north | Edible young shoots. Reeds for instruments, fencing |

Source: Data from Anderson, 1993.

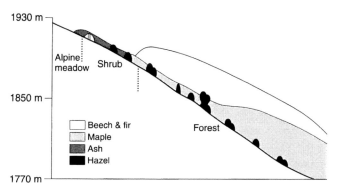

**Figure 3.4** Gradual upper timberline of fir-beech forest on the southern slope of Assara Mountain, West Caucasus Range. Note how the height of the trees falls with altitude and different species take over (adapted from Armand, 1992).

## Box 3.2   Cloud forests

These forests are entirely dependent upon moisture captured from clouds and thus their distribution is sharply defined by the altitude at which condensation nuclei form. They occur in the western Andes, Ruwenzoris (2000–3500 m), Indonesia, Malaysia, Hawaii (as low as 100 m), Papua New Guinea and the Philippines. As saturated clouds rise up the mountain ranges from the oceans, for example on the west side of the Ecuadorian Andes (1500–3500 m), when they reach dew point the drops condense on the surfaces of the leaves, capturing 500–10 000 mm moisture per year. The dense canopy helps to maintain a humid atmosphere during sunshine hours.

These forests are extremely rich in species numbers and densities. They are complex assemblages of trees, epiphytes (plants which feed off trees), climbers and several layers of forest canopy. Characteristic endemic forms abound, and there are species of insects, birds and mammals such as the howler monkey and jaguar (Ecuador) and gorilla which occur only in cloud forests of a particular region. Above and below the cloud forest very arid conditions often occur, making the forests stand out, a feature which has earned them the name of 'ceja (eyebrow) de la montana' in Colombia, Ecuador and northern Peru. Protection of these forests is vital as once they are cleared, the mechanism for water capture has been removed, the water balance of the area deteriorates and restoration is very difficult.

*Sources*: Aldrich *et al.*, 1997; Hamilton *et al.*, 1997; Long, 1995; Parker and Carr, 1992.

## The timberline

The upper limit of forests is defined by the timberline, which is the most distinctive ecotone on many mountains. This may variously represent the species limit, forest limit or tree limit. The species limit represents the upper

limit that a species can tolerate and survive. The forest limit represents the
upper limit of forest growth, which may not be located at the environmental
limit of the species in it. The tree limit is the physical upper limit of tree
growth. The timberline may be a discrete and sharply defined line between
forest and grassland or tundra. Alternatively, it may be diffuse, with gradual
changes from upright, recognizable forest into increasingly open woodland
with scattered trees which have a dwarf form and are stunted and twisted,
shaped by the prevailing winds. Figure 3.4 shows an example of a diffuse
timberline. Such trees are termed *krummholz*. Forests may have upper or lower
timberlines; in arid regions where the lowlands are dry and mountains relatively wet, lower timberlines mark the lower altitude of sufficient moisture to
support tree growth.

The altitudes of timberlines vary considerably. They may be near sea level
in polar regions where the permanent snowline is depressed. In Britain the
timberline would naturally occur around 650–800 m, but only remnants survive now. Timberlines rise towards the tropics at a rate of about 100 m per
degree of latitude to 30° north and south when the altitude levels out between
there and the Equator. The highest occur in the Andes – 5000 m in the
northern Chilean Andes and 4700 m in Tibet, although this may reflect the
altitude of the mountain masses on which they occur. Coastal mountains tend
to have lower timberlines, as do the aspects facing prevailing winds; in both
these cases there is more abundant moisture.

The cause of the timberline is much debated in the literature. In the
temperate latitudes, the timberline more or less coincides with the 10°C summer
isotherm, indicating that temperature is a critical control. It may be caused
by the limits of tolerances of species to long winters, short summer growth,
diurnal temperature changes and cold. Locally, the height of the timberline
also varies – in temperate latitudes it may be higher on ridges as trees do not
survive as well in valleys where there is significant cold air drainage and snow
lies late. In the tropics, however, diurnal temperature ranges are reduced compared with exposed ridges, offering more sheltered and favourable conditions
(Troll, 1968). In the case of cloud forests the limits of tree growth are determined by the altitude of condensation of fog so these forests have clear upper
and lower timberlines. In semi-arid subtropical mountains lower limits of
forests may also be determined by the availability of moisture; the surrounding
lowlands are often too dry to support tree growth and so only at altitudes
where orographic rainfall creates wetter conditions can trees grow. Tree growth
is also limited by soil development and by competition between species which
may prevent or limit regeneration of trees. Grazing is also an important factor
in the prevention of forest regeneration. In the central European Alps, treelines
are higher on igneous rocks than on calcareous ones, suggesting that moisture
or nutrients are local factors controlling their altitude.

Other factors which contribute to timberlines include snow, which can
smother and retard new growth; wind, causing damage and desiccation, and
solar radiation – particularly the intensity and the higher UV component at
higher altitudes. The thick cuticles of evergreen trees are more able to tolerate

these effects than broadleaved species. The success of reproduction of trees is also a factor. Growth is slow at high altitude, and production of seeds rare (Holtmeier 1994). Conifers may reproduce by layering; the down-sweeping lower branches in contact with the ground take root and in time become separate trees. Although low temperatures are probably the most significant factor in determining treelines, early or late frosts, snow cover, nutrient availability, soils, etc. all contribute.

In most forests the timberline is artificially lowered – or raised, in the case of lower timberlines – as a result of human exploitation. Grazing of upland pastures prevents regeneration of young trees, whilst cultivation of lower slopes also prevents tree growth. There may also be relict stands of trees surviving at unusually high or low altitudes due to climatic change and the survival of small areas of forest from previous climatic conditions. Some such trees may be very ancient, as for the bristlecone pine (*Pinus aristata* var. *longaevea*) in the Sierra Nevada, California, Snake Range, Nevada, Rockies, Colorado and New Mexico. Individuals of this species are over 4000 years old (Brunstein and Yamaguchi, 1992). The very thin rings of the trunk testify to very slow growth, although this has not always been the case. Such long-lived trees provide useful sources of information about climate, as well as a means of calibration for radiocarbon dating in archaeology.

## Mountain fauna

Like the flora of mountain regions, the fauna is diverse and often rich in endemic species. Some species are closely associated with particular ecological assemblages and conditions – for example, the cloud forest gorillas. This adds weight to the argument for conserving forest and other floral resources as a means of conserving its fauna. Many mammal species have an important symbolic value both for local cultures and as a global focus of conservation and symbol of wilderness, such as the ibex (*Capra ibex*) in the Karakoram, the snow leopard in the Himalayas, the legendary yeti and other species.

Altitude has an important influence on the distribution, density and type of faunal species. Animals are more mobile than plants so tend to show less clear zonal distributions, unless closely associated with a particular ecological niche. Adaptations are both physiological and behavioural. They need to adapt not only to temperature, exposure and, in temperate latitudes, seasonal snow cover, but also to reduced air pressure and oxygen availability. Physiological adaptations include seasonal change of coat or plumage – the arctic hare and ptarmigan, for example, turn white in winter to blend in with the snow. Coats are thicker in winter and change of coats is attuned to day length. In aseasonal areas, mammals tend to have thick coats, with bald areas under the stomach or between the legs. This enables them to keep reasonably cool during hot days, but stay warm curled up at night. Other adaptations include an ability to slow the heart rate and for extremities to survive with reduced blood flow; this prevents excess heat being lost and occurs in animals that hibernate in winter. Rapid maturation of young ensures they reach self-sufficiency and a reasonable size before the winter arrives. Lung capacity and haemoglobin levels in the blood are enhanced to cope with oxygen deficiency. Reptiles face problems in being unable to regulate their own temperature. Lizards, snakes and other creatures abound in hot areas, but need to shelter at night and during the winter, and may share burrows and caves with other creatures in the winter. Their metabolic rate slows down by around 60 per cent to a minimum maintenance level, as does that of mammals, in order for them to keep ticking over until spring arrives. Flies (*Diptera*), butterflies (*Lepidoptera*) and beetles (*Coleoptera*) are all relatively hardy and springtails (*Colembola*) have been found in the nival zone above 6000 m (Franz, 1979).

Behavioural adaptations include migration; some species remain at the same altitude all year whilst others migrate to lower altitudes in the winter. This is particularly common in birds and insects that can fly and thus cover large distances relatively easily. Animals which remain in winter may hibernate, or store food or fat reserves to live off whilst food is scarce. Hollows in scree slopes and caves are important hibernation sites for invertebrate and vertebrate animals. In the spring herbivorous birds arrive first, followed by others once the insects have migrated. Rodents and scavengers also arrive earlier than carnivorous mammals which need to wait until a food supply has already arrived before they can survive! Many insects remain at the same altitude all year, hibernating in various ways in the winter – for example, larvae may be hidden in bark and roots, and can tolerate freezing. Some 50 per cent of wingless insects remain at the upper timberline in the Himalayas (Jeník, 1997) whilst this rises to 60 per cent at the snowline. This reflects the greater distances involved in migration, which is not possible for non-flying species.

Animals are relatively catholic – i.e. they can survive in a range of niches and reflect adaptations of lowland species. However, others are highly specific to their niches. There tends to be a reduction in total numbers, species range and species diversity with altitude. In the Swiss Alps there were 96 species of birds in the coniferous forest, 27 in the shrub zone and only 8 in the high alpine zone. This reflects the diversity of food supplies and shelter, which are much greater in forest environments (Hamilton *et al.*, 1997). In the cloud forest of Mount Kilimanjaro in Africa, 600 species occurred at 2000–2800 m, 300 species in the montane forest at 2800–3500 m, 150 in the upper montane zone at 3300–4200 m and only 43 above this in the alpine desert. Overall in cloud forests some 10 per cent of birds are restricted to this ecosystem. Research into the survival of insects such as butterflies indicates that, in fact, spatial heterogeneity of environment is important as ecosystems tend to be fragmented and each small area is not large enough to continually sustain populations. The mobility of species is therefore critical to their survival (Boggs and Murphy, 1997). The number of species of insects increases with the complexity of the habitat mosaic (Haslett, 1997a).

The effective adaptations of many animals to mountain environments have also made them useful to humans. Many traditional societies relied on livestock for a substantial part of their livelihoods and their use of local species has made this strategy very effective. In the Himalayas, for example, the yak

(*Bos grunniens*) is adapted to survive at altitudes of over 3000 m and suffers heat stress below this. It is able to live all year round at high altitude. The domesticated yak is smaller than its wild counterpart, but makes a good beast of burden across the mountain passes along trade routes; the versatility of the animal enabled the tradition of portering of the Sherpa people to remain an important part of their livelihoods. The yak has been successfully cross-bred with domestic cattle to produce various hybrids such as the *zomo* (a cross between a male yak and female lowland cow) (Bishop, 1998). The males are usually sterile, but the females are fertile so they can be backcrossed with cattle to produce *pamu*. The *zomo* is able to survive at altitudes of 2100–3660 m where pastures exist and dairying enterprises may thus be undertaken. Because the *zomo* can calve every year (the yak only breeds every two years) this has advantages in herd management and milk production (see Chapter 9). As cattle cannot survive higher than 2000 m, or yak below 3000 m, these hybrids occupy an important ecological niche, making the high pastures productive for human use.

Other Himalayan species include mountain sheep and goats, such as the fat-tailed sheep, which can store water in its tail (rather like a camel in its hump) and thus survive arid conditions; *Capra ibex*, which is a goat of symbolic importance in the Hindu Kush – Karakoram region; and Kashmir goats which are renowned for their fine hair. In the European Alps the angora, chamois and bouquetin are all species of goat adapted to the high summits of 1000–3000 m altitude, spending the summer on the high pastures and glaciers and taking refuge in the forests in winter. They produce fine, soft wool and leather which has many uses and is a high-value product.

The Andes are best known for their indigenous camelids, the llama and alpaca (*Llama* spp.), which are adapted to survive the diurnal extremes of temperature of the high puna and paramo grasslands. Llama prefer the drier areas although alpaca are reputedly more manageable. Both can carry loads at high altitude and produce hair for local weaving. Increasingly, alpaca hair is valued on the international clothing markets, producing a source of income for farmers.

## Environmental change and mountain ecosystems

In explanations of their evolution and response to changing environmental conditions, mountain ecosystems have frequently been likened to island ecosystems. Island biogeography is concerned with the colonization and change of isolated land areas and mountain peaks are similar in that they too can be isolated, surrounded by areas of substantially different environmental conditions which act as effectively as the sea as a barrier to migration between peaks.

The colonization of mountain peaks takes place by migration and adaptation of lowland species, and by the adaptation of indigenous species to changing conditions. The alignment of mountain ranges is important in determining the pattern of migration throughout the world. The Himalayan range is

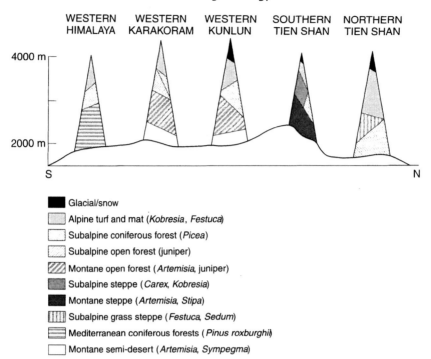

**Figure 3.5** Sequence of vertical and horizontal changes in vegetation across the Himalayas. The vegetation closely reflects changing humidity; in humid areas coniferous forests and alpine turf dominates, and in arid regions semidesert scrub with discontinuous cover occurs (adapted from Richter *et al.*, 1999).

orientated roughly east–west, which facilitates migration across longitude, as the mountain range would be a barrier to movements north–south. However, climatic conditions are so variable, getting drier westwards, for example, that characteristic floras exist for subregions, reflecting local speciation (Figure 3.5). Thus the flora has distinctive eastern and western elements, which have as close affinities to the adjacent lowlands as to the adjacent mountains. The fact that the Andes run north–south enables migration of species across latitude rather than longitude, with the flora reflecting a predominately southern hemisphere influence (Figure 3.6).

Changes in climate, such as the Quaternary glacial–interglacial cycles, caused shifts in the distribution of the main types of flora and their associated fauna. The expansion of icecaps and glaciers, and the lowering of the altitude of snowlines, pushed the alpine tundra and grasslands and forests down slope. Pollen evidence from lakes in mountain regions shows an alternation of grassland and forest vegetation, reflecting these oscillating warm and cold conditions. The types of species which occur in the pollen record give an indication of how far the shift was; in many mountain regions the treeline was lowered by about 500 m per 3°C fall in temperature (Peters and Darling, 1985) during the last glacial maximum. Flenley (1979) and van der Hammen (1974)

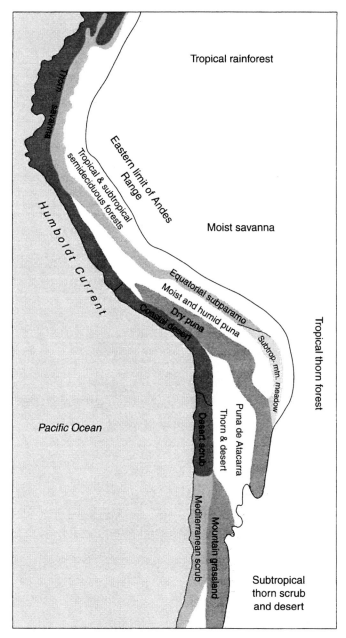

**Figure 3.6**  Simplified pattern of vegetation zones in the Andes. In the central belt there is a humid–arid gradient running west–east, whilst from north to south the gradient is from an aseasonal humid Equatorial climate to a seasonally dry Mediterranean climate (adapted from Troll, 1968).

Mountain geoecology

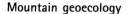

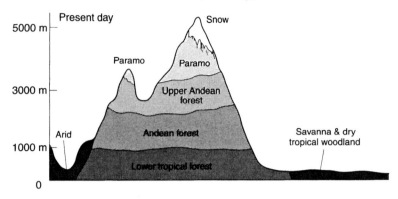

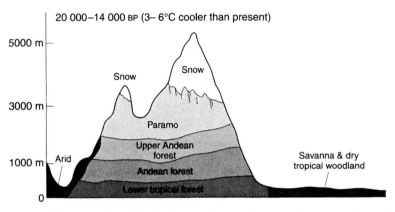

**Figure 3.7** Differences in the altitude of vegetation zones between the present day and the last glacial maximum for the Eastern Cordillera of the Andes (after van der Hammen, 1974).

estimate a shift of around 1000–1500 m in the Central and Southern Andes. As vegetation zones move down slope, they are able to expand and merge with neighbouring floras, allowing a mixing and commonality to be established (Figure 3.7). When the climate warms again and the zones more up slope, however, the higher-altitude assemblages become more isolated and also more constrained in space, as there is less land surface available. This may cause extinction of some species which are unable to compete, or move fast enough. With time, isolated floras may develop their own distinctive characteristics and endemism. The recolonization by vegetation of volcanoes such as Krakatoa, which erupted in 1886, provides useful evidence of the processes of soil formation and development of vegetation mosaics with time.

The reconstruction of past vegetation responses to the Quaternary and more recent climate change has contributed to the prediction of the response of mountain vegetation to future climatic change, given a global increase in carbon dioxide and rising temperature (Box 3.3, Figure 3.8).

CURRENT CLIMATE                          +3.5°C/+10% precipitation

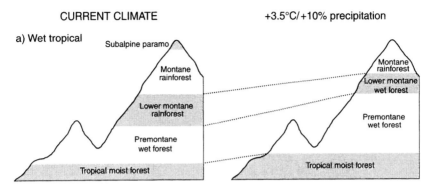

a) Wet tropical

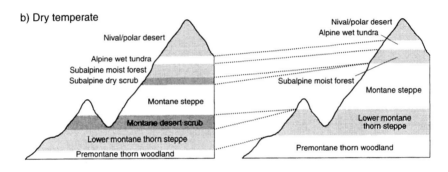

b) Dry temperate

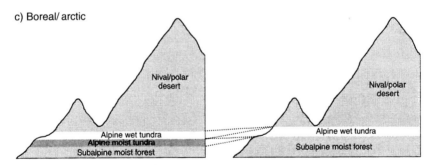

c) Boreal/ arctic

**Figure 3.8** Comparison between present-day vegetation zones and predicted future zone locations for three sites, and for a +3.5°C and +10 per cent precipitation scenario. In both the dry temperate and the boreal/arctic locations vegetation assemblages are lost (adapted from Halpin, 1994).

Local conditions are very important in determining the vegetation, and present assemblages do not always have past analogues for comparison, so it is difficult to predict changes on the scale of a whole mountain. A rise in $CO_2$ would enhance the rate of photosynthesis and raise the productivity of

## Box 3.3   Predicted changes in vegetation due to global warming

The changes in zone shifts due to global warming will vary according to latitude, as climate change has different effects globally. One model of these effects showed the potential differences between wet tropical (Costa Rica), dry temperate (US Sierra Nevada) and boreal/arctic (central Alaska) regions. The simulations were based on a +3.5°C and +10 per cent precipitation scenario. The predicted changes are shown in Figure 3.8. In the boreal/arctic region, the greatest changes occurred at lower altitudes, where zones were compressed between expanding forest moving into areas where permafrost had melted. and persistent higher-altitude snow fed by the rise in precipitation, much of which would fall as snow.

In the dry temperate region the greatest changes were observed. At high altitudes there is a contraction of the nival zone, and loss of the dry scrub due to warmer, wetter conditions. At lower altitudes the expanding lower montane thorn steppe has moved up slope and the desert scrub is lost.

In the wet tropical site, the nival zone contracts and the lower forest assemblages expand up slope. Note that the upper rain forest has reduced in area as there is less land at higher altitudes. This has implications for biodiversity conservation (Chapter 11).

*Source*: Halpin, 1994.

mountain vegetation, if other factors such as moisture and nutrients are not limiting factors. Change in moisture availability is likely to be important, particularly in summer where it may coincide with a higher temperature, creating much hotter and drier conditions. This may change the vegetation assemblage.

Increasingly, a landscape ecology approach is taken to study changes in future vegetation (Haslett, 1997a). This encompasses the change in vegetation, soils, local climate, etc. as a whole, rather than linking vegetation response directly to a given climatic scenario. The response in space and time is highly variable (Fischlin and Gyalistras, 1997), with species such as annual grasses and insects responding rapidly but forest composition and location taking decades or longer to respond. Thus, although the movement of the timberline up and down a mountain has been taken as a sensitive and useful indicator of past climate change, it is of limited value in studying future change, especially on timescales of less than a decade (Kupfer and Cairns, 1996). Limitations such as soil development and slow rates of movement and the effect of human activity on the natural regeneration of forests mean that it is difficult to study the movement of the timberline in a useful sense. However, studies have indicated that there has been an increased recruitment or recovery of trees at or above the timberline in mountains throughout North America and Canada, as well as in Russia and New Zealand. These changes are attributed to decreased snowfall and warmer temperatures (summarized in Graumlich, 1994).

Mountain vegetation and, by extension, fauna, is sensitive to environmental change, however unpredictable its response. Given the complex mosaic of life on any mountain slope, there is constant change in relationships between species. Some will respond quickly and colonize aggressively, for example on a landslide scar, whereas others may be shaded out or regenerate slowly. Pollution of the atmosphere has been a significant factor in vegetation change in recent decades. Several studies of the effects of acid deposition on coniferous forests in Europe have shown that the pattern of dieback follows the distribution of clouds (see Chapter 2) and also local winds, which serve to accumulate pollutants in pockets of the landscape, causing dieback. At the upper timberline, where many species are at or near their ecological tolerances, they are more sensitive to such effects and unless action is taken to actively replant, regeneration is slow and fails due to a continuation of the pollution effect. Pollution is an increasing problem in many mountain areas, especially with the improvement of road networks and increase in motor traffic – some 150 million people now cross the Alps for example (Denniston, 1995)

The projected environmental changes for many areas, such as the Mediterranean, coastal ranges of California, the Blue Mountains of Australia, Mount Kenya and the Alps include warmer, drier summers. This, together with a reduction in the tradition for collecting deadwood and keeping forest floors relatively clear, means that there is a greater likelihood of forest fires (Le Houérou, 1987; Beniston *et al.*, 1995). Some of these regions are located close to major population centres (as in California and Australia, Spain, Italy and France), which presents a potential major hazard. Major forest fires in the 1990s in California have already demonstrated this. In the Mediterranean Basin alone, the number of fires each year has almost doubled since the 1970s to 50 000 per year, affecting 500 000 ha (Alexandrian *et al.*, 1999). Most (55–77 per cent) are of unknown cause and only 1–5 per cent can be attributed to natural causes. Major efforts in public education and imposition of penalties are adopted by most affected governments, together with controlled burning in order to reduce the effects of fires.

The mobility of fauna enables them to cope with changing conditions but whilst species populations can survive extreme events of low frequency, a mean change in conditions may result in extinction (Haslett, 1997b). However, the prospects for mountain fauna and flora in the future are much more endangered by the expansion of human activity than by climatic change at present. Many mountain regions have extensive grasslands maintained by grazing and elsewhere the increasing introduction of new varieties of crops decreases the reliance upon local species, which may die out as a result. In recent years, however, it has been increasingly recognized that the native forms of many crops need protection as the whole world may benefit from them in the future. This also applies to medicinal plants and other elements of the mountain ecosystem and provides a strong argument for conservation of the rich biodiversity of mountains (see Chapter 11).

## Key points

- The concept of geoecology encompasses not only fauna and flora but also the soil and local environmental conditions, giving a holistic approach.
- There is a generalized succession from forest to grassland to snow with increasing altitude, but great variations in type of vegetation and species occur.
- Soils are often poor, thin and stony. Where fine material accumulates in loessic soils or those based on glacial till, they may be prone to waterlogging. Coniferous forests and acidic rocks maintain acidic soils, whilst deciduous forests and calcareous rocks support more alkaline brown soils. Soils on slopes are prone to erosion.
- Grasslands and meadows occur in a variety of forms and are often floristically rich. Many adaptations have arisen to cope with poor nutrients, temperature fluctuations and short growing seasons. Hence there is a high degree of endemism.
- Forests are also highly variable in form and species. They are important economically, ecologically and culturally. Special types of forest, such as cloud forests, occur. The diversity of species and their usefulness as well as their associated fauna and stabilizing function make them an important conservation target.
- Animal populations have adapted physiologically and behaviourally to cope with mountain environments. Different adaptations arise in the tropics compared with temperate mountains. Some species are endemic to specific niches. Others have a high cultural importance and attract conservation initiatives.
- Climatic change induces shifts in the altitudinal zones of mountain ecosystems. Past analogues go some way to assist in the prediction of the response of mountain ecosystems to changing climate but there is still great uncertainty on a local to regional scale due to the complexity of controlling factors.

Part 2

## Mountain people and cultures

# Introduction

Part 2 approaches mountains from a different angle – that of the people who live there, and who shape and are shaped by their unique environment. It is somewhat artificial to separate totally the human and physical environment of mountain regions as there is considerable interaction between the two spheres. However, in order to complete the background understanding required to comprehend these interactions fully, it is important to explore the aspects of the human environment which are most significant in this respect. These aspects are the nature of the people themselves (their culture and physiology), their socio-political and economic systems (which have a bearing on the exploitation of resources) and the demographic patterns of mountain populations (which highlight coping strategies in the context of limited resources as well as trends in population change).

The underlying cultural and social environment of mountain populations and their patterns of demographic change are closely allied to physical environmental constraints and opportunities. These factors represent checks and balances in the controlling of resource use which is examined in Part 3. The concepts of allegiance and honour in these communities give them a strong awareness of what is acceptable behaviour and a sense of identity where challenges between communities arise, but they also give a coherence to the social structure across wider mountain landscapes, creating a social mosaic superimposed on the physical one.

Despite the wide variety of physical conditions and the range from unpopulated to densely urban populated mountain landscapes, a number of common characteristics of mountain communities have been identified by several workers (Rhoades and Thompson, 1975; Brush, 1976a; Guillet, 1984), based on comparisons between the major mountain systems of the world. Whilst these common characteristics may provide a basis for the identification of a 'mountain culture', they have limitations as such general traits mask an enormous diversity of forms and variations. These common traits represent a shared repertoire of practices, indigenous cultural knowledge and local management systems which contribute to the resistance and persistence of these communities (Stevens, 1993).

The common characteristics can be broadly divided into two groups – agro-ecological adaptations to the physical environment (see Chapter 5) and indigenous management practices (Chapter 7).

1. Agricultural adaptations:
   - Mixed, integrated, diverse agro-pastoral economy
   - Mainly subsistence
   - Exploitation of a range of environmental zones/niches
   - Vertical and horizontal trade/exchange networks
   - Migration on permanent/temporary basis
2. Indigenous management systems:
   - Common control of critical/low-productivity/multiple-use resources
   - Individual ownership of terraced plots
   - Citizenship determining rights of access to resources
   - Tribal/village council/committee to oversee decision-making
   - Terraces, manuring, soil conservation strategies
   - Irrigation and water allocation systems

Both groups of characteristics represent deliberate strategies for risk avoidance and optimum resource use. The importance of interaction up and down valley by trade and exchange relationships, and the obligations and reciprocity systems within tribe, clan and household maintain a means of ensuring access to resources. The responsibility for resources, livelihood and survival strategies rests not only with individuals, but also with households, villages and the wider community. This brings an assumption of harmony and egalitarianism, but as some of the examples below and in Chapter 7 illustrate, it may mask feuds, intolerances and imbalances between adjacent groups. Communal co-operation, responsibility and risk avoidance strategies, which are enforced and maintained within the strong cultural context, are critical to survival. But the persistence of communities which appear to have survived on a 'knife edge' for centuries bears testimony to the efficacy of these arrangements, despite underlying tensions.

# Chapter 4

# Mountain culture

Culture is a broad and elusive concept encompassing beliefs, language, behaviour and attitudes of societies. There is a close interaction between culture, particularly ritual and belief systems, and livelihood activities: cultures are closely attuned to their environment. Cultural norms constitute a framework shaping the activities of individuals and communities and forming a basis for control of both society and environment (Rosman and Rubel, 1995). All societies have some sort of cultural identity, although this is rarely 'pure', but incorporates ideas, language and other aspects from different sources. At one level, a single 'culture' may thus appear to be an accumulation of fragments from different sources, but the coalescence of these fragments produces something which is unique and identifiable and becomes a culture in its own right. Such is the case of mountain cultures; whilst elements can be identified as specifically 'mountain' in character, it is still debatable whether there is such a thing as a 'mountain culture' as distinct from all others.

There has been a traditional perception of mountains as areas of isolation and refuge, with cultural forms surviving unchanged for millennia. However, it is increasingly appreciated that cultural change is the norm, with populations assimilating and adapting to change, incorporating new ideas and traditions even if they have escaped wholesale subjugation by invading forces. In this chapter, the different facets of what constitutes a mountain culture are explored and the concepts of refuge, isolation and change considered. The role and pattern of settlements will also be examined.

## The influence of culture

The belief system of a community is a fundamental part of the traditional cultural framework, and impinges upon almost all areas of social and economic activity. Many regulations concerning the use of resources are manifest in the context of beliefs in the form, for example, of taboos on hunting and land use. Ethnicity is also an important aspect determining identity, and is linked to beliefs, and often expressed in the more visible forms of culture. Ethnicity may be reinforced by caste systems, or equivalent systems of hierarchical organization, although Guha (1989) shows how often such systems may be more loosely interpreted and more flexibly upheld in mountain regions than in adjacent lowlands. This arises partly from the diversity of other cultural influences and histories and the often only superficial dominance of external ruling authorities. In addition, within the traditional livelihood strategies of mountain regions, the needs for labour and cooperation within the mountain environment may make some facets of caste and religious systems inappropriate; for example, the strict determination of labour divisions dictated by caste and the incomplete seclusion of women in nominally Islamic cultures such as the High Atlas Berbers. However, cultural traditions do dictate labour divisions in many societies and this can create obstacles in development initiatives if not understood. For instance, in Nepal certain tasks in caring for *zomo* are for women or men only (Bishop, 1998) and in Pakistan, women are not permitted on the high mountain areas, limiting their search for firewood to lower altitudes (Hewitt, 1989).

Mountain populations are characterized by great cultural diversity, manifest in arts, language, dress and livelihood traditions as well as beliefs (Figure 4.1). This cultural diversity is now increasingly recognized by the wider world as an important aspect of mountain culture which is under threat from modernization

---

### Box 4.1   The concept of caste: the Newar of Kathmandu

The Newar people have occupied the Kathmandu valley since prehistoric times. They are not of a single origin but have developed into a culturally distinct people. There was a transition from Buddhism to Hinduism in the fourteenth century. The Newar include elements of both religions, although originally the fourteenth-century King Jayastiti Malla determined which religion was responsible for the welfare of each caste. However, others contend that this division evolved from the fifth century as Hinduism was incorporated into a Buddhist society. The caste system has been modified by successive invasions and political changes; the Newar are still banned from military service as a result of a defeat in the eighteenth century when they were downgraded. The caste system has not been embedded in legislation since 1963 but socio-economically it determines each Nepali's activities.

Each Newar village comprises a range of castes as each has a specific function – agriculture, land management, buying and selling of different products, as well as ritual, educational and political roles. Older settlements on trade routes tended to have a stricter division of labour and more rigid caste hierarchy. This arises from the more varied labour requirements but also possibly from the external influence of traders who may unconsciously have prompted the observance of ritual more strictly by the Newar than in the more remote areas. Other taboos are not observed in remoter areas, partly as a result of the necessity of making a living with whoever and whatever is available, which may not allow the observance of all rules.

*Source*: Müller-Böker, 1988.

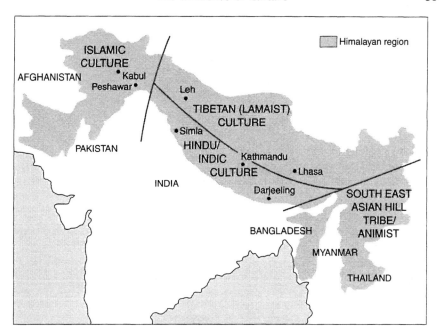

**Figure 4.1** Regions of the Himalayas showing the dominant cultural influence in different areas (adapted from Karan, 1987).

and is worthy of recognition and revitalization. The identity of many mountain populations is closely tied up with their cultural character, and the weakening of the cultural context in which livelihoods are acted out reduces social cohesion, and often causes an imbalance in activities in terms of equality within the community and sustainability in the environment.

The UN has initiated a Declaration (UN, 1994) on the recognition of the rights and celebration of the uniqueness of indigenous cultures. Although the Treaty was not ratified, its existence does clarify the value of indigenous cultures and go some way towards the protection of such people's rights to self-determination; i.e. their right to evolve in the future as they have in the past and without coercion and exploitation. These issues are increasingly important as mountain (and indeed other) areas are targeted by multinational, external commercial exploitation. A particular example is that of 'biopiracy' and the exploitation of indigenous knowledge of plants and their uses in ways which do not benefit the indigenous peoples.

Mountain communities were previously considered bastions of traditional culture, unchanged for centuries, and thus became the target of anthropologists seeking to study the unique, almost primitive nature of such communities. Certainly, the isolation and marginality of mountain communities contributes significantly to their resilience and survival, and to the development of individual cultural expressions (Stevens, 1993; Mehta, 1995), but they are no longer regarded as the untouched remnants of ancient cultures. Although isolation may operate to protect many mountain populations from wholesale takeover by invading external powers, reinforcing their resilience, it does not prevent them from adopting selected elements from other cultures which suit them politically, economically or socially. The location of many communities along trade routes, for example, made exposure to external cultural influences the norm rather than an unusual event.

Mountain communities, therefore, have not only been isolated, but have changed continually over time, as the Balti example in Box 4.2 demonstrates. This dynamism is also critical to survival, as flexibility in changing circumstances enables communities to adapt to and cope with changing physical, socio-political or economic conditions. Many elements of mountain cultures are shared with other traditional societies; their uniqueness lies in how they are expressed. Thus just as geomorphological and climatic processes are not unique to mountain environments, but the combination of processes and their rate and intensity make mountains distinctive; where cultural traits are concerned, the distinctiveness of mountain environments lies in the expression of cultural traits appropriate to their physical environment and sustainable livelihood activities. Thus, whilst the physical environment is the 'theatre' of operating a

---

### Box 4.2  Cultural complexity: the Balti people of the Karakoram

The Balti people of Hunza, Pakistan, despite being geographically isolated and highly dependent upon their extreme environment, have a culture which is an amalgam of many diverse elements:

> '. . . the Baltis are devout Muslims, mostly Shiites, converted in the 15th to 16th centuries from Buddhism. They use much of any cash they earn to send menfolk on pilgrimage to Mecca and Iran. In villages of . . . 50–100 folk, you will find two or three who read and speak Persian and Arabic. Most can converse in Urdu, the language of Pakistan. Many have . . . English and Italian . . . Their own language, Balti, is Tibetan, not Indo-European, but with a Persian script.'

In addition, external historical factors have been important, including Islam and colonial rule. The Hindu Rajput rule conscripted men and taxed the region in the beginnings of the continuing Kashmiri problem. Trade routes through the Karakoram were important sources of revenue but disrupted by the British, Russian and Chinese in the 'Great Game' played out in the region. There were kinship links in China, as well as opportunities of labouring on roads and railways for the British in India and, in the present day, some work in the Persian Gulf. The Karakoram Highway became an important agent of economic integration in the twentieth century (Plate 4.1).

*Source*: Hewitt (1988: 17–18).

*(continued)*

*(continued)*

**Plate 4.1** The construction of the Karakoram Highway has changed the economy of many northern Pakistani villages. Trade between the mountains and adjacent areas has increased and cash crops such as potatoes can be traded for flour and grain, reflecting the integration of the mountains with the national economy. Heavily laden lorries (of which this is a modest example!) ply the route as long as the roads remain open.

subsistence lifestyle, the people are not limited only to the stage and wings, although this may be as far as many outsiders choose to look. In other words, it is the external observer who constructs the stage, but fails to observe that, in reality, the extent of activity and interaction with environment and other peoples is limitless in time and space.

## Belief systems

In traditional societies the dominant form of belief systems comprises a combination of animist beliefs and ancestor worship. This may be overlain or replaced by mainstream religions such as Buddhism, Christianity or Islam. However, it is common to find that they exist alongside each other. Because of the close intertwining of beliefs with subsistence practices, it is not easy for communities to shed their old beliefs, as this undermines their security, identity and stability. Hence the phenomenon of more 'dilute' or at least less 'fundamental' following of mainstream orthodox religions mentioned above.

## Box 4.3   The ritual of shifting cultivation

In northern Thailand, many rituals determine the lives of the hill tribes. Most maintain an active interest in the spirit world despite conversions to Buddhism and Christianity. Many different hill tribes have coexisted in close proximity but each has retained its own unique culture and identity which amalgamate themes from animist and mainstream religions and which shape their activities and attitudes to resources.

The Akha, for example, shift location fairly frequently (about every decade), and the choice of a new site is determined by rituals involving the placing of rice grains in a clearing. If these move after a given time, the spirits are considered disturbed and a new site is sought. Likewise, in clearing the land for cultivation, stumps of trees must not be cleared, as these are where the 'Grandmother' sits. These stumps also have an additional significance in aiding the recovery of forest after abandonment. The most important spirit is that of the Rice Spirit, guardian of the staple crop. The Rice Spirit is invoked at rites of passage such as birth, marriage and death as well as in sickness.

*Sources*: Lewis and Lewis, 1984; Lewis, 1992; Bragg, 1992; Anderson, 1993.

A common element in many animist belief systems is that of the spirit world and the need for people to placate, sustain or otherwise interact with spirits. In many cultures, such as those of the Zinacantecos of Mexico (Vogt, 1990) and the hill tribes of northern Thailand (Lewis and Lewis, 1984), these spirits are thought to be the 'owners' of various resources, such as forest, land, rice, maize and water. The spirits are therefore consulted and rituals precede the clearing of land, ploughing, planting, harvest, etc. Failure of harvests, or other misfortunes, are considered the result of angered spirits and the shamans or other priests are called in to perform various ceremonies to put the matter right.

A second element of these belief systems common throughout many animist/ancestor rituals is that of the soul. In Mexico, the Zinacantecos believe in both the inner, personal soul, and a spiritual alter ego in the form of an animal. In northern Thailand, the soul is sacred to individuals but the Karen tribe also have a ceremony of tying the umbilical cord of a baby to a particular tree. In both cases, individuals and nature are closely related spiritually, which is critical in maintaining an attitude of worship, or at least deep respect, for their resources. Thus in the Karen's case, the tree and the person are closely identified, and the viability of the tree is critical to the health and prosperity of the individual. Such trees cannot be cut, and great emotional upsets can arise from government acquisition of land with such trees, particularly where the population are excluded or the trees scheduled for clearance.

Illness and lack of prosperity are attributed to angered spirits, or lost souls. In Thailand, tribes have elaborate soul-calling ceremonies for sick persons, where priests or shamans seek to reunite the soul with the body. As this is to some degree in the power of the soul/spirits, it leads to a pragmatic, if not fatalistic, attitude to illness, death and misfortune. The power of the shamans and priests can be great, especially if particular rituals, such as soul calling and curing, or rituals determining new sites for occupation or clearance, lie in their hands.

Where mainstream religions predominate, as in the Buddhist cultures of Tibet, Nepal and other parts of the Himalayas, the monastic institutions may perform both ritual and secular functions, such as settling disputes, collecting taxes and renting land. Here the temple may form the ceremonial centre of the village, as at Melemchi, (see *Settlement*, page 98). In Mexico, the descendants of the ancient Maya have largely converted to Catholicism, although they retain strong rituals and beliefs associated with spirits and lords of water, land, etc. They have numerous sacred sites, and observe rituals throughout the year, although many of these have been aligned with the Catholic calendar. Thus they celebrate Easter and Christmas as well as participating in rainmaking ceremonies. They traditionally have a ceremonial centre which served a number of outlying hamlets. This still operates, although it is now dominated by a church and government administrative services.

The agricultural year is punctuated by ritual and ceremony. Leaders may be the local ruler, priest or their representatives; in Hunza, the Mir (local ruler) used to cut the first sod and pick the first grain, signalling the time for ploughing and harvest for the whole community (Biddulph, 1880; Lorimer, 1938). Dates for ploughing, clearing, planting, or harvest may be set by a religious calendar, marked by saints' days, or by dates considered auspicious by shamans or priests. This applies equally to pastoral activity – the dates and event of sending cattle up to the alp in Switzerland, for example, is marked by a saint's day, and by ceremonies to invoke good fortune. At Oukaimeden in the High Atlas of Morocco, the dates when reserved pastures are officially opened were set by a saint in the fifteenth century, and remain in force today (see Chapter 7).

This carefully controlled pattern of activity prevents abuse of resources, and ensures a degree of social coherence vital to the effective functioning of the community 'organism'. This is strengthened by codes of honour which strongly dictate behaviour. The Basques have such a concept, called *indarra*, which bridges secular and sacred aspects of life. Tribal feuds are strongly determined by codes of honour and the need for redress to preserve honour of tribe, name or individual, as is the case in Middle Eastern mountain regions and parts of eastern Europe.

The Muslim code of conduct is also important, dictating the status and protection of women and resulting in gendered space. For example, a natural hot spring in Chutrun in the Karakoram is an important social centre. Throughout the day male and female uses of the spring alternate and the two sexes do not mix socially. A development initiative to construct a women's bathing house, though held up as an example of unusual consideration of women's needs, is in fact interpreted as entirely dictated by the needs of males from outside the community.

The increasing use of a jeep road has made this spring more visible to outsiders and in order for the visitors not to be embarrassed by seeing the women, they built the bathhouse. In previous times the spring was screened from the tracks leading to the village, and this 'high-handed intervention' is blamed for breaking down the intricate social structures in the village (Hewitt, 1998).

## Other aspects of culture

Cultural identity is also expressed in the form of arts and crafts, architecture, language and music, ornamentation and dress. These serve as external forms of identity, by which other tribes may know where they belong. Ceremonial and ritual events are generally marked by music, dancing and traditional clothing. Such clothing may be special forms of everyday dress.

Language is an important source of community identity. The relative isolation of many valleys during the past enabled local forms of language to evolve into distinctive dialects. These reflected the language of origin of many populations, such as Persian in the case of the Balti people, and the modifications by later influences – for instance, Arabic and more recently French on the Berber dialects. These different dialects are not often mutually intelligible: the three dialects of Berber – Chleuh in the western Atlas, Braber in the High and Middle Atlas and Riffian Berber in the north – have evolved from similar origins with different modifying influences (Camps, 1987). The nomadic Tuaregs, who are pastoral Berbers, speak Tamachek, which is different again, reflecting a much weaker Arabic influence than in the mountain Berber languages (Brett and Fentress, 1996). Many mountain dialects and languages are not written, and often only translated to paper by visiting anthropologists. In the Hunza valley there are some four languages and two dialects within a relatively constrained area. These include Burushaski, which is spoken by a few villages and has no connections with other languages, whilst Shina is spoken by adjacent villages. A third language, Wakhi, is spoken only in Upper Hunza by the Wakhi people (Felmy, 1997; Biddulph, 1880; Lorimer, 1938). The traditional languages are used in rituals and spoken by the older people and particularly the women.

In a similar way to language, dress is an important part of traditional life. In the hill tribes of Thailand, each tribe and subtribe can be identified by its dress and ornaments. Some remain relatively plain, whilst others, such as the Lisu, are highly competitive on an individual level, vying for the most elaborate ornamentation (Lewis and Lewis, 1984). Some motifs may have spiritual significance or be representative of life stages. Dress may differ in style between children and mature adults, between married and unmarried women and widows, and between men and women. In addition to dress, ornaments, particularly on women, are culturally significant. In cultures where women's wealth is in the form of dowry, a proportion of their personal wealth is worn as silver (often) or other metal jewellery. Again, some patterns may serve as tribal identifiers. Tattoos, particularly on men's bodies, or women's faces, as in the case of some Berber women, identify that woman as belonging to a specific tribe or village. In northern Thailand, the wearing of traditional dress, and the embroidering of such clothing has been revitalized by the tourist industry, which demands such cultural artefacts.

## Cultural change

Culture is constantly adapting to external and internal influences. These changes reflect the cyclicity which we examined in the physical environmental context; some changes are only temporary in response to periodic shortages in resources. Permanent change may arise from the incorporation of refugees, from invasion, or from adaptation to changing political, social or economic circumstances, which may affect the balance of power and modify the social norms. In the process of change, there is the risk, or reality, of the 'loss' of culture, knowledge and indigenous characteristics. This process is one of great concern to many development and anthropological workers, for example, who decry the loss of indigenous culture. However, this supports the conception of a static culture, not recognizing that the transformations taking place are simply a contemporary example of what has been going on throughout history.

The application of cultural theory to mountain communities (Price and Thompson, 1997) illustrates the cyclicity of cultural change, which resembles the concepts of cyclicity described at the beginning of Part 1 with respect to the physical environment. Price and Thompson propose four 'myths of nature' which describe different relationships between humans and their environment and response to change (see Box 4.4). These responses may be on the scale of individuals or communities. They are not restricted culturally or environmentally to any one response or 'myth' but may move between these different states, demonstrating different degrees of adaptability or coping, with different societal or environmental outcomes. This application of cultural theory attempts to broaden the understanding of outsiders such as development practitioners in order to help them propose 'appropriate' development strategies and to understand the often unpredictable response of indigenous people to development initiatives (Funnell and Parish, 1999).

The loss of indigenous knowledge is important for a number of reasons. It is increasingly recognized that the 'wisdom' of people who have been born and bred to a lifestyle making use of marginal environments to the maximum potential, can be eroded or lost. In the Hunza valley, Pakistan, for example, the children of economic migrants have been educated in an urban environment and have not had the same opportunities to learn the local irrigation systems, history and traditions of their villages of origin. This makes it more difficult for them to be reintegrated into their traditional society and, as a result, many do not return. However, they bring new and different forms of knowledge from a different educational background, which provides new opportunities for economic diversity, such as networks with urban markets which can help sustain local livelihoods (Parish, 1999). In other cases, the success of economic migrants can supply finance for electrification, rebuilding mosques and road construction. This is the case with migrants from the

---

## Box 4.4   Cultural theory and coping strategies

Price and Thompson propose four conditions of attitude and existence to describe mountain peoples' coping strategies and behaviour:

1. 'Nature is benign', where an attitude prevails that catastrophe is impossible and justifies an *individualistic, laissez-faire* approach.
2. 'Nature is ephemeral', where any disturbance to nature may bring collapse and an attitude of 'treading lightly' upon the earth is necessary. This requires an *egalitarian precautionary* approach.
3. 'Nature is tolerant', where there are recognized limits which nature can tolerate before collapse occurs, and it is in the *collective* interest to stay within these limits, as reflected in hierarchical approaches of statutory *regulation* and sustainable development.
4. 'Nature is capricious' is an *individualistic fatalistic* approach where it is seen as irrelevant and a waste of time to attempt to do anything as efforts will achieve nothing.

It is not difficult to apply some of these categories to individual situations in mountain regions, or to see why applying 3 might clash with a particular population's tradition of 1 or 4. It is also shown that societies can move from one category to another – from 2 or 3 into 4, for example, as a result of population growth and demand for resources, coupled with external controls hindering their traditional methods of management. Such is the case in the history of Nepal's forest (see Chapter 10).

*Source*: Price and Thompson, 1997.

---

Tafraout region of the Anti Atlas, Morocco, who have been highly successful in the grocery trade. Cooperatives of migrants have obtained financing from Electricité de France for the electrification of a number of villages (see also Chapter 6).

However, the introduction of new economic strategies often clashes with traditional cultural mores. In the Basque homeland of the Pyrenees the status of people undertaking different economic activities represented the social stratification of society. Shop owners who handled money were traditionally viewed as of lower status in the social hierarchy than farmers who worked with their hands. However, as shopkeeping proved increasingly lucrative due to the opening up of new markets, and the work was recognized by the younger generation as less arduous and dirty, such employment became more attractive. Thus the younger generation regarded shopkeeping as preferable employment to farming, creating a reversal of the social hierarchy (Greenwood, 1975). An inversion of status is reflected in many other regions, for instance, in the development of tourism activities in the Himalayan regions as compared with traditional farming.

People belonging to a particular caste are usually associated with particular activities (warriors, priests, merchants) and there are certain activities and associations which are traditionally forbidden (for example, intermarriage with lower castes). Thus, caste determines social status, occupation and where individuals may live, but each caste is interdependent with others. Modernization, a change in the economic base (such as the growth of the wealth of merchants and hostel owners) may affect the distribution of economic power – for example, the growth in demand for meat by hotels in Kathmandu has benefited lower castes preferentially, as higher castes are constrained by tradition from engaging in this activity and therefore cannot benefit. However, it has been shown that milk is now sold by 'impure' castes, which was inconceivable even in recent times (Müller-Böker, 1988). In addition, new jobs, such as those in the services and industries, lie outside the traditional system. They can loosen rigid caste boundaries: school teachers can now come from any caste, whereas in the past teaching was the role of the priests. There is also the economic and social aspect to ritual activity which means that those who cannot now afford to be so lavish and punctilious are reducing their expenditures in this area.

In the current climate of modernization and penetration of Western tourism, many traditional crafts, and even some festivals and ceremonies, have been revitalized or preserved by production for an outside audience. This is a controversial matter, for although it may help to perpetuate some forms of craft work, such as basket weaving, wood carving and textiles, it may also serve to belittle the cultural significance of special ceremonies and cause them to be performed to order, rather than at spiritually significant times of the year (see Chapter 12). In addition, some of the traditional motifs used in weaving, carpet making and embroidery are lost, and replaced by designs demanded by the tourist market. Thus the knowledge of the significance of these motifs is eroded (Kalter, 1991) and some believe that their spiritual power is also weakened or lost.

## Physiological adaptations of human populations

An important cultural aspect of mountain populations is their ability to live at high altitudes. The ways in which the human body adapts to cope with reduced oxygen conditions (hypoxia – the partial pressure of oxygen at 5800 m is about half that of sea level) and extreme cold have been of interest to the medical profession for many years. There have been many myths circulating about the extreme longevity of humans in remote mountain regions, but many of these have been overturned as they often reflect the fact that hard labour outdoors in such regions 'weathers' faces beyond their years, giving the appearance of great age. This is compounded by the fact that many are unable to give an accurate date of birth, but assign their age in relation to particular events in their region – and such memories can be very long!

However, there are certain identifiable physiological characteristics of mountain people which enable them to cope with high-altitude life. It is possible for lowlanders to adapt – the process of acclimatization to high mountain conditions is well known to mountaineers and is a critical component of their training as mountain sickness can kill. Mountain sickness results from changes in the air composition with altitude – reduced oxygen availability being the most acute. It causes nausea, drowsiness, insomnia and faintness and the effects begin to be felt at around 2500 m (Pawson and Jest, 1978). The body adjusts by producing more red blood cells (haemocytes) in order to carry more oxygen around the body, and by increasing the heart size and activity. The lung capacity of Sherpas and Andean Indians has been shown to be proportionately greater than that of lowlanders, which increases the intake of air into the body, thus again compensating for its lower oxygen content. These regions have been of particular interest as they have substantial populations living at much higher altitudes than elsewhere in the world.

There are considerable debates surrounding the issue of stature. Mountain people are often smaller than average but this may be due to poor nutrition and genetic traits rather than a particular adaptation to mountain environments. The hardiness of individuals – walking in snow in bare feet, for example – may well be an ability developed of necessity, much as a desert dweller can stand the high temperatures of sand with bare feet, without any particular physiological difference to the blood or nerve supply to the extremities. Other areas of physiology which have been studied by anthropologists and the medical profession include the fertility of women (Abelson, 1973). Fertility can be reduced for a number of reasons (see Chapter 5) and diet and time of year as well as general stress of physical labour can delay puberty and reduce the fertility of women without having to be attributed to mountain environments *per se* (Goldstein *et al.*, 1983). However, it has been shown that the combination of factors including hypoxia, physical exertion and poorer nutrition is likely to affect general health and fertility (Frisancho, 1993).

## Settlement

Settlements in mountain regions range from scattered farmsteads to major cities. As populations moved into mountain areas, cleared land and settled, villages and hamlets grew up, with larger regional centres at points of passes, valley confluences and other communication points. Other settlements developed as a result of specific trade routes, such as those crossing the Himalayas, until they were closed as a result of the formalization of political boundaries. Others sprang up on pilgrim routes, as in the Indian Himalayas. Yet others arose as a central focus of ritual or religious significance, due to the presence of a sacred site, shrine or ceremonial centre. Thus settlements evolved out of function as well as on the basis of their location.

Settlements in mountains did not emerge in isolation, however, and links between different regions in the mountains, as well as with the lowlands, have always been important. Interaction with other settlements may arise from the function of stopping places along trade routes, or the need to exchange goods necessary for survival. In Mexico, several outlying hamlets are served by a central ceremonial complex – in former times, during the Mayan period, this

## Box 4.5   Andean city development

There has been a history of high-altitude settlement in the Andes stretching back 14 000 years. The first inhabitants were hunter gatherers who had temporary shelters in many regions between Chile and Patagonia. Concentrations of early transhumance settlements occurred at 2700–3300 m in the central Andes. Advanced societies from 900 BC to the Spanish conquest in AD 1532 centred first on Chavin in north Peru up to 200 BC, then the Tiahuanaco Empire AD 600–1000 and then, from 1438, the Inca, based on Cuzco in south Peru. In between these empires there were periods of transition – just before the Incas, the region was divided up into political states which were incorporated into the Incan administration. It could not expand into the lowland forest regions as the political and subsistence system was too different to incorporate effectively, so the empire remained centred in the highlands, reaching around six million people at its zenith.

After the Spanish conquest, the states were quickly incorporated into the new administration; land was taken over and many Indians forced to work in the metal mines, which became the biggest employer. Replacement of traditional with European grain staples, which were less productive at high altitude, and the spread of alien diseases caused the native Indian population to fall to around one million by about 1800. In the next 20 years Chile, Colombia, Ecuador, Peru and Bolivia all achieved independence. Political power was transferred to landowners and the Catholic Church. Now, 80 per cent of Peru's population of 22.5 million are Indian, and 50 per cent rural dwellers. In Bolivia, 80 per cent of the population live above 2500 m altitude. The region around Bogata is the most densely populated high altitude region of the Andes and the city (2600 m altitude) has nearly half a million inhabitants and Quito in Ecuador, at 2800 m, has a population of 1.3 million.

*Sources*: Cooper, 1946; Guhl, 1968; Pendle, 1967; Steward and Faron, 1959.

was a temple, palace and meeting place, but since the conquest the centre has been dominated by a church, and additional administrative, legal and political functions for the region are based there.

The construction of roads, industrial activity and changes in political boundaries have profound effects upon settlements. Staging posts and animal supply centres may fall into disuse, or change to become truck stops. Alternatively, the closure of trade routes by the establishment of political boundaries may remove an essential part of the livelihood of many populations, such as the Sherpas of Nepal, for whom trade was an essential part of the economy. Whilst the Sherpas have been able to benefit from the tourist industry, providing portering and guiding services for the growing trekking market, others have not been so fortunate, and their fortunes have declined. Abandonment of settlement by depopulation may also occur, as in parts of the European Alps when easier living and more attractive opportunities occurred in the lowlands urban centres, echoing the example of the Basque country.

Tourism and its related infrastructure have a great impact on the form of mountain settlement. The development of winter sports facilities and summer trekking has resulted in the construction of hotels, ski facilities, and associated roads, sewage systems, etc. This has often occupied many areas of the valley previously left clear, or taken over productive terraces (informants in Hunza referred to the cultivation and herding of tourists as their new source of livelihood). In the former case, land was often left clear for good reason, such as the knowledge of avalanche tracks or flood levels. The construction of new buildings on these lands has had tragic consequences as in recent years severe floods and unusually severe avalanches (see Chapter 1) have resulted in great damage and loss of life.

In the past, literature has concentrated on the concept of mountains as refuges for persecuted minorities fleeing the invasion and subjection of overpowering political forces. The Berbers are thought to have retreated to the highlands to escape the Arabs, as are the Druze in Syria. This is reinforced by the past, and contemporary sources of many claims for independence by mountain-based minorities (Basque, Andorra, Montenegro) and current locations of guerrilla warfare and unsubjugated peoples. Certainly, mountain people had a reputation for being fierce warriors, as the Balkan, Greek and other conflicts

---

## Box 4.6  A Balkan tribal refuge society

The Dinaric mountains of Montenegro are karstic and so have little surface water. Their apparent barrenness and steep topography have offered sanctuary to the Montenegrins, who were able to maintain their autonomy throughout successive centuries of invasion. They have been able to use the mountain environment to their advantage, supporting their own populations and resisting the efforts of outsiders to invade and conquer. They used guerrilla tactics, and their intimate knowledge of the landscape, its hiding places and locations of food and water gave them considerable advantages over the Ottoman armies in the eighteenth and nineteenth centuries. They differed in this behaviour from the Serbians to the north and the Albanians to the south, whose regions were equally mountainous. This difference is attributed to the richer pickings in these latter zones for the Ottomans and the relatively less suitable environments for true guerrilla tactics compared with Montenegro. Thus Montenegro provided a natural mountain fortress which remained impenetrable until the economic and political forces of the modern world established the area as the smallest Yugoslav state. The recent fragmentation of the former Yugoslavia is, to a large extent, the latest episode in a history of continuing tribal friction.

*Source*: Boehm, 1984.

have testified. However, there are also reasons to consider that such populations did not so much flee to the mountains – although some undoubtedly did – but deliberately chose to live there. This might arise from the comparative advantages of climate (compared to arid desert lowlands), or because of the greater security and defensive potential of the mountain landscape. Men who know their landscape well can control large armies with small forces, as the history of warfare from classical times onwards has documented (see Box 4.6).

In the High Atlas, the rise and fall of the dominance of different tribes or 'houses' (e.g. Glaoua, Goundafa), was still central to the politics of Morocco right up until Independence in 1956. Blood feuds existed between the tribes, and the central administration often operated to play one off against another (graphically described by Maxwell, 1966). On the other hand, whilst the Arabs and French relied on the allegiance of particular dominant tribes, these often proved defiant. At times of particular stress, the blood feuds between tribes would be put on one side, and a temporary alliance of united tribes operated against the invading forces. Even today, in parts of Yemen, where blood feuds are still rife, the assurance of safe passage through tribal territory is a matter of honour.

The nature and location of settlement in mountains also reflects the environmental niches exploited, social organization and the physical configuration of the landscape. Lauer (1993) identified three types of cultural adaptation reflected in settlement – compact, extended and archipelago (Figure 4.2). Allan (1986) in his critique of verticality refers to the impact of road construction on the isolation of settlement and development. He argues that where roads, such as the trans-Alpine and Himalayan routes, were constructed, mobility permitted development to occur at a faster pace than at settlements some distance from the road. Thus there was a breakdown of the traditional model of increasing isolation with altitude, and also of the degree of primitiveness/tradition. Certainly, the concept of isolation has been rethought in recent years as our understanding of the interdependence of mountain communities over large distances has grown.

Settlement can also be defined with respect to agro-ecological function. The distinction between permanent and temporary settlement is a fundamental difference reflecting the livelihood strategies of mountain populations. Permanent settlement in mountain landscapes represents areas of dwellings and community facilities, such as meeting houses and ceremonial or ritual centres. The settlements may be compact, as in some of the defended hilltop towns of Yemen and southern Morocco, or dispersed throughout the fields belonging to the village, as at Melemchi, Nepal (Bishop, 1998) and to some extent in the Hunza valley, Pakistan – particularly in recent years where infilling has occurred through the construction of hotels and new houses on terraces. The former offer more effective defence against invasion, which was a common characteristic of the Bedouin and Berber (both later Arab) lands. In the latter case, security was offered by the feudal overlord (in Melemchi, the Chini Lama, and in Hunza the Mir, and, more recently, centralized national government and relative political stability). Permanent settlements tend to be located

Vertical access patterns

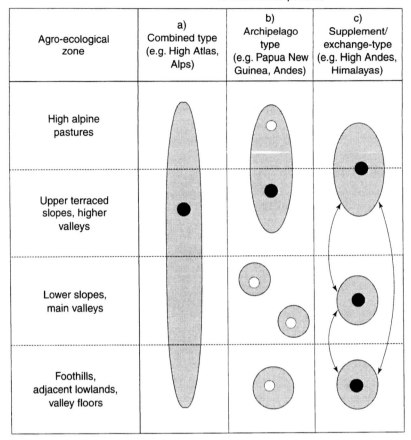

| Agro-ecological zone | a) Combined type (e.g. High Atlas, Alps) | b) Archipelago type (e.g. Papua New Guinea, Andes) | c) Supplement/ exchange-type (e.g. High Andes, Himalayas) |
|---|---|---|---|
| High alpine pastures | | | |
| Upper terraced slopes, higher valleys | | | |
| Lower slopes, main valleys | | | |
| Foothills, adjacent lowlands, valley floors | | | |

● Permanent settlement or marketplace

○ Temporary or seasonal settlement

◉ Zone of activity

**Figure 4.2**  Three patterns of settlement in mountain regions. In a) the valley is sufficiently small to allow the same community to access all the different resource zones. In b) the distance is greater, so subsidiary settlements are developed and occupied seasonally at different altitudes. In c) the distances mean that different communities occupy the different zones and are specialists in a particular type of subsistence. In order to meet the shortfalls in other needs, trading and exchange networks are established between different altitudinal groups. Although this model was developed for the Andes, here it has been adapted to reflect other mountains (adapted from Lauer, 1993).

on higher ground – in Yemen they are deliberately situated on rocky promontories to conserve productive soil (Swagman, 1988), and in Morocco, on slopes above the valley floor where defence and protection from seasonal flooding were important factors in site selection. In Southeast Asia, the animist traditions have a series of rituals associated with site selection, involving the consultation of spirits and omens as to the suitability of a particular site.

Other forms of settlement include temporary dwellings – temporary both in terms of time (for example, huts on high summer pastures which are only used for part of the year) and space (as for the shifting cultivators of Southeast Asia, where villages are relocated at regular intervals). In the former case, the huts are similar to the main dwelling, but simpler, though no less substantial. In the European Alps, summer dwellings were associated with hay barns where fodder could be stored once cut from the meadows, for use in winter. In the High Atlas, the possession of a summer dwelling and cattle enclosure was a prerequisite for maintaining rights of access to reserved pasture lands (see Chapter 7). Likewise in Nepal, the ownership of a *Gode*, comprising an area of grazing land and associated buildings, was a necessity for *zomo* husbandry.

At the other end of the scale from permanent dwellings are the tents of pastoral nomads. Whilst these may follow ancient, redefined routes, their tents leave no enduring physical evidence, although they may be located at the same site each year. Such dwellers are only present seasonally, such as the nomads of the Middle Atlas who move up onto the summer pastures and move continually across tribally defined territories. These people are the remnants of a previously much larger population, many of whom have been permanently settled by the French and by later independent Moroccan governments.

Houses tend to be of robust construction, usually being built of stone, which endures. However, in the southern Atlas mountains houses are traditionally constructed of mudbrick – a factor arising from its ease of working and transport and its greater availability compared with suitable stone (Plate 4.2). On the northern flanks, where the climate is considerably wetter, adobe is still used, but at higher altitudes such as at Imlil, stone prevails. The architecture varies regionally according to available materials, symbolic and spiritual trends and customs, and climate. For example, wood is used in relative abundance in the Swat valley of Pakistan, whereas over the watershed in the more arid Hunza and Chitral regions, timber is rare and adobe and stone were the traditional materials, now replaced largely by tin roofs and breeze blocks. In northern Thailand, teak planks and bamboo predominate, with wooden shingles on the roof, reflecting both the abundance of both these materials, and the relatively temporary nature of buildings (Plate 4.3).

In many cases, houses comprise adjacent human and animal dwellings. This is particularly true of the middle and high latitudes, where during the winter months cattle are stall-fed in the villages whilst the pastures are closed by snow. During the summer they are moved up to distant pastures. This combination results in many households owning both permanent village houses and summer huts – Lauer's 'combined' form of adaptation. The arrangement of the dwelling often consists of a lower storey for animals and an upper storey for

**Plate 4.2**   Square, flat-roofed houses are common to many mountain areas. They may be constructed of stones and/or mudbrick, depending on local materials. The flat roofs are important extensions to the living space, where grain and fruits are dried, fodder stored, and washing, cooking and spinning take place – and even sleeping in summer. From the ground these houses appear very enclosed and tightly packed, which makes the added dimension of the rooftops important. Notice the satellite dish in the lower part of the frame; TVs and videos are increasingly common and although this village has electricity, many others use car batteries as the power source instead.

humans or a hay barn. Such an arrangement is typical of the European Alps, Scandinavia and the High Atlas. In Nepal, it is also common to find the human part of the house in the centre, with outbuildings built onto it for the animals. Both the in–out and up–down arrangements allow humans to benefit from the animals' warmth. In Thailand, several of the hill tribes build their houses on stilts. The lower part is often left open, although still used for animals and other storage, but in some newer households, the lower portion is also boarded up and items such as motorbikes or small cottage industry facilities may be stored instead of, or as well as, animals.

In many cases, the house comprises a single room, with fireplace, and sometimes separate partitioned areas for bedrooms – for the parents, for children of different sexes, or for guests. Other areas may be separate for cooking. Rice pounders may be present in all or only some houses. A granary or rice bin is an important feature of mountain homes, where the foodstore is kept close to the living quarters. In other villages, such as in Morocco, a separate,

**Plate 4.3**  In northern Thailand, houses are constructed of wood, which is the most abundant building material. The houses are raised to give storage space below and reduce damp. The balconies outside each house serve similar functions to the rooftops in Plate 4.2. Notice the plank ladders for access. The roofs are covered in plastic and tin which have replaced the traditional banana leaves.

communal granary may exist. Many barns and granaries are built on stilts or staddle stones, to permit free flow of air underneath, to reduce mould, to keep the food above the level of wet or frosty ground, and to reduce infestation by rats. In other areas, tubers may be dried, as in the Andes, or buried, or stored in separate buildings.

In more recent years, many traditional types of building have been replaced by cement and concrete buildings with tin roofs. These are less aesthetically pleasing (to some eyes), but more durable. A traditional shingle roof may need replacing every ten years or so (Bishop, 1998), or a thatch more frequently, so investing in a tin roof not only confers status (by conspicuous wealth) but also represents a longer-term saving. In areas of shifting cultivators, enforced settlement (through lack of land to move to, or as a result of government decree) has enabled the population to take advantage of modern methods of construction. Whereas before the continual movement of the village every few years meant it made no sense to construct such durable buildings, the transition to permanent settlement changes this perception. Though often constructed of modern materials, the new buildings tend to resemble traditional architectural forms, even if more ostentatiously as well as substantially built. Such details include the incorporation of forged grilles at the windows, and shutters as well as glass (Morocco), painting traditional motifs on door posts and central pillars that once were carved in wood (Swat), and still building on stilts with traditional open and closed porches (Thailand).

## Key points

- Culture is a broad concept encompassing language, beliefs, societal structure, power and gender, dress, music, arts, etc. It is the combination of these elements which gives any one society its cultural identity and affinities.
- Mountain populations cannot be assumed to have lived in isolation and their cultural traditions are drawn from a wide variety of sources.
- Belief systems are important in shaping the attitudes of individuals and communities to each other and to their environment. Belief systems can operate a controlling mechanism of resource use and act as as a framework of power relations in communities.
- Language and dress as well as other elements of culture can vary from community to community, giving rise to a highly varied mosaic of cultural traditions within relatively small areas.
- Cultural change has been the norm in many communities as elements are assimilated and lost with time and with changes in external influences. The process of modernization in recent years has eroded some hierarchical social structures, such as caste, and changed the traditional balance of wealth and power.
- Patterns of settlement in mountains have arisen in response to agricultural adaptations, trading opportunities and external interventions such as colonialism and invasion.
- Both the style of house construction and the location of the settlement have changed as a result of modernization. This has resulted in improvements in health but has also placed some areas at greater risk of natural hazards.

# Chapter 5

# The socio-political environment of mountain communities

The 'socio-political environment' of mountain communities refers to the distribution of power and the control of access to resources, and of the exchange networks that operate between communities. In mountain environments the distribution of resources to all who are entitled to access, and the management of those resources for the benefit of all is critical. It represents not only some form of sharing of the means of subsistence but also a sharing of the risk and of the consequences of failure of one or other aspect of the resource base. This sharing may not be equitable, or even democratic, but is embedded in various forms of power structures within traditional societies, with interactions occurring at all scales from individuals to whole tribes of mountain regions.

In traditional societies, power structures are often culturally determined. Aspects such as gender and caste play significant roles in determining access to resources and patterns of exchange between groups. In addition, trade and reciprocal exchange arrangements between different specialist groups in regions operate to form a mutually supporting complex of communities. Indigenous institutions of governance and law have been targeted as a focus for mobilization and implementation of development projects, increasingly so since the advent of local participation as a development mantra. However, in the development process, traditional or quasi-traditional institutions may be coopted and given powers which disrupt or modify the existing balances of power, with potentially disastrous results both for the local community and for the development project.

This chapter will consider first the processes at the individual and household level; second, those at the community or valley level, and finally the processes of interaction between mountain communities and the wider world, particularly with respect to development. Many processes may operate at all levels, and thus are not mutually exclusive. The controls on the processes may in part be determined by cultural values such as those described in the previous chapter.

## Individual/household-scale organization

The household is the fundamental unit of production and consumption, and of social organization and gender and labour divisions (Fricke, 1984; Zurick, 1986). Households, whilst being relatively self-sufficient, are not independent, as exchange networks of labour, foodstuffs and other aspects of livelihoods are fundamental to their survival. It is this interdependence, in the form of reciprocal exchanges, which contributes not only to survival but to a strong sense of identity, entitlement and mutual benefit. In the Basque region of the Pyrenees the extended family has always been the basic unit of society rather than the village (Ott, 1992).

The household may consist of a joint or multiple family or a nuclear or extended family, with local custom determining the pattern of inheritance, sibling roles and opportunities for setting up new households. Miller (1984) documents households comprising two families in the High Atlas, and elsewhere,

---

## Box 5.1    High Atlas social hierarchy

Imlil is situated at about 1700 m in the Rereya valley in the northern flanks of the High Atlas. It lies at the end of a made road and serves as a starting point for High Atlas trekking parties. Its relative accessibility has made it a greater focus of study and development than other villages further up the valley. The social structure comprises groups of households belonging to a particular lineage and affiliated to clans of a particular tribe. Tribes are further grouped into confederations (Figure 5.1).

Each clan in the area is heavily reliant on its own territory for food and fodder, but only 12 per cent of the food is produced locally. Bottled gas is increasingly becoming a key fuel source. The existence of the road means that this is easily available. Both combined (extended) and nuclear households exist – the latter being about 50 per cent more common. Daily activities are centred around the provision and use of fuel, fodder and fertilizer, which form the basis of the subsistence economy. Fuel comprises firewood and LPG as well as animal and human labour expended to produce food and fodder. The fertilizer from the animals is carefully collected and distributed on cultivated land. The household is the basic unit of production and consumption, but certain resources, such as grazing and water, are communal and thus interactions between clans occur on the basis of shared access to these. Whilst households can dictate what happens on their own private land, changes of use or access to common resources requires negotiation within the clan, or in some cases between tribes if encroachment onto other tribal property is at issue. In the Middle Atlas, some tribes may have rights to water livestock on another tribe's property, but they must carefully observe the rules which dictate the route taken and ensure that visiting flocks and herds are not allowed to stray or graze. This particular situation arises because of the limestone bedrock which means that surface water is relatively rare.

*Sources*: Miller, 1984; Dougherty, 1994; Funnell and Parish, 1995; Parish and Funnell, 1996; Bencherifa and Johnson, 1990.

*(continued)*

*(continued)*

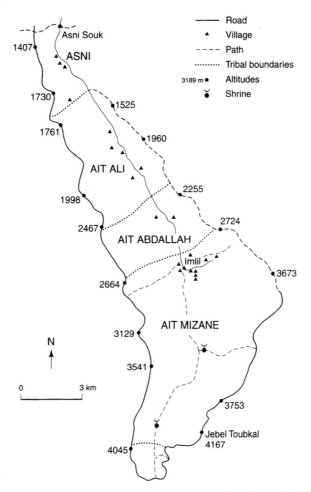

**Figure 5.1**   Map showing the tribal divisions in the Rereya valley. Although the different groups are concentrated in particular zones, they interact frequently as a larger confederation (adapted from Miller, 1984).

as in the Alps (Netting, 1981), some households comprise several siblings, some or all of whom may be married. Where this arrangement arises the family estate is maintained as a whole.

There are regional differences in inheritance which affect the pattern of household development. There is a balance between division of land between siblings and retaining sufficient land and access to resources for survival. Estates may be divided up between siblings (partible inheritance) or passed to one sibling (impartible inheritance). In the former case, the land is divided up, not always equally, between heirs. Although this limits the amount of land per sibling, new estates are formed on marriage by the amalgamation of spouses' properties. This results in excessive fragmentation of estates, producing tiny plots of questionable productive potential, but in fact it represents a good and fair way of redistributing land amongst individuals. It also redistributes different qualities of land so that any one household may have plots scattered through-out different altitudinal zones, which ensures access to all zones and to re-sources such as common grazing and water, as well as distribution of risk. This represents household-scale vertical organization. Impartible inheritance systems maintain the family estate. Such estates often have established access to differ-ent zones, with a number of land holdings located throughout the valley. These patterns of inheritance have implications for demographic changes (see Chapter 6).

Other advantages of scattered plots lie in the highly diverse mountain environment. Variations in cropping potential can occur on a plot scale, so having dispersed plots means the range of crops and cropping conditions can be distributed amongst the households. It also helps in the management of labour, as small plots will tend to ripen sequentially rather than all at once, thus smoothing out the labour demands at critical times such as harvest. Labour management in the household is an important part of subsistence economies. In mountain regions the economy comprises a mix of cultivated crops, livestock, and use of forest and other resources. These may be supple-mented by trading (see below) or by migration to external sources of employ-ment (see Chapter 7).

The nature of the physical environment means that cultivation tends to occur on terraced slopes in the valleys, with hay production and grazing on valley floors when they are not flooded. On higher-altitude meadows, grazing takes place by transhumance, where herds and flocks remain at high altitude for several months during the summer. This pattern of activity requires not only sufficient numbers of animals to make it worthwhile (though numbers are limited by the need to provide winter fodder for them on relatively scarce land), but also the available labour to tend them. In some areas, the young males of the household may be absent for the summer on the high pastures, as in the past in the European Alps and still today in the High Atlas. Elsewhere, permanent herding communities occupy the high pastures and trade with settled agriculturalists lower down, as is the case of the Hindu farmers and Sherpa herders in Nepal and the llama herders of the High Andes. Herders in Afghanistan and western Iran have a similar practice (Balland, 1988; Bradburd, 1996; Browman, 1990; Stevens, 1993). In cases where there are too few animals to make a small household warrant the loss of active labour all summer, the animals may be rented out to herders, or combined with the herds and flocks of others in the care of a member of another household, who is paid in kind for caring for the amalgamated flock. Instances of this are recorded in the High Atlas (Gilles et al., 1986).

In the same way that herds may be pooled, amalgamated or rented, labour is also pooled and exchanged for certain tasks. This strategy is a critical part of community survival. Each household may have reciprocal arrangements with other families, related or not, to share labour during times of high demand. This operates in addition to the traditional pooling of labour for communal tasks such as irrigation channel maintenance and terrace construction. The principle of reciprocity, where lending labour time comes with an assurance that it will be returned when needed, is a fundamental characteristic of the successful operation of traditional mountain economies. It maintains stability by enforcing community allegiance. In recent years, with the penetration of capitalism into these areas, labour is increasingly exchanged for cash; i.e. it becomes paid labour. This change breaks the bonds of reciprocity, and may serve to promote the fortunes of a subjugated underclass, but may also reduce stability and community solidarity by introducing elements of individual wealth and power rather than community-based systems.

Another feature of social organization at a household level is the distinction between gender roles. Whilst this may be enforced by wider cultural or religious tradition, it operates at a household level in determining who is responsible for different areas of work. In the Atlas and Karakoram, women look after household livestock – cows and poultry – but not the sheep and goats, which are tended by men. This is because the flocks are an important source of wealth and status, and thus their care is a man's work. It also arises from the fact that women in these societies are relatively secluded and so not able to be absent for many months in summer pastures. Hewitt (1989) also reports that women are excluded from these areas as they are 'impure' (due to their child-bearing roles) whereas high pastures are often sacred and 'pure'.

In Nepal, however, couples work together in the *gode*, tending yak and crossbreeds (Bishop, 1998). Women also tend to be responsible for tasks such as weeding and planting, vegetable production, and, importantly, collecting fuelwood, water and fodder. Men tend to plough and construct the roads, terraces and irrigation systems. Traditionally, in Melemchi, Nepal, only men could irrigate, but this task is taken over by women where there are no available men in the household to do it (Bishop, 1989, 1998). Thus some tasks are interchangeable, and if men are not there, women can take the task on. Other tasks, which may include marketing produce, handling money and hiring labour, may be considered dishonourable to women and they are not permitted to do them, as in the Karakoram and Atlas.

In Thailand the division of labour is much less obvious, as the Buddhist and animist beliefs of the hill tribes do not restrict women in the same way as the Muslim societies. Thus, in Thailand, women may be itinerant peddlers and market traders, whereas in Morocco the women in certain areas of high emigration have had to be creative in their handling of money and labour in order avoid being ostracized, as it is considered inappropriate for women to handle money and to hire labour (de Haas, 1996) (see also Chapter 6).

## Community/intra-community-(valley-) scale organization _____

Households are grouped together to form communities which may be very close-knit and clearly defined, or loosely connected except during specific

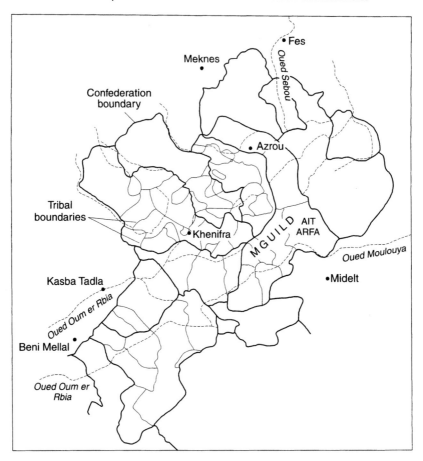

**Figure 5.2** Part of a tribal map of Morocco, showing tribes and confederations in part of the Middle Atlas (adapted from the Naval Intelligence Division, 1942).

ritual events or during times of crisis. The connection between households may be based on kinship, either by extended blood relationships which ensure cooperation between the different branches of the family by a system of obligations (as in Mexico), or it may be based on the claim of descent from a common lineage in the past, as in the case of the stem families of the European Alps. A modification of this involves the claim of common social descent, although not common lineage – i.e. by long-established occupation of a shared geographical space, as in the Ait Mizane of the High Atlas, Morocco (Miller, 1984). Here the clan and confederation form a hierarchy of cohesive social units (Figure 5.2).

The sense of community is usually very strong, forged by the need to share resources, including labour, and concern for the good of the community may take precedence over that of the individual. Cooperative linkages may extend beyond the immediate family or lineage group to include others within the wider community whose identity and sense of belonging are derived from residence (Dresch, 1989). Elsewhere, as in the Alps, labour is drawn almost entirely from the family, and only specific duties such as pasture maintenance may be communal (Netting, 1981). Thus the primary allegiance varies both in time (according to seasonal duties) and between different cultural groups – it is based on the family in many European mountain communities, such as the Sarakatsani shepherds of Greece, and the Alpine villages of Switzerland, where honour is upheld in the family. In other groups, such as the Lisu in Thailand, a clear distinction exists between the obligations and rights of sharing within and beyond the close family; whilst reciprocal exchange involving carefully repaid 'debts' operates outside the family, within it everyone has a right to a share in all resources, from clothing and food to tools and labour (Durrenberger, 1983c).

Ethnicity is another basis for the identification of communities, as in the case of the hill tribes of Thailand, where different villages comprise people from different ethnic groups, with their own lifestyles, rituals, language, etc. and with relatively little interaction between them on a daily basis. In parts of the Himalayan system, communities are also defined by religion, as in the case of the Shia and Ismaili of the Hunza valley. Cultural identity therefore defines the group of people, or community, who cooperate in the use and management of resources.

The affiliation of the individuals and households within the community may be tightly controlled, or be insignificant in daily life, but be called upon during times of stress or conflict. Under these conditions, informal management systems may become formalized or otherwise made visible in order to establish more clearly each community's claim.

In the past, authority within the community often lay in the hands of a hereditary leader who may have been a feudal lord or ruler (as in the case of the Mir of Hunza, for example). Alternatively, it may have lain with a monastic or similar institution, as in Nepal, or power may have been held by a group of elected officers with defined duties, such as priest, headman or forest guardian, or by an elected or hereditary body of 'tribal elders' forming a tribal or village council. This is common in many parts of the Arab world, such as Morocco in the past, and Yemen, where tribal power and affinity are still important. In many cases, in the Swiss Alps, for example, there is no record of a hierarchical authority (Netting, 1981), whereas in Nepal the village of Melemchi rests under the authority of the Chini Lama (a hereditary lama who has almost feudal authority over a designated area) as a result of a land grant by the king in 1847 (Bishop, 1998). The authority of the Chini Lama and equivalent hierarchical rulers is absolute, and their role in institutional control of social and economic resources is discussed in Chapter 7.

## Traditional communities and resource management

The traditional or indigenous livelihood of mountain people encompasses a variety of practices and institutions concerned with the management of the diverse but limited resource base on which they depend. These institutions are of critical importance to sustainable living, as they determine who has access to which resources, when and in what quantity. The critical resources in question are primarily those of land, water, grazing and forests. These will be considered below, after a brief examination of the nature of wealth and power within a community, which are closely associated with ownership of and access to resources.

The concepts of identity and survival are drawn from these institutional structures: identity because belonging to a particular group determines in many cases whether access to particular resources is permitted or denied, and survival because without careful management of resources which are in limited supply or slow to regenerate, sustainable livelihoods are short-lived. The survival of mountain communities for millennia, and the flexibility of their management structures are testimony to the efficacy of indigenous institutions. It is only in more recent years that these institutions have increasingly been unable to cope with the much greater pressures of modernization and capitalist penetration. The integration of societies into the wider national and international economies has changed the nature of the objectives and operating mechanisms of the societies and many have been replaced either actively or passively by central government institutional structures and control.

The interdependence of communities is an important part of effective resource management. In small valleys, such as the Rereya valley in the High Atlas, the communities occupying the valley are able to make use of all the resources. However, in much larger valleys such as those in the Karakoram, the distances are much greater and the existence of other groups means that there is a greater occurrence of interdependent and usually ethnically distinct communities who specialize in an area of the traditional agro-economy, but who exchange produce, resulting in a symbiosis and a mutual meeting of needs (see Figure 4.2, page 102).

In northern Thailand, several distinct hill tribe peoples coexist. They occupy different niches and each is relatively independent. However, there is an exchange of goods, particularly (until recently) between opium from above 1400 m and rice from lower altitudes. Interethnic relationships between these tribes are complex. Trade between tribes may favour certain tribes above others for honesty and integrity, in the same way as other cultures interact with certain castes. The Lahu may take on Karen workers as labourers during the opium harvest, but regard them as unreliable and dishonest, with frequent disputes concerning payment, and the addiction of many workers to opium, which prevents them from working effectively (Dessaint and Dessaint, 1992). The Karen in return consider the Lahu wasteful (by using swidden cultivation instead of settled farming) and dislike them for their prosperity arising from opium earnings. The hill tribes generally prefer trading with Chinese rather than northern Thai, partly because of their greater knowledge of the Chinese language, but also due to a distrust of the Thai arising from their encroachment on hill tribe lands and association with centralized control from Bangkok.

Markets have an important role in the economic and social exchanges between communities (Bromley, 1974). Periodic markets which occur on a

## Box 5.2    Pastoral–sedentary exchange strategies: Afghanistan

Nomadism has increased during the last century in Afghanistan at a time when it has been declining elsewhere. The Gujars originally followed a seasonal cycle of winters on the Indus plains and summers on southern slopes of the Himalayas. However, the expansion of irrigated agriculture during the second half of the nineteenth century forced them to either settle or seek new pastures in Afghanistan where they have been increasing at least up until the 1960s. They occupied the alpine pastures of Nuristan and the areas bordering Pakistan. The Gujars are limited to short-range intra-mountain transhumance with little association with the adjacent plains. This is compared with the long-distance movements of most Afghan nomads.

The Gujars' position in the region is precarious as they arrived well after existing settled Nuristani and considerable friction, often involving the military, occurs between different groups, jealous of guarding the scarce resources. As the Gujars did not receive official sanction for occupation and do not have advocates in Kabul or access to trade networks, they remain at the mercy of local extortion through grazing fees. This contrasts with other nomadic groups such as the Pashtun who share mutual needs with local Hazara cultivators and thus benefit from a wary interdependence. The Pashtun have similar political advocates as the Hazara and it is in their common interest to remain 'on speaking terms', whereas the Gujars are considered by most other groups as foreign interlopers and so remain less privileged in their connections and prospects. Nomadism even within this small region is a highly variable undertaking, intertwined with local political and social conditions.

*Sources*: Balland, 1988.

regular basis, often once a week or maybe at longer intervals, are important aspects of the economic life of mountain people. In the High Atlas each confederation holds a weekly souk at the same location where goods and produce are exchanged (or now purchased and sold) and the news of the day shared. Different confederations hold their markets on different days so itinerant traders can move from souk to souk. Despite the rise in motor transport the souks remain lively; at Asni (the souk of the Rereya valley) in the High Atlas and Azrou in the Middle Atlas, the fodder is an important part of the spring produce (Plate 5.1). In Yemen, there are many more weekly markets than the population might warrant, reflecting their importance as a medium of distribution – for smuggled goods also (Schweizer, 1984, 1985). Here, each market is held on tribal territory and it is a point of honour to ensure that no conflicts arise on market days, otherwise the market may be discontinued. This could make life very difficult for people who would have to travel further to trade.

Within these markets, different produce is grouped together – meat, vegetables, dry goods, household items or livestock. Some markets have become

**Plate 5.1**    A periodic market or *souk* at Azrou in the Middle Atlas of Morocco. There is a mixture of permanent and temporary buildings and shelters. This picture was taken in the spring, when fodder and grain would have accounted for the greater part of purchases.

more specialized, especially with the development of road transport. In Yemen the sale of *Qat*, which is chewed as a mild narcotic, must occur every day and in all towns there is a lively *Qat* market daily. In fact, the growing market for *Qat* and luxury goods smuggled in from the Gulf States has increased market activity (Swagman, 1988). These markets are much less altered by modern tourist activity than those, for instance in Morocco, where large numbers of souvenir sellers haunt the souks of the Ourika and adjacent valleys.

Activity is concentrated in the mornings. Trading may be gendered, according to local custom – in Thailand there is less restriction on women's activities in markets than in Muslim countries, for example. In recent years these markets have also expanded their service role, especially with the extension of government health and education schemes into rural areas. Thus, the barbers, storytellers, garagistes and dentists have been supplemented by pharmacies and sometimes development extension workers and rural credit facilities. The meeting point provided by the markets is therefore a crucial part of the functioning of mountain life.

Interdependent and reciprocal relations may only arise during times of hardship. In Papua New Guinea, pig farmers descend with their pigs to stay with lower-altitude peoples in times when successive frost has destroyed their potato crops. They come down once failure is established and famine looms, with some returning immediately to replant. They stay until the harvest is ready, when they return. The pigs form a source of payment for their stay (Allen, 1988).

## Institutions

The power and control over resources in the traditional mountain community may rest in the hands of an individual, who is an absolute ruler, or in a group of commonly recognized leaders, forming a council or committee. In the former case, the ruler may be a prince or monarch of a mountain state, as in the Himalayan kingdoms in the past.

In some communities, the role of the monarch was effectively in the hands of a local religious institution such as a Buddhist monastery. As discussed in Chapter 5, belief systems are central to the culture of mountain populations, and the power of the shaman or priest could be considerable in determining what land was cleared, or whether certain decisions or activities were compatible with their belief systems. In the Buddhist communities of Nepal and Tibet, the importance of the monastery may go beyond being the focus of religious activity and security. These monasteries not only contained significant numbers of monks and nuns from the local community but could also own significant proportions of the land of the community. This is also the case for land owned by mosques in the Arab world. Such land might be rented

---

### Box 5.3   The role of the Mir of Hunza

In Hunza in the Karakoram of Pakistan, the Mir held feudal authority over his lands and people until his deposition in 1974, when princely states in the Indian Himalayas were abolished and integrated into wider society. These rulers held absolute power, and generally owned most, if not all of the land. In Hunza, land was rented to tenants and payment of rent (usually in the form of a proportion of the crops) supported the Mir and his household. In return for their allegiance, they could expect the Mir to represent their cause to regional governments (Gilgit), to negotiate the terms of trade for those caravans passing through their territory, and to organize and initiate land clearances, irrigation development and road construction. The Mir needed to count on his people's support in order to present a façade of stability, prosperity and strength to the outside world. The Mir also had an important ritual significance, and during the year he had to conduct ceremonies to mark the first ploughing, planting and harvest. Following his deposition, a power 'vacuum' developed in Hunza, in which development organizations were able to operate to provide a new focus for social control and economic development.

*Sources*: Kreutzmann, 1988, 1993a; Sidky, 1993.

or sharecropped, and in exchange for their religious duties, such foundations might also perform the decision-making and conflict resolution roles normally undertaken by the ruler or elected council.

The third main controlling institution in traditional societies is that of the council of elders. This may comprise a group of (usually) men who represent the heads of well-established and recognized households or lineages. Such positions may be hereditary in communities with well-established stem families (as in parts of the Swiss Alps; Netting, 1981). These individuals and families may become very powerful, such as the shaykhs of the mountain communities in Yemen. They may derive a significant proportion of their income from dispute resolution and trade revenues (Swagman, 1988). Under the process of modernization many of these shaykhs have been replaced by government-elected representatives and their power base has been eroded. Elsewhere, the council of elders may comprise a group of elected individuals; the Karen in northern Thailand have modified their traditional organizations as a result of integration into the national government administration. There are now both local and formal elected groups who are responsible for managing forests. The traditional system elected 26 elders for life to sit on a council. In addition to this are new committees established by the government, which are composed of temporary positions. These committees make decisions on forest management in response to changing needs and conditions. Both groups work together to reduce conflict (Tan-Kim-Yong, 1997).

---

**Box 5.4   Moroccan tribal councils**

In Morocco, the council of elected elders is the *jema'a*. Social organization is by confederations of tribes, with each tribe occupying adjacent areas of land in the same valley. Thus the Ait Mizane occupy the Rereya valley in the High Atlas. Such territories are delimited by watersheds and topographic features. The *jema'a* is a tribal council with responsibilities for adjudication and settlement of disputes regarding land, water, grazing rights and violations of these. It is also responsible for providing contributions towards projects funded by a number of tribes (roads, mosques and other facilities) as well as organizing initiatives within its own territory. The *jema'a* is concerned to uphold the honour of the tribe and community and to maintain its status with regard to other tribes. Its authority is accepted by the whole community, to which it is held accountable. However, with the integration of mountain areas into the wider economy and political system, the *jema'a* is now largely defunct as it is overtaken by these new systems. As the new balance of wealth lies increasingly with the younger generation of economic migrants who do not necessarily have a place in the *jema'a*, it is bypassed in favour of modern political and administrative institutions emerging from central government.

*Sources*: Miller, 1984; Parish and Funnell, 1996.

In the normal course of events, the tribal council or its equivalent may not impinge on the daily lives of individuals other than by operating to maintain the familiar cycle of livelihood activities. However, when crises threaten, such as disputes over resources with adjacent tribes or communities, blood feuds, or conflict with external authorities (colonial, central government, etc.), identity is emphasized and affiliations become polarized to support a particular point of view. Under such pressures, informal, nominally accepted forms of control, such as denying or granting access to resources, may become less flexible and more formalized in order to clarify the claims of the tribe to its opponents. Certainly, this has been the case in northern Thailand where incursion by Hmong immigrants onto Karen lands caused them to formalize their own institutions in order to draw support from external government institutions (Walker, 1992). In Yemen, tribal allegiance may be called upon in such cases, although it is generally irrelevant to daily life.

Under modernization pressures, communities which have adapted effectively to incorporate new ideas, technologies and crops and to cope with immigration and emigration and changing environmental conditions, are suddenly faced with much deeper, fundamental and powerful forces of change which are exogenous to the community and are too great to cope with. For example, the rate and nature of transition from a subsistence to a cash economy, which breaks down the ethic of communality in favour of privatization, cause disruption and marginalization of traditional institutions. Several references have been made above to communities in transition. Stevens (1993) illustrates how, in the Khumbu Himalaya of Nepal, such indigenous institutions are no longer able to cope. In other cases, traditional community management structures have been adopted as the basis for participatory development initiatives, as with the case of the AKRSP in Hunza (see below, page 122).

## Common property resources

A central theme of resource management in mountain communities is the concept of common property resources (CPR); that is, resources managed for the good of a designated community rather than for the individual. Ownership of resources is not seen in the same light as in the materialist West, where absolute ownership is considered the foundation of the capitalist economy, and the amassing of wealth for individuals or discrete groups is the norm (Gray, 1995). In many mountain societies, property (land, livestock, buildings and rights of access) may be individual or communal, and in the latter case, communality may have a variety of different definitions. There is often a mixture of ownership, with some land being held in common, and some privately. Critical resources such as water, multiple-use resources (forests), or extensive, lower-productivity resources (grazing lands) tend to be communally managed to ensure an equitable (i.e. on the basis of need rather than of absolute equality) distribution between entitled users.

In order to have access to these communal resources, users need to hold some sort of entitlement. This may be established by residence, lineage or membership. It is the nature of the entitlements and of the type of ownership of the resource which determines the resource base available to individuals (Leach *et al.*, 1997). The well-known story of the 'Tragedy of the Commons' (Hardin, 1968) where uncontrolled overuse causes destruction, is often upheld, but in fact it is often the case that enclosure and privatization cause the destruction, as will be shown in Chapter 10 in the context of mountain forest resources.

There is a large literature on CPR and its significance to indigenous societies, especially with the more recent growing recognition of the importance and effectiveness of many of these strategies in sustainable living in marginal environments (see, for example, Berks, 1989; Bromley, 1991; Ostrom, 1991). Communal ownership and management require the existence of appropriate institutions, and in the current era of participatory development these institutions form a starting point for many initiatives (Ostrom, 1985, 1991). This is not without its problems, however (see Part IV).

## Mountain communities and the wider world

The role of communities in survival and adaptation strategies within the mountain context has been examined above, but these communities also have extensive links with the wider world, through their interaction with national agencies, state government and international development and commercial exploitation (colonial influences). Through these encounters, institutions of resistance and of development have emerged, and, with the rise of the participatory school of development, these institutions have proved vital to the successful implementation of appropriate development options.

As discussed above, communities are complex networks of interdependent households and individuals, with various structures which control power, resource entitlement and access. Such structures have the capacity to mobilize the whole community to action in order to achieve mutually beneficial and shared objectives, be it warfare or infrastructure construction. The 'community' of action may not comprise the whole population of an area, rather it varies according to need or issue – it may be a women's group cooperating on a task, or a group of elders representing the wider community at a regional or national level. These communities of action have been increasingly recognized as power bases both for resistance and for encompassing change.

Communities of resistance have been long associated with mountain areas. This stems from the idea of mountains as refuges from invading forces. This has been the case in the Atlas, where Arabs pushed the Berbers into their mountain strongholds. However, the Berber strongholds resisted domination up until the time of the French, and in fact used the mountains as a base for controlling the rest of the country for periods. In Thailand, the hill tribes are increasingly marginalized by the expansion of lowland Thai onto their lands. Where lands have been left to regenerate and the swidden cultivators have moved on, these areas are perceived as empty and thus open for settlement by other tribes. They are also excluded from areas of national park and commercial

## Box 5.5    The Basques of the Pyrenees

The Basque people are of unknown origin but were certainly living in the western Pyrenees before the time of the Iberians. They have their own language, Euskera, which, like many other mountain languages, has no known association. In the thirteenth century the Basques were renowned for their expertise in whaling, first in the Bay of Biscay and then further afield.

The Basque homeland has never been united but always divided between France and Spain. On the Spanish side, General Franco abolished the use of the Basque language in 1937 but in more recent years there have been strong demands to become independent like neighbouring Andorra. Andorra, in the eastern Pyrenees, is one of Europe's smallest countries, speaking Catalan. Its economy since the development of roads in the 1930s and tourism in the 1950s has become almost totally dependent on tourism.

The Basque nationalist party ETA (*Euskad Ta Azkatasurra* meaning 'Basque Homeland and Freedom') was founded in the later nineteenth century and has been involved in attacks of considerable violence on tourist resorts and car bombing of important people. ETA split into a violent faction, composed of fanatical Basque nationalists, and a socialist faction, for whom Basque nationalism is secondary. Since 1980 the Spanish Basques have had limited self-rule, with their own government, language and control over issues such as education as well as representation in the Spanish national government. However, on the French side, there is no likelihood of independence as the French authorities fear that greater autonomy would be associated with the type of violence seen on the Spanish side. However, administratively there would be advantages in a single planning Département. Even in Spain, sympathy for the continuing resistance to non-Basque exploitation is undermined by the continuing violence. The French Pays Basques have seen peaceful protests against the overriding central government investment in tourism at the expense of any other activities, but it is probable that many cultural aspects of the Basques are likely to be maintained through tourism, although a wider economic base is sought.

*Source*: Astrain *et al.*, 1997.

exploitation of forests, or plantations which were so common throughout the colonial era. Most mountain communities are skilled in guerrilla warfare – for example the Balkan people of Montenegro (Boehm, 1984; see Box 4.6), Afghanistan (Balland, 1988) and in many other areas (Box 5.5).

In more recent times, resistance has come from the mobilization of groups affected by various developments. Chipko, the movement to protect forests in the Himalaya, and the Campesino rebellions in the Andes both represent the effective mobilization of affected people. The movements resulted in reforms at national level of forest laws and land tenure respectively. Chipko arose from the exploitation of forests vital to the agro-economy of the local population, by commercial logging firms. It was characterized by Gandhian non-violent resistance, with protestors hugging the trees to prevent them being felled. The Campesino revolts in the Andes resulted in extensive land reform comprising the parcelling up of the haciendas from the colonial era to be given over to individuals and to cooperatives of peasants to farm for themselves.

Communities are therefore seen both as critical to stability in mountain populations and also as a catalyst for change. Development practitioners have increasingly espoused the concept of 'participatory' development, where individuals and various groups are targeted and encouraged to be active facilitators of their own change, rather than passive consumers of it. Indeed, it is generally held that without participation, development cannot be effective. However, it is also the case that people cannot be made to participate. The leadership and power structures of the village or household are important, as by tradition they lead the way, and their approbation of a development will help others to follow suit. Leaders also act as mediators between external organizations and their people. Interventions for development and change may not come from external sources, but the role of an intermediary who enables the mountain community to negotiate non-exploitative relationships with wider markets can be invaluable. Such was the case in the Ecuadorian and Bolivian Andes where priests and university professors as well as European volunteers and donor agencies acted as effective facilitators, enabling development opportunities to be taken up successfully (Bebbington, 1996b).

Nevertheless, such groups need to participate and to buy into the opportunities. There may be a number of reasons preventing them from doing so. In Hunza, the Aga Khan Rural Support Programme (AKRSP) has mounted many development programmes in the Ismaili villages of Hunza, with schools, health care and other facilities. This was done through the formation of village organizations of individuals who put forward proposals for development to the AKRSP. However, the village of Ganesh which, like neighbouring Nagar communities over the river, is Shiite and distrusts the Ismaili benevolence, has been unable to agree on, or has been disinclined to bid for funds for projects. Instead, they have acquired funds from the initiatives of individual entrepreneurs exploiting the tourism potential of the valley. It is difficult to assess which route promotes truly sustainable, community action in the long term (Parish, 1999).

Whilst development initiatives are seen as having to capture local participation through existing networks and organizations and institutions, they can create problems. In Hunza the establishment of different village organizations under the AKRSP, and in Ganesh the financial power of the entrepreneurs, begins to shift the traditional power balance, creating conflict. In this case, though, it is interesting to note that the success of the AKRSP lies in part in the filling of the power vacuum left by the deposition of the Mirs in 1974 and the abolishment of princely states. Elsewhere, as in Nepal, for example, there is a tendency for additional layers to be added to existing power hierarchies, and for powers of action to be given to new offices. These may be filled by individuals who thus attain more power than they would under the traditional system. As they are appointed by the state – and particularly if they are foreign to the area – they become separate from the community, as the formidable barrier of state autarchy intervenes.

The capture and mobilization of institutions is thus not without problems in the development field. The physical character of mountains, together with the complex relationships between different groups and their environment present additional problems to the development of such regions. These include the physical remoteness of many mountain regions from many national capitals, and the likelihood of relatively poor returns on investment by industry due to the lack of local markets and the higher costs involved in production, which discourages many private investors. It is not enough for them to invest in such regions purely for altruistic reasons of supporting marginal economies; this trait is all too pervasive in the current capitalist globalization trend. The particularities of development in mountain regions are considered in more detail in Chapter 10.

## Key points

- The socio-political environment refers to the social organization and community structures which form the framework of activity. It encompasses issues of power and authority, gender and resource management.
- The household is the basic unit of social organization, production and consumption. It provides the labour and manages it to provide the fuel, fodder and food necessary for its maintenance.
- Households may be grouped by lineage or affiliation into larger communities such as villages or clans. These groups may be further amalgamated in a hierarchy of organization. This interconnectedness adds some stability, as well as being a cause of friction and competition in the wider society.
- Power may be held by hereditary individuals, such as rulers or priests, or by elected or hereditary groups of elders. These make up the local institutions which lie at the root of social organization.
- Some communities have considerable freedom of participation in decision-making, whilst others are bound by feudal-like systems in which there are obligations and expectations on both sides.
- Institutional structures embracing people on many different levels, and the distribution of power are both important in managing access and use of natural resources, and in resolving disputes regarding grazing, forest and water.
- Mountain communities are increasingly being integrated into the wider world economy. Development initiatives can upset the local balances of power, both directly by investing power in different areas of the social hierarchy, and indirectly by the effects of capitalism and shifts in wealth distributions. These initiatives have both positive and negative effects on traditional structures.

# Chapter 6

# Mountain populations and demographic change

The structure of and changes in mountain populations have, like their anthropogenic character, been the target of early studies of mountain communities. Mountain communities were thought to be sufficiently remote for early demographers to consider them as essentially 'closed' (having little connection with external populations) and dependent upon their own finite resource base. Thus such communities became the focus of studies, such as those of Malthus in the eighteenth century, who wanted to understand the dynamics of population control within such finite constraints.

The concept of environmental determinism, prominent in later nineteenth- and early twentieth-century work, emerged, propounding environmental controls, in particular food resources, and the adaptation to the mountain environment as the main determinant of population dynamics (Ratzel, 1882; Semple, 1923). This was superceded by studies concerned with other controls on population growth and a growing understanding of the complex checks and balances in operation. These checks and balances involved cultural and social constraints, such as marriage and inheritance practices; and physical constraints such as seasonal labour demands, diet and health which affected reproduction rates.

The traditional alpine demographic model has therefore evolved into a greater recognition of a social and cultural framework which responds not only to physical resource constraints on food, diet and livestock support, but also to external opportunities afforded by different types of migration, and to changes in economic conditions. These are considered below: first the changing conceptions of demographic change in mountains, including the roles of social and cultural values in controlling growth, and second in the consideration of migration as a conscious strategy for survival and for economic progress.

## Models of demographic change

Population models seek to illustrate the relationships between the two components of population change: fertility (birth rate) and mortality (death rate). Other components are also important: nuptiality (marriage) affects the death rate, numbers of households and the ways these households appear within the community. Migration may be on a temporary, permanent, or a cyclic (e.g. seasonal) basis and represents net gains and losses in population. A 'Traditional Alpine Model' (TAM) of population emerged from the early studies mentioned above, based primarily on data from the European Alps. It is only in more recent years that sufficient data from other mountain regions have been available for comparative analysis. This traditional model refers to the status of populations before the Second World War, which has been taken by many writers as a watershed between the '*ancien régime*' and the modern world.

The TAM comprised high rates of legitimate fertility, low nuptiality and relatively high age at marriage, resulting in relatively low, constrained growth. The norm for the Alps used to be 8–10 persons in a family – now perhaps only 2–3 children are born. The high age at marriage resulted from the need to own a home before setting up a family – usually arising from inheritance following the death of a parent. Connected to this was the fact that as relatively few marriages occurred, there were high rates of celibacy amongst adults. This arose from the practice of impartible inheritance (i.e. the estate remains intact) usually going to the eldest son (primogeniture). This practice limits the number of potential households to the number of existing estates, and often meant that only one sibling of a family married, thus reproducing rather than increasing the family. The model corresponds to the 'Western European' model of Wolf (1970). Variations on these patterns exist throughout many mountain regions. In Nepal, the youngest son inherits the family house and is expected to care for his parents in their old age. Elder sons can set up their own households, however. Other communities pass the estate through the female line, such as the Karen in Thailand.

However, to apply a single model of demographic processes to the whole Alps is to oversimplify reality; Wolf (1970) described an alternative model, the 'Mediterranean' type (described below) to account for differences not so much in the demographic trends as in the controls in operation. Furthermore, Viazzo and Albera (1986) made a detailed study of five regions of the European Alps, including locations in Switzerland, Austria and the western French Alps and proposed three variations on the Traditional Alpine Model:

- the TAM with low illegitimate fertility, typical of the Swiss canton of Ticino in the 1870s
- the TAM with high illegitimate fertility typical of the western Alps in the nineteenth century
- a pattern of very high legitimate fertility, low nuptiality and low illegitimacy as in the Swiss Canton of Uri, 1870s

These variations were attributed to the coexistence of different economic and social circumstances in each region. High illegitimate fertility tended to be associated with regions where impartible inheritance prevented many adults from getting married and setting up their own households. Where low rates of marriage occurred, they might be associated with either impartible inheritance, or due to emigration of adult males for economic reasons, which would reduce both marriage and birth rates.

## Box 6.1    Törbel: two centuries of change

Netting's detailed study of the village of Törbel in the Valais canton in the Swiss Alps is an important one. The pattern of change (Figure 6.1) shows a stable condition until 1770, after which there is a general increase to a peak around 1950. Mortality fell from 30/1000 in the eighteenth century to 20/1000 in the nineteenth. Birth rates fell dramatically in the 1830s, 1860s and 1890s, during which time there were also increases in deaths, creating periods of decline within the general trend. It is these data which have given rise to much detailed conjecture regarding the causes of the rises and falls in population, in relation to climate, food, disease and economic change. Until 1890, there appears to be a link between fluctuations in births and deaths: they both follow the same pattern with the latter lagging slightly behind births. At the beginning of the twentieth century the breaking of this link was thought to be due to a distinct increase in births in each family, arising from marriages taking place much earlier than previously. Marriage was linked to inheritance, as it was necessary to have access to land or livestock to support a family, but whilst this held true for many areas of the region, in Törbel partible inheritance occurred, allowing more siblings to marry. However, Netting also shows that estates were not necessarily actually broken up, but existed with multiple households, remaining more or less intact. Total fertility rose as a result of increased longevity of adults, and also reduced infant mortality.

*Source*: Netting, 1981.

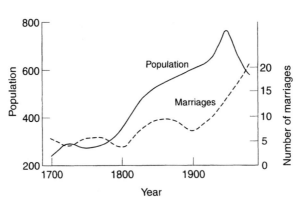

**Figure 6.1**    Population changes in Törbel in the Swiss Alps for the period 1700–1974. Numbers of marriages are also shown (data from Netting, 1981).

## Cultural constraints on demographic growth

Further research in the European Alps revealed significant differences in the inheritance patterns, ages at marriage and family structures even in adjacent valleys. For example, Viazzo and Albera (1986) examined distinctive differences in adjacent towns of St Felix and Tret in the Swiss Alps. The former corresponded to Wolf's (1970) 'Western European' pattern of impartible inheritance; stem families; fewer later marriages; high legitimate fertility; high celibacy. Low growth is thus maintained by the fact that there are few opportunities to set up productive households, and the family line and estate are kept intact.

The second village, Tret, conformed to a 'Mediterranean' pattern of partible inheritance; multiple, extended or joint families; low legitimate fertility; low celibacy. Population growth was controlled by social and religious institutions which controlled sexual activity, but did not prevent marriage of several siblings. The family estate was divided equally between siblings, but reamalgamated by the marriages between households, which could call upon an extended network of resources to survive.

Both the models demonstrate similar patterns of overall low growth, but different ways of controlling population growth sufficiently to keep within available resources. Later marriages in St Felix occurred when the adults were in their late twenties, thus reducing the number of reproductive years per couple. This mechanism is status-driven, with the need to retain the estate intact, and the pressure to do so increasing the acceptance and expectation of many adults that they would never marry. In the case of Tret, marriages might occur in the early twenties, but the expectation was that there would be no, or few children, indicating a degree of sexual abstinence or other birth control. The joining of families allowed a larger pool of labour, and made up for the fragmentation of the estates upon the death of the parent.

Several factors are therefore operating to control population growth in these cases and are applicable to many mountain regions. The two cases above illustrate the interconnected nature of religion (such as Catholicism and its view of illegitimacy) and social conditions governing the nature of inheritance (Box 6.2) in restricting population growth.

Inheritance systems not only controlled the population indirectly, but also maintained the honour, status and wealth of established family lines, and stratified social hierarchies. Citizenship and property rights tend to be associated with household ownership or membership. Those without such rights could not survive and were therefore physically excluded from their subsistence base. Impartible inheritance patterns also acted as a mechanism for maintaining honour, status and prestige, and land became an object of status rather than reflecting need. Thus a wealthy household with few people owned more land than it needed. This might be controlled by renting out excess land and by the fact that a small household might lack sufficient labour unless they could buy in extra workers. This aspect has increased as a result of cash cropping – for example, the employment of Karen tribesmen by Lisu and Lahu farmers who are growing lucrative cash crops instead of opium.

Other controls on marital fertility are associated with diet and with the agricultural calendar, where the physical separation of families had a significant influence, including the separation of spouses as a result of different labour demands for part of the year, particularly summer. Whilst the men were up on the alp with the cattle, their wives would be working the lower fields. This

## Box 6.2    Demography and theology in Nepal

In Nepal, Fürer-Haimendorf distinguished between the Hindu settled 'cautious cultivators' and the Sherpas, who were Buddhist and 'adventurous traders'. Whilst this distinction is now less clear than it might have been in the past, it is an interesting example of cultural distinctions. The Hindus followed patterns of partible inheritance, similar in many ways to the 'Mediterranean' pattern of demography, whilst the Buddhists had a system of impartible inheritance, with the youngest son inheriting on the understanding that he looked after the parents. The Hindu farmers lived on the lower-altitude zones where it was possible to cultivate land and engage in settled agricultural economies. They were viewed as more conservative in their attitude to change when compared with the relative openness of the Sherpas to experimenting with alternative sources of income. However, the farmers were well adapted to their environment, and used every available piece of land to best effect, but the populousness and intensity of cultivation gave an impression of overuse and crowding. The Hindus and the Sherpas coexisted and each benefited from trade and exchange with each other, supplying shortfalls in each other's production economy. The Sherpas, however, limited by less productive land, and less amenable to settled cultivation, needed to supplement their predominantly pastoral economy with other sources of income. Their physiological adaptation to high altitudes enabled them to become guides, traders and later tourist and mountaineering guides and porters, and they formed links both over the mountain and down the valley. Several Sherpa siblings normally entered the monastery and thus became celibates, and whilst they might contribute labour, or were supported by the community, they did not contribute to demographic growth, whereas all the Hindu siblings worked the land, however small the plot available to them might be.

*Sources*: Fürer-Haimendorf, 1964, 1975.

applies to many societies where an element of transhumant pastoralism is part of the economy. Netting (1981) noted that most of the registered births occurred in the winter, which conveniently coincided with times when the family was confined to the house or village by snows. This implies conception in the spring, before separation during the summer.

## Health and nutritional constraints

Conception in the spring months was probably facilitated by the fact that this represented the time of year when the lean, monotonous, carbohydrate-based diet of winter began again to be supplemented by fresh green vegetables. Individuals would have had the chance to rest as activities were constrained during the winter months. The boost in the quality of nutrition is known to be significant in revitalizing women's fertility. Not for nothing is the 'spring itch' seen in most living, reproducing things. In addition to this, various spring festivals, particularly religious ones, tended to mark the spring months with holidays and feast days, when work would be laid aside, and revelry indulged in. Thus the social calendar and the quality of diet are important factors in controlling women's reproductive capacity in traditional mountain societies. In addition, the long nursing (up to three years) of infants, combined with poor nutrition and hard labour, would also reduce fertility.

It is frequently proposed that the introduction of the potato significantly enhanced the quality and quantity of food to each household in the Alps, Himalayas and elsewhere. This would have enhanced women's fertility and contributed to stronger, healthier babies, and lower infant mortality. The introduction of the potato and other New World crops such as maize certainly had an impact on the agricultural subsistence base of many mountain areas but the nineteenth-century idea that it was directly responsible for a population explosion is debated. The potato, originating from the Andes, is well suited to cultivation in poor soils and harsh climates (Zuckerman, 1998). The Andean peoples had some 230 varieties to choose from, and as the tuber is 80 per cent water, they were able to develop a freeze-drying process using the diurnal fluctuations in temperature. The resultant meal could be kept for some 10 years, but was rapidly reconstituted and made edible. The tuber has a high nutrient value, providing all the essential nutrients except calcium and vitamins A and D, which could be derived from milk and dairy products. Moreover, it could support many more people per unit area of cultivation than maize, wheat or soyabeans. The replacement of tubers with cereal grains by the Spanish in the Andes was one cause of the population decline. Cereals did less well at high altitudes and represented poorer nutrient yield per unit area.

The potato arrived in Spain around 1570 and had spread to most of Europe by 1600, although mainly as a garden crop (McNeill, 1992). It was adopted cautiously by many populations. In the Swiss Alps, it was taken up relatively rapidly, in order to meet shortfalls in food supplies. The expansion of cultivation could support higher populations and may have enabled families to have more children (Netting, 1981; Wiegandt, 1977). This may have contributed to the unprecedented growth of populations in the Swiss Alps in the late nineteenth century. However, in the Austrian Alps the crop was only taken up on a small scale. This is attributed to discouragement by local lowland elites and to the fact that the traditional coping strategy here in times of famine was emigration rather than starvation (Viazzo, 1989). Thus the potato could maintain higher populations and provide greater nutrition and intensified production without undermining the traditional economic base (Pfister, 1986). The endemic in-stability of mountain environments, where cold wet summers inevitably led to failure of grain and hay harvests, could be circumvented by the potato, which is more resistant to poor climates (Pfister, 1983). Thus it could introduce greater flexibility and resilience into the traditional system. However, its distribution is limited to wetter areas of northern Europe rather than drier southern Europe and North Africa, where it requires irrigation (McNeill, 1992).

In the Nepal Himalayas, the potato seems to have arrived in the mid-nineteenth century, at a similar time to the start of its cultivation in European mountains, by several possible sources – Darjeeling through the British colonists, from Tibet by Sherpa trading routes, or from Kathmandu, where it seems to have been cultivated in the eighteenth century (Fürer-Haimendorf, 1964, 1975; Hooker, 1969; Stevens, 1993). It has had an important economic impact and some claim that the Sherpas were pastoral nomads prior to its arrival and have settled since (Bjønness, 1980). However, again it was taken up gradually. Some villages were highly resistant to it and banned it from their communities' lands, despite its greater yields and resistance to disease. The potato has, since then, become an important staple and a response to the need for intensification, rather than a perpetrator of it. New knowledge, beliefs and festival cycles needed to emerge alongside the new crop for it to develop its own niche in the cultural framework of the community.

General health is another factor affecting population growth, particularly the effect of epidemics. The plagues, which in some periods coincided with climatic cooling, brought about devastation of the population, abandonment of land and lack of labour to tend and harvest crops. In other regions, the introduction of modern health care, and particularly the eradication of diseases such as malaria have had substantial effects on population growth rates. Indeed, it is this that has traditionally been blamed for the population growth and assumed consequent land degradation envisaged in models of Himalayan degradation and collapse (see Chapter 10). In parts of the Andes the indigenous population was decimated because of its lack of resistance to European diseases such as measles (Figure 6.2).

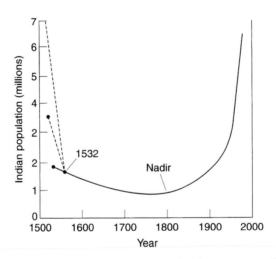

**Figure 6.2** Population of Andean Indians in Peru and Bolivia, 1500–1970. Figures before the Spanish Conquest are estimated from archaeological studies of pre-Conquest settlements. (Data from Kubler, 1952 and Salzano, 1968; after Stone, 1992).

The complex interaction of environmental, economic and cultural constraints, which we have seen in operation in the Alps and Himalayas, is, like any finely tuned model, subject to instability. The balance between population growth and resource availability is one which, both directly and indirectly, has been critical to limiting population in mountain regions. This balance has been disturbed throughout history, by both human action and environmental change (see the LIA, Chapter 2).

## Other factors

Changes in employment and education opportunities have reduced the dependence of many mountain communities upon the land to support their populations. The opening up of alternative employment in industries (for example cottage weaving) and particularly tourism (see Chapter 12) has brought new income and thereby the means to import any shortfalls in local food, fuel and fodder production. Alternative livelihood strategies can therefore play an important role in determining the population that can be supported in a given area.

Reforms to inheritance laws have also had an effect. Despite the Swiss government making it a legal requirement to divide estates equally between siblings, many families agreed informally to maintain the estate as a whole, and honour their older traditions. Thus, although the reform might make it possible for many more new households to be established, in fact the cultural tradition is maintained and can still operate to constrain population growth. However, the new opportunities in tourism in particular have reduced the necessity for individual households to possess or have access to land in several environmental niches and thus the rationale behind the maintaining of the estate as a whole is weakened. For example, a family might make a going concern of winter and/or summer tourism by only having access to high pastures, whereas in the past this was but one element of the subsistence economy. Population may grow in these areas if farm buildings which have been abandoned or converted are taken up by outsiders as second homes (see below).

In conclusion, it is difficult to separate environmental from cultural and economic controls on demographic change, and whilst the present intellectual climate has moved away from pure environmental determinism, the role of the physical conditions of mountain life cannot be ignored. The concept of carrying capacity, or the optimum number of people a finite resource base can support has largely been criticized (Zimmerer, 1994; Leach et al., 1997) as a simple reading of the physical environment cannot predict the pattern of population change, or the opportunities or strategies of the resident population. As we shall see in Chapter 9, it is true that economic pressures can overcome environmental constraints by opportunity or necessity, but the physical structure of the landscape remains the fundamental framework upon which all these activities take place.

## Alternative strategies for population control: migration

Migration has always played an important role in the subsistence strategy of highland economies. However, it has more recently become an issue of concern

on a national scale. This is because of the resultant depopulation and abandonment of areas – for example, high alpine pastures and villages in the European
Alps. Elsewhere, households have become highly dependent on migrant income
rather than agriculture, as in Yemen (Vogel, 1988a) and the Anti Atlas. In addition
to this, national governments are concerned about the impact of migrants on
urban centres and the social and economic problems of their integration into
society. Three broad 'types' of migration strategy are discussed below.

The traditional mountain economy is based on cultivation, pastoralism and
sylviculture; the pastoralism element may comprise some form of transhumance
or nomadism. For significant parts of the year, the shepherds or animal tenders
are absent from the main homestead on an annual cycle of temporary migration.
This is usually up to higher pastures in the summer, returning to the village,
or moving to lowland pastures during the winter. The degree of attachment to
the sedentary community varies – in the High Atlas of Morocco, members of
the household may only be away seasonally.

For the Sarakatsani shepherds, who occupy the western limits of the Pindus
mountains in Greece, close to the Albanian border, the migration is a permanent
movement of ethnically distinct households and their flocks, which occurs in a
continuous cycle throughout the year (Campbell, 1964). The pattern of continual movement is closer to true nomadism than a seasonal transhumance
between fixed bases, occurring at fixed times of the year. During the summer
the community is more closely knit as they move across the high pastures, but
in winter the shepherds and their families must return to the lowlands and are
dispersed across the coastal plains over a much wider area. They do not have
grazing rights in the lowlands as they do in the uplands, so their search for
winter fodder is a difficult business, especially with the extension of developments and privatization of land in the lowlands. They have access to the high
pastures because the local settled farming population, the Zagori, derived most
of their wealth in the eighteenth into the nineteenth centuries from remittances from emigrant members of their households. The émigré's local wife
stayed at his parents' home, until he could afford to establish his own household, eventually retiring there. Thus, as the source of subsistence was not
based on the land, these farmers could safely leave the high pastures to be used
by the Sarakatsani shepherds. This way of life is primarily threatened by the
disappearance of the winter pastures. This pattern is reflected elsewhere such
as in the Middle and High Atlas of Morocco (see Box 6.3).

Transhumance and nomadism permit the seasonally available high pastures
to be effectively used, and in the tropical mountains, a flexible rotation system
prevents overgrazing on high grasslands which are subject to diurnal rather
than seasonal fluctuations. The success of this strategy depends upon sufficient
labour within each productive unit for some to be spared during the summer
(see Chapter 7) or on cooperative labour exchanges or herd-sharing between
households. Transhumance is a common feature of traditional mountain livelihoods and continues in many areas, including Spain (Garcia-Gonzalez et al.,
1990; Perez and Saez, 1990), the American West (Rinschede, 1988) and the
Middle East (Shoup, 1990), where it occurs seasonally.

## Box 6.3    Morocco Middle Atlas – changing pastoral strategies

Pastoralism in Morocco does not occur in a 'pure' form; i.e. it is part of a more extensive collection of activities which make up the people's livelihoods. The mixture of herding and farming spreads the risk of failure of one or the other much more widely. In the past, low population densities and slow growth rates favoured pastoralism, but in the present day not only are the populations much larger, and thus less easy to move and sustain on the move, but encroachments onto traditionally accessible rangelands have affected the pattern of seasonal mobility. Access to markets has encouraged settlement and investment within reach of the markets. But efforts to intensify the pastoral production system have generally failed.

The Ait Arfa tribe (shown on the confederation map, Figure 5.2, page 109) kept flocks of sheep on the limestone rangelands of the Middle Atlas. Precipitation is higher in the mountains, making them a more favourable location than the adjacent lowlands, despite the relative lack of surface water. Seasonal movements were dictated by the availability of pasture and water, and these were supplemented by areas of cultivation in the Gigou valley. The effect of colonial French rule was significant. Large areas of winter lowland grazing were closed, which meant that the Ait Arfa were effectively cut off from their kin on the Atlantic coastlands, making it difficult to move their flocks between winter and summer grazing lands. As a result, some tribesmen were able to maintain rights of access to the lowlands or even buy their land, but many others could not. Therefore, the upland settlements became increasingly important and cultivation increased. This encroached further onto common grazing lands, restricting pastoralism still further. Introducing new techniques, especially the concept of privatized grazing lands with strict rotations, did not sit well with the traditional extensive system, and as a result many innovations failed. This was partly the result of intermittent droughts which meant that specific pasture areas with carrying capacities calculated on the basis of average conditions were not able to support stock in poor years. This reflects the difficulties of applying modern scientific theory to reality.

*Sources*: Bencherifa and Johnson, 1990, 1991.

Another form of permanent migration is that of shifting cultivation, as practised by the hill tribes of Thailand. Whole villages remain at one location for a period of a few years, clearing the forest and growing crops until the natural fertility of the soil is exhausted and the family moves on to a new location. After their departure, the forest regenerates and under conditions of relative land abundance, the site may not be reoccupied for a century or more (Walker, 1992). Under the present political and demographic conditions, however, population increase has resulted in a scarcity of sites to move to, and thus longer use of each plot and a faster return to a regenerating one.

Politically, the imposition of national boundaries has made it difficult for populations to move freely in some areas which may fall on either side of a boundary – the Karen people of Southeast Asia, for example, exist in Myanmar, Thailand and Laos and it is not possible for them legally to move freely across these boundaries. In other cases the establishment of a national park, conservation area or the enclosure of land for commercial logging concerns restricts movement also. The restriction of free movement has also occurred elsewhere, where the settlement of nomadic populations has occurred and has modified the traditional livelihood substantially, and not always successfully as in the Middle Atlas of Morocco (see Box 6.3).

*Temporary* migration from the household may also occur for economic reasons. As the impact of a capitalist, cash-based economy has penetrated mountain areas, both the need and the opportunities for earning cash as an addition to household income have grown. During the relatively slack months following harvest of their own crops, which always occurs before that of the lowlands, farmers from high valleys, such as the northern High Atlas, and the Himalayas of Nepal, migrate down valley to earn cash harvesting lowland crops. Fruit picking is particularly common, as this is less amenable to mechanization than most lowland cultivation, and thus still requires large numbers of manual workers. For the most part it tends to be the young males who migrate, especially in cultures which restrict the activities of women. Other forms of migration for cash earnings include the weekly or slightly longer-term

---

## Box 6.4   Andean coffee strategies

Coffee farmers in the Andes have developed a pattern of activity which permits wealth to be accumulated during the period when a couple are married but before children appear. They cultivate coffee as a cash crop on leased lands at lower altitudes whilst they can spare the labour. This enables them to accumulate cash, to make up for lean years when crops are poor, or when the woman of the house is pregnant (and thus less economically productive), and for the time when the children need to be educated or otherwise supported. This mechanism worked well, enabling lower land to be used extensively, thus not exhausted, by spreading the risk over many different options, and by using labour when available. It is, however, at the mercy of several external factors, such as the prices commanded on the coffee market, competition from the large haciendas, and the eviction from land seen by officials as abandoned. This lack of security of tenure prevents a more full exploitation of this cash cropping, leading, in turn, to the taking over of this land by commercial enterprises. An inherent distrust of cooperatives has also prevented farmers in a similar situation joining forces to pool resources and maintain production on combined plots of land.

*Source*: Collins, 1983.

migration to work during the tourist season, or in a nearby urban centre, returning at the end of the season, or at weekends.

A second form of temporary migration is that related to crop failure, and occurs at a particular time during the household's life cycle. This is illustrated by the case of coffee farmers in the Andes, described above. Such migration occurs in different forms in many mountain regions. For example, Allen (1988) describes the migration down valley of Papua New Guinea highlanders following a series of failures of their potato crops. They maintain exchange networks with lower-altitude tribes for just this eventuality, moving down with their pigs to support them, returning briefly to plant new crops, and then remaining in the lower villages until they are ready for harvesting.

Finally, there is permanent or at least long-term economic migration, consisting of several years away (even two or three decades), with occasional visits home. Migrant workers from the mountains of Morocco are common in Europe, particularly in Italy, the Netherlands and France and those from Yemen in the Gulf States (see below). In Morocco and Hunza (and elsewhere), having a migrant worker in the household can add substantially to the income of the family, allowing improvements in agriculture and houses, education of children, and a fall-back in times of crop failure (Plate 6.1).

**Plate 6.1** A Black Lahu man in northern Thailand. In many villages the emigration of young males may mean that changes occur in gender roles to ensure sufficient labour exists to do all the subsistence work. This man is minding a grandson. The fighting cockerel is a treasured possession.

## Box 6.5   Local entrepreneurs: changing societal traditions

In the Hunza valley, Pakistan, entrepreneurs of one village, Ganesh, have established highly lucrative tourism businesses, based on the hiring of jeeps and guides for mountain trekking. Adjacent villages, particularly Karimabad, have an apparent monopoly on hotel accommodation. One family in Ganesh has done particularly well; the lineage has a long tradition of being prominent in decision-making, and now influences the pattern of development. This family can now afford a house and office in the capital, Islamabad, and educates its children there. Like other children from such families, they become distanced from their village peers, and learn a different set of knowledge from them. Whilst the village-based children learn more of the local traditions – how the irrigation systems work in particular – their city-based peers have an advantage in securing opportunities in a different world. The cultural clash comes when these sons try to maintain their position or prominence in a village to which they are no longer native. Therefore, whilst economic migration has positive benefits in endowing this village with schools, hospitals, mosques, etc., it has distanced some of its children by the different opportunities they have received and by their initiation into a different world. The father of this family has tried to keep his sons in touch with the village, showing them their ancestors' graves and making sure they participate in traditional ceremonies such as Eid with their village peers. However, they are conspicuous by their new, clean clothing and sophisticated and well-fed appearance, a distinction which goes much deeper than mere looks.

*Source*: Parish, 1999 and unpublished observations.

There are a number of development and economic problems associated with migrant labour, including the economic dependence of the remaining families on remittance income, which, as in the case of Yemen, can be brought to a swift end, thus undermining their resource base. On the other hand, the influx of these earnings has done much to improve the standard of living in many communities. In Morocco, a consortium of migrant labourers originating from the same tribe and valley pooled their resources to provide electricity for their home valley (here the tribal allegiance enabled the wealth of many people to be combined in a form of obligation to their community). Likewise, funds can be used for schools, mosques, roads, etc., which all confer status and prestige upon the donor. Income is also invested in new houses and luxury items such as televisions (run off car batteries if electricity is lacking), videos, cars and other forms of conspicuous consumption. The degree to which income is invested in agricultural infrastructure varies, and it is common to find in some areas almost negligible cultivation, whilst the houses demonstrate relatively luxurious living.

In northern Pakistan, two worlds exist in many village communities. One is inhabited by older men, women and children, who remain in the mountains,

## Box 6.6    Emigrant economy of Yemen

The economics of remittance income is dramatically demonstrated in the case of Yemen. During the 1960s and 1970s, the boom years of the oil industry, large numbers of Yemenis migrated to the Gulf States, particularly Saudi Arabia, to work in the construction of the developing infrastructure of these countries. They were able to call upon high incomes (compared with those in Yemen), and sent much of this money back to their families. There, it went to improve the standards of living of the family, increasing their purchasing power, and thus permitting the purchase of imported foodstuffs with the consequent reduction in agricultural activity.

Two issues arose – first the cultivation and consumption of Qat instead of subsistence crops. This required more irrigation water, and thus disturbed the traditional pattern of water rights (see Chapter 8). Second, much of the money bypassed the national economy, so that individual wealth could be substantial in a country where national wealth was very low. Thus any development of infrastructure, or enterprises such as commercial development could only arise from individuals, not from the state. This high dependence on remittance income was terminated as a result of Yemen and Saudi Arabia's different views of the Gulf War. As a consequence of this, Yemeni workers were thrown out of the Gulf, and returned to find a collapsed agricultural system, few economic opportunities and an agricultural base which seemed to consist almost entirely of Qat. In more recent years, there has been resurgence in agricultural development, as Yemen has sought to find its feet after the external mainstay of its private economy had collapsed.

*Sources*: Swanson, 1985; Al-Kasir, 1985; El-Daher and Geissler, 1990.

maintaining the subsistence production needed to support them. The other world of the younger men exists outside the valley where greater numbers go as wage labour, taking industrial jobs in the cities or portering in the mountains. The men become dependent on the women for subsistence but without raising their status. As many development initiatives are directed at male activities, the women do not benefit from them. The partnership between family workers turns into a master–servant relationship, and family values and the economic and political significance of marriage all become hollow and marginalized (Hewitt, 1999).

## Permanent outmigration and foreign immigration

Many mountain areas of the developed world are suffering considerable outmigration on a permanent basis as families decide to move to urban areas in search of employment. This arises in part from the economic marginality of the areas; the greater costs of production, limited yields and limited potential for mechanization mean that these areas appear substantially economically disadvantaged. The attraction of employment in the growing service sector in many European countries is greater than that of remaining in the mountains, despite subsidies and development capital investments by the EU. The results of abandonment include the collapse of terraces, the reversion of pastures to scrub and resultant increases in forest fires. It is difficult to rebuild such landscapes, and the alternatives to traditional agriculture are limited, but there are areas which have capitalized on the growing market for specialized products such as organic and trademarked items. Whilst this is a successful counter to emigration, the greatest alternative economies lie in tourism (Chapter 12).

In the Swiss Alps, concerted efforts have been made to reverse the outmigration from the mountains. Grants are made to support traditional agriculture, particularly the use of the high alp, where the maintenance of meadows and pasturing is an essential part of the character of the Alps, sought by the tourist industry. Financial incentives in the form of reduced taxes, transport subsidies, etc. also help to make life more attractive for permanent residents, even those not involved in farming. This has helped to ease the problem of too many second and holiday homes, and the consequent demise of local services as a result of a primarily non-resident population. Certainly, the impact of 'amenity migration' is a significant force for change in mountain regions (Price et al., 1997).

Education is an important factor in social mobility. In traditional societies such as those of the Swiss Alps up to the 1950s there was little incentive for children to be educated. Their labour was needed on the farm and they focused on practical skills rather than academic study. The ties of the family, farm and church shaped their lives. The lack of alternatives meant that there was no reason to change. However, with the improvement in road networks and increases in construction jobs this began to change and investment in education from the 1960s onwards became an increasing agent of betterment, though the transition from practical to academic is slow (Friedl, 1984).

Parallels exist with outmigration from the Appalachian mountains, though the pattern of change shows differences. The distances from the mountains to cities such as Detroit, Cincinatti, Cleveland and Chicago were too great for frequent trips home. However, workers returned about once a month and eventually retired to their mountain villages. Population in the Appalachians is much more dispersed and communications less good, compared with the densely populated and nuclear settlement pattern of the Alps. The effect of the Appalachian Regional Commission set up in the 1960s was very variable, with some areas developing education and alternative employment and others changing little. Appalachia was also much more dependent on coal and timber as the basis of its economy and the farming tradition was much weaker than in the Alps. Another difference is the setting up of communities in the cities of kin, which acted as ports of entry for emigrants from the mountains. They could retain their own cultures, blended less with the city and thus had stronger links with their home. In the Alps there tended to be a division between rural and city; emigrants blended into the city life and did not retain their mountain culture so strongly, though they still returned in retirement (Friedl, 1984).

*Amenity migration* is the movement of lowlanders and outsiders to the mountain regions, into mountain communities in search of natural beauty, tranquillity and raw nature. As mountains have long held a fascination for many people, the growing wealth and leisure time available to urban elites is now penetrating traditional mountain economies. This is often met with disfavour by the old-established residents, as these incomers do not have the local knowledge of the community and its environment, and are seen as rich parasites with excessive consumption habits. However, others, including national and regional administrations, can view this as a means of rejuvenating and effectively using mountain regions which they perceive as having little else to offer.

Whilst a resident population can manage and look after the landscape, visitors, whether permanent or seasonal settlers, can bring in substantial reserves of capital. The restoration of abandoned buildings and land can provide employment, especially as either the taste of the new residents or the planning regulations of the area may dictate that local techniques and materials are

---

## Box 6.7    Santa Fe: the effects of amenity migration

The effects of the development of amenity migration in Santa Fe, New Mexico, are typical of the global trend. The attraction lies in both the cultural traditions of the Native American, Hispanic and Western American peoples, and the natural environment, including the cooler summer climate, which is more pleasant than the desert lowlands. Land is taken from other uses, including tourism, as the establishment of migrant communities is exclusive. One aspect of this is the marginalization of the local Hispanic community from the historic core and periphery due to high property values and taxes.

The activities of the migrants are primarily knowledge-based: intellectual establishments such as colleges and museums as well as alternative learning centres specializing in spiritual and philosophical traditions predominate. A survey in 1988 found 50 such establishments and by 1995 the number had increased. The retail concerns (575 in 1988) were heavily dominated by restaurants (200), clothing and art galleries (150), serving a resident population of 64 000. Few of these offered suitable employment to the local population who lack the skills and are unlikely to develop them in this context. Although economic rejuvenation has occurred the price has been a hijacking of the local culture and environment. Such rejuvenation can also bring exploitation of the locals and their knowledge (where local knowledge of plant resources is exploited for external uses, for example) but also environmental protection initiatives. Thus there are positive and negative sides of such developments.

*Source*: Moss, 1987, 1994.

used. Some of these traditions originated in the colonial era, such as the summer retreats of the British in the Indian Himalaya. This now occurs in many areas including the Phillippines, northern Thailand, the Indian Himalayas, the High Atlas, Morocco, Mexico, the European Alps and Canadian Rockies. Investment in the upgrading of access roads and of the quality of buildings and their facilities is important in some areas, but in others the attraction lies in the remoteness and 'primitive' nature of the living conditions.

## Key points

- Demographic change is driven by changes in birth and death rates and in patterns of migration.
- In mountain societies, there are both environmental and cultural constraints on demographic change. Environmental constraints include diet and nutrition, the effects of high altitude (hypoxia) and the harsh physical life.
- Cultural constraints include controls on household formation through inheritance, migration as a method of coping with overpopulation and separation of spouses due to the requirements of labour in traditional subsistence livelihoods.
- The Traditional Alpine Model of low overall growth has a number of variations; the two most important are the 'Western European' model of impartible inheritance, stem families and nuclear households; and the 'Mediterranean' model of joint households and partible inheritance.
- Migration is an important factor in mountain demography. It may take the form of permanent or seasonal nomadism or transhumance, where communities or parts of communities are on the move for all or part of the year.
- A second strategy is a temporary migration to cope with harvest failure; moving to another area for a time, or revisiting other land holdings.
- A third strategy is a well-established one in mountain communities where usually young males take work elsewhere and send remittances back to their families. Some mountain populations can be heavily dependent on such remittances, which can support the family and provide facilities for the community but can also act as a disincentive to investment and to maintain agricultural activity.
- A final type of migration is that of amenity migration where people move in from outside the mountain area, attracted by the natural or cultural environment. They can rejuvenate the economy but may also marginalize the local populations as they are much more wealthy and introduce new skills and unsuitable employment.

Part 3

Mountain resources and resource use

# Introduction

Having considered both the physical and socio-cultural environments of mountains and their communities, we now turn to an examination of the use of available resources, both environmental and social, and how they are managed. It is here that the interaction between human and physical environments is particularly close. An understanding of the way that resources are viewed and used, and how they fit together into the pattern of agro-economic livelihoods, is important in order to identify the nature of the problems and potentials of mountain development in Part 4.

The two most important resources are land (Chapter 7) and water (Chapter 8). Each chapter looks at the nature of its resource and how it is allocated, managed and used. Forest resources are included in Chapter 7 as they are closely associated, but not synonymous, with land. The traditional agro-economy, which is discussed in Chapter 9, examines how all the elements of the physical and social environments fit together, forming the basis of the subsistence livelihood. In many mountain areas, the traditional livelihood has undergone significant changes and the search for sustainable rural livelihoods is a central focus of development and change. In some cases the agricultural economy has adapted, in others the basis has shifted to alternatives, such as mining, industry or tourism. The conflicts over control of resources such as land, forests and water are critical issues in the management of mountains.

As a result of the diverse nature of the physical environment and the various hierarchies of socio-economic activity from individual to regional, the themes which characterizes mountain resources and their use are diversity and flexibility. This diversity is reflected in land, for example, in terms of the altitudinal variation in the quality of land and the variations in its potential use. We have already seen in Part 2 the strategies of inheritance which seek to maintain access to a broad range of resources, including hay meadow, upland and lowland grazing, cultivable land and forests. It is the same with water resources, with the additional dimension of downstream users and contested developments such as dams, which benefit lowland urban elites, whilst threatening to bypass the indigenous mountain dwellers. Patterns of ownership and rights of access, of allocation and entitlement, are issues where traditionally the greatest disputes have occurred, and still do in many mountain societies.

The traditional mountain subsistence livelihood is largely based on cultivation of staple crops together with some fruit and vegetables, the keeping of livestock and the use of forest resources for fuel, timber and fodder in particular, although there are many variations on this theme, as summarized in Table Pt 3.1. This reliance on a variety of resources is a risk-spreading strategy: if one element fails there are others to fall back on. The social structure of the society with its elements of individual activity and communal responsibility allows each household a degree of flexibility and self-determination whilst at the same time sharing the responsibility for critical resources such as water and the disproportionate burden of lower productivity resources such as grazing. Thus the social structure is an important framework in which livelihoods are played out. Of course, many regions have undergone substantial change, with varying impacts on the economy, social and political structures and also on the physical environment.

**Table Pt 3.1**  Generalized models of different types of mountain agriculture.

| Type | Characteristics | Location | Sustainability issues |
|---|---|---|---|
| Eurasian mixed agro-pastoralism | Single population exploitation. Mixed land tenure system. Local varieties of crops and animals. Strong communal controls. High level of indigenous knowledge. Historically isolated. | Locally small populations in steep valleys. European Alps, Eurasian mountains, Himalayas, inner China, Mexico, some Andean regions. | Loss of indigenous knowledge. Loss of local land control. Degradation of forests. Government intervention. Tourism impacts. |
| Andean mixed agro-pastoralism | Single population exploitation. Community-based tenure and rights. Hardy local crops and livestock. Strong communal controls. Cultivator transhumance. High levels of indigenous knowledge. | Full range exploitation from low- to highlands. Tropical system: Andes (Ecuador to Bolivia), East Africa. | Loss of indigenous knowledge and local controls. Shifting cultivation and small plantations on colonized hills. Replacement of local varieties with imported variants; genetic erosion. |
| Pyrenean symbiotic agro-pastoralism | Dual population exploitation. Communal pastures, may be rented to outside groups. Local hardy species. Cultivation and long-distance transhumance. Strong symbiotic tie between cultivators and herders. Highly specialized indigenous knowledge. | Full range tropical or subtropical exploitation. Commonest traditional form occurs from Spain to the Himalayas, western Andean slopes. Rare today. | Breakdown of symbiotic tie between cultivators and pastoralists. Government intervention to settle herders. Upland overexploitation leading to degradation due to lack of social controls. |

(*continued*)

Table Pt 3,1   *(cont'd)*

| Type | Characteristics | Location | Sustainability issues |
|---|---|---|---|
| Andean specialized zone exploitation | Multiple population exploitation. Strong trade/exchange links between zones. Land use and tenure specific to each zone. Highly developed community exchange systems. Specific zonal crops, tools and livestock. Political struggles between groups of each zone. | Full range exploitation of tropical mountains. Andes, Eurasia – Afghanistan, Central Africa – Rwanda-Burundi, where zones are reversed – lowland grazing and upland cropping. | Complex issues: High zone depopulation. Middle zone conversion to cash cropping if markets available. Breakdown of trade and exchange networks. Loss of indigenous knowledge. Loss of biodiversity. Decay of irrigation systems. Political conflict leading to overexploitation. |
| Himalayan complex stratified exploitation | Ethnostratified with mixed zonal specialization and intra-zonal exploitation. Complementarity of zones (e.g. seed potato production from highland bought by lowland). Political dominance greatest in lower zones. Strong exchange networks; specialized traders. | Zonal specialization and ethnic stratification loosely aligned. Karakoram – Himalaya, Andes. | Complex issues as above. Complexity means that main issues become hidden. Population pressure in middle hills, colonization projects in lowlands causing deforestation, loss of biodiversity, etc. |
| North American commercial agricultural exploitation | Commercial goals of entrepreneurs. Transhumance movement of livestock by road. No cultivation. Government and public land used. Direct links to external markets. | Western USA, former USSR. | Overgrazing and erosion. Uncontrolled deforestation. Introduction of improved grass species – loss of local varieties. 'Scientific' approaches – carrying capacity, etc. Elimination of wildlife predators. |
| Southeast Asian near urban market farming | Vegetable production, commercially orientated. Intensive cultivation. Few or no livestock. Modern technologies and hybrid varieties. High inputs. | Occurs in all areas adjacent to urban centres. Southeast Asia, South Pacific, Tropical Latin America (1000– 2500 m). | Deforestation for new land. Heavy erosion. High pesticide use and pollution problems. Displacement of local varieties. Displacement of local people by high land values. |

*Source*: After Rhoades, 1992.

# Chapter 7

## Resources: land

This chapter looks first at the concepts of ownership of land and the strategies employed in modifying the landscape to increase its productivity – particularly physical engineering technology such as terracing. It then considers the cultivation of the land – coping with the physical environment by improvement of soil fertilization, planting strategies, etc. which minimize the negative and maximize the positive effects of the mountain climate. The chapter goes on to look at pastures and forests as particular examples of the indigenous strategies employed to control use of these resources.

### Ownership

Property is the basis of wealth in many traditional societies, and especially in marginal environments such as mountains where the availability of good land is at a premium. Property includes land and livestock as well as items of personal wealth, such as the heavy silver jewellery comprising a woman's dowry in Morocco and elsewhere. The importance of land to households has been mentioned in Chapter 6 in the context of inheritance strategies, but it also operates on the level of the tribe. In Yemen (Swagman, 1988), Thailand (Tan-Kim-Yong, 1997), Nepal (Stevens, 1993), the Swiss Alps (Netting, 1981), Morocco (Miller, 1984) and elsewhere, complex regulations exist concerning the prevention of sale of land to anyone external to the immediate community. Land might be offered for sale to a kin group first, or at most, to others within the tribe, but stringent efforts are made to prevent its being lost to the village. Such an event is seen as a weakening of the power and status, as well as the resource basis, of the community. In any case, it is only in more recent years that the concept of buying and selling land has evolved.

Land certainly changed hands, but by mechanisms which did not involve cash or exchange. In many mountain communities, land is not considered to be owned absolutely and thus individuals do not necessarily hold legal title to land as much as an entitlement to cultivate it or to have access to it. Communities tended to hold land and their entitlement to it was recognized by neighbouring communities. In Mexico the Zinacantecos, descendants of the Maya, believe in the 'Earth Owner' (*Yahaval Balamil*) who owns livestock and water, controls weather and all the products of the earth, which are gifts to the people. During times of hardship or misfortune, compensation offerings must

## Box 7.1    Shifting ownership in Thailand

In northern Thailand, shifting cultivation is practised by the hill tribes (Plate 7.1). They only consider that they own land temporarily, i.e. during the time when it is actually cultivated. Land is reallocated by the local headman according to need and available labour. Cultivation occurs on a plot for one or two years after which it is left fallow for up to 12 years – or longer for the Karen people – and maybe for 80 years at higher altitudes, where recovery takes longer. The land therefore reverts to ownership by the spirits and deities. The requirement of central government that all households should register their land is incompatible with this situation – for how can they claim permanent ownership of physically static plots when the natural course of events is always to rotate land? Confusion also arises as a result of the government claiming ownership of land which it sees as unused, and which the hill tribes see as 'theirs in waiting'. Modern pressures on land and enclosure by the state and logging companies means they must cultivate their plots for longer and return more quickly.

*Sources*: England, 1997; Hirsch, 1997.

**Plate 7.1**    Shifting cultivation in northern Thailand. Increasingly, steep land is cleared and cultivated for longer due to growing populations and restrictions of tribal movement. Clearance of these plots is surrounded by ritual to ensure good fortune.

be presented to make amends (Vogt, 1990). In Ethiopia, the products of the land are considered to be a gift from God, who retains ownership of all the earth. Thus the people have no real concept of a market economy. Travellers should be made welcome and share the bounty of the earth – it is considered inappropriate to make money, or even to buy and sell food (Wolde-Mariam, 1991). The introduction of a cash-based market economy therefore is at odds with Ethiopian cultural perceptions.

In some cases, registration of land by communities has made it possible for their claims to be upheld against adjacent tribal incursions, and has enabled their traditional rights and practices to be continued and recognized as legal by the authorities. This does, however, bring the penalty of taxation, although this may not be significantly different from the taxes previously claimed by the local rulers. When the Mir of Hunza was deposed, his land was reallocated by the state to tenants and sharecroppers on the basis of which land each household had traditionally cultivated over several generations (Kreutzmann, 1988).

A number of different patterns of ownership may be recognized – whether this is considered to be absolute or in the context of stewardship:

• Privately owned land belonging to households or individuals. This includes most cultivated, highly productive terraced land, which is prized and confers wealth and status on the owner, e.g. lower terraces of richer soils and irrigated rice paddies. Private land can be claimed by clearance and improvement; in Morocco this is known as 'zina'. In Yemen, this rarely occurs, as such land is generally not worth improving, being of such poor quality. Elsewhere, the labour needed to construct new terraces is such that it becomes a communal or at least joint venture between households, each of whom benefits. Private land may be rented or sharecropped.
• Privately owned but communally managed land. This may actually belong to individuals, but communal decisions control which crops are grown, where and when. This applies in parts of the tropical High Andes where year-round cultivation occurs, and strict rotations are orchestrated to minimize disease i.e. to ensure seven years between potato crops on any one plot, the length of time needed to prevent nematode infestation (Sarmiento et al., 1993) and to ensure sufficient time of fallow for the land to recover. Private land may also revert to being common seasonally or for longer periods; for example, in Nepal, after the grain harvest land becomes open-access for grazing, and thereby common, until the time for planting again. The owner benefits from manuring, whilst the shepherd gains grazing. The opening of fallow land for grazing is a common feature of the pastoralist economy (see Chapter 9). Strict rules apply in many communities, such as in Nepal, concerning the prevention of animals straying onto growing crops when grazing adjacent fallow land (Bishop, 1998).
• Village common lands. True common lands are owned by the community and have associated with them rights of access or use by designated people, and exclusion for others. Such land often comprises extensive rangelands or various kinds of forest. Regulations are enforced to restrict numbers of

flocks, animals, forest products, etc., thus controlling the pressure of use. Such land is often territorially defined, by watersheds, or topographic features. Other lands nominally do not belong to any particular tribe, although several may claim rights to it. In the Andes, the transition to a market economy has caused an increase in demand for cultivable land over common grazing land, and the latter has been parcelled up in some cases, and distributed out to entitled households. Thus the concept of common property and common responsibility has been eroded in such cases.

• Reserved forests and grazing lands. These form a particular category of common lands, where the regulations of use may be strictly enforced, or the land preserved for specific purposes. In sacred forests, individual trees or the forest as a whole have particular ritual or spiritual significance and felling, gathering or hunting may be prohibited. Such is the case in northern Thailand where individuals are identified with trees at birth. An umbilical cord ceremony ensures a close affinity between people and forest. Other examples include the sacred groves of the ancient Mediterranean, where the remnants of former forest remain (for instance, the cedar groves surrounding shrines in Morocco) and the Lama forests of the Himalayas, which are preserved for spiritual reasons. In the case of grazing lands, the concept of set dates between which entitled people are permitted to graze under strictly enforced conditions is discussed below.

The pattern of mountain land ownership is therefore complex. In two villages in the Indian Himalayas, for example, ten categories of land use with different associated rights were identified, including three kinds of forest, four kinds of grazing, and other categories (Berkes *et al.*, 1998). The varied physical environment produces lands of different qualities suited to different uses. The functions of different zones – pasture, forest, cultivated land – are managed in different ways in order to minimize risk and maximize potential productivity, both on an individual terrace and on the scale of tracts of communal land managed for the benefit of all entitled users. The effectiveness of these systems is a major factor contributing to the resilience of mountain communities over the centuries. In more recent years, however, due to population growth and external intervention – particularly the penetration of the capitalist market economy into mountain regions – these systems have been disrupted in many areas.

Land ownership and tenure is a serious issue in many mountain regions. In Chapter 5 the idea of communities of resistance was mentioned; the Chipko movement concerned with their rights of use of forests and the Andean Campesino movement with their calls for land reform are typical. In Chapter 10 the differences in the conception of property and of litigation between traditional and modern societies is explored in the context of the problems of development. In Guatemala, Ecuador and Peru 90 per cent of the farms are too small to sustain a living and farmers may have 20 dispersed fields to manage. However, 80 per cent of the land in Chile, and 50 per cent in Colombia, Ecuador and Guatemala is held by large landowners and reform in the 1950s and 1960s has been incomplete. Reallocation of land as a result of colonization

## Box 7.2  Appalachian poverty in the midst of wealth

Landownership is fundamental to the local culture and dictates power relations within the community. Two periods of transformation have occurred; in
the 1890s and the 1980s. The region has an image of poverty but is in fact
very rich in mineral wealth. The investment of UK funds in the 1890s in
economic development, especially of the rich coal and iron resources, meant
that much of the land was bought up cheaply from illiterate farmers who did
not appreciate the mineral wealth. Within three years towns of 5000 or more
incomers grew, along with mining, tunnels and roads, etc. Where previously
55 000 pieces of land were owned by 33 000 people, land became concentrated in the hands of coal 'barons', so 1 per cent of the owners held over
50 per cent of the land. Some 85 per cent of the mineral wealth and 72 per cent
of the land was held by absentee landlords.

In the 1950s and 1960s the decline in the coal industry and massive emigration of the rural population, who were landless and unemployed, led to the
region being targeted by a War on Poverty campaign. However, the mines
were effectively replaced by the oil companies, so the situation did not change
greatly, with a growing gap between rich and poor. Much of the wealth is
exported, whilst strip mining and other environmental impacts remain. Many
of the local people are separated from both land and hope of a living there.

*Source*: Cole, 1986; Gaventa, 1984.

is also an issue; in the Mount Meru region of northern Tanzania land was
requisitioned by the German colonists in the late nineteenth century, and its
redistribution by the British in the early twentieth century did not favour local
African populations. For example, the reallocation of coffee plantations was
very slow and thus the system continued to marginalize indigenous users who
had already long-standing disputes over rights of ownership and use (Spear,
1997).

## Management of land

The management of cultivated land concerns not only ownership and access
issues but also the physical management of hill slopes to develop and maintain
flat land suitable for cultivation. Terracing of slopes is a ubiquitous adaptation
in most mountain areas where settled cultivation takes place, and in many
areas of considerable antiquity (Plate 7.2). The construction of terraces requires the cooperation of the whole community. Terraces may be individually
owned, but they cannot be constructed individually or by individual families.
Terrace building is a precise art; the dimensions of step and riser (Figure 7.1),
the number, the surface angle, and whether they are designed to drain or to
hold water are all aspects which are involved in their construction. The steep

**Plate 7.2** Terraces in Hunza. Each of these plots has been prepared for potatoes. The wall running across the picture demarcates adjacent village lands. Notice the shade cast by the trees; on these south-facing slopes trees may only be planted where the shadow falls on the same owner's terraces, not on a neighbour's.

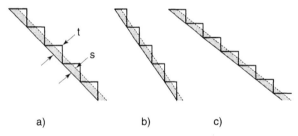

a)                    b)                    c)

**Figure 7.1** Relationship between slope angle, soil depth (s) and terrace height (t). The height of the terrace perpendicular to the slope is toughly twice the soil depth (Figure a)). Figures b) and c) show the differences in terrace size arising from different slope angles but similar soil depths (after Moldenhauer and Hudson, 1988).

flights of terraces so well known on mountain slopes not only represent a tremendous physical effort, but also a detailed knowledge and understanding of the climate, slope geometry and surface sediment characteristics.

Methods of construction have not changed substantially over time, although some mechanization is applied in the present day. In the past it did not involve surveying instruments common to Western planners. Instead, the people had an inherent 'feel' for what the land can support, which is still the case in some traditional areas today. There are only rough estimates of the labour required for terrace construction; Treacey (1989) estimates 610 worker days are needed to construct 3 ha terrace in Peru; Guillet (1987) 40 days for 0.03 ha walled terraces, and Clark (1986) 1320 days for 1 ha in Eastern Bhutan. Thus, a team of 10 would take four or five months to construct a hectare of terraces, assuming no delays. This could be accomplished during slack seasons. However, as terraces are so vital to livelihoods, providing the only extensive areas of flat land, construction tends to occur on greater scales where land permits, and as several families would gain land, the pool of labour available is not only larger, but allows for individuals to come and go as other duties require.

The construction of terraces takes place working upslope in a series of steps (Figure 7.2). They are usually built up on the slope rather than being cut into the slope itself. A stone wall or earth bund is constructed, and backfilled to form a level surface which can be cultivated. The height of the riser and size of terrace area depends on slope geometry: walls are higher and terraces narrower on steeper slopes. Considerable care is needed to maintain terraces, as failure of one wall has knock-on effects both up- and down-slope. Major collapses require substantial labour investment to repair, and usually involve loss of land for considerable periods until repair is effected, which may be delayed until the slope appears stable. Small breaches are repaired as they occur, but larger breaches and failures need to stabilize first. Mitigation measures include modifications of the basic flat terrace in order to hold or to shed water, together with effective drainage mechanisms to channel water to minimize damage.

In Nepal, two main types of terrace occur: *khet*, which have a bund to retain water to enable paddy rice to be grown, and *bari*, which are rainfed terraces gently sloping outward to shed water. *Khet* occur on less steep slopes which can bear the weight of saturated terraces without collapsing. Steeper slopes have *bari* terraces, which need to shed rainwater to prevent collapse. Thus *bari* terraces are reserved for rainfed cultivation whilst *khet* terraces are irrigated. However, in recent years, the need for intensification has resulted in steeper land being used for irrigated rice paddy, with the result that small-scale failures are an increasing problem, taking up farmers' time and resources (Gerrard and Gardner, 2000).

The creation of terraces can stabilize slopes and protect against erosion, so long as they are maintained. In the Mediterranean Basin, large-scale abandonment of terraces as a result of the resurgence of pastoralism following the collapse of the Roman Empire resulted in the deposition of sediments in valleys throughout the region. This did, however, coincide with a cooler and wetter climatic phase so it is difficult to separate these causal factors. In the Peruvian Andes and in Polynesia land degradation has also been shown to be of some antiquity (Erickson, 1992; Kirch *et al.*, 1992).

Other engineering practices include erosion control of slopes and modification of slopes to channel irrigation water (see Chapter 8). Monsoon rainfall and storms may be very heavy and quickly exceed infiltration capacities. On terraced slopes, therefore, channelling of water off terraces is an integral part

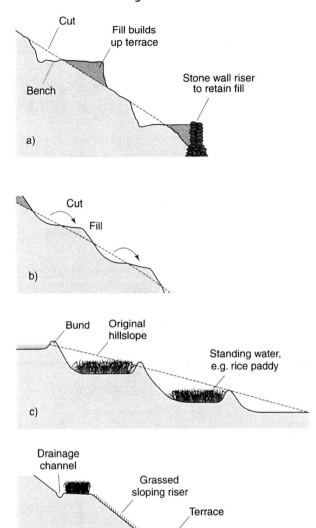

**Figure 7.2**    Construction of two types of terrace: a) bench terraces with walled riser; b) cut and fill terraces with sloping risers. Two common types of terrace are shown: c) a wet or paddy rice terrace (Nepali *khet*) with bunds designed to retain water – these can only be constructed on shallower slopes; d) a dry or rainfed terrace (Nepali *bari*) with sloping risers and channels at the foot to drain water off the surface – these occur on steeper slopes (adapted from Barrow, 1999).

**Plate 7.3** Stone check dam crossing a gully in the Anti Atlas, Morocco. In the foreground is a collapsed dam. The semi-permeable nature of the construction helps the wall stand up against floodwater, but such dams need regular maintenance.

of terrace construction. Unterraced slopes are particularly vulnerable to surface erosion, though, especially if unvegetated. Deep gullies and sheetwash may occur. Gully erosion may be controlled by a sequence of check dams. These are walls built across the gully and serve to trap sediment, and to break the flow down-slope (Plate 7.3). Eventually, the area behind the check dam may be infilled, forming a series of 'pseudo' terraces, which may be cultivated. However, building concrete and other impermeable constructions across gullies causes problems of erosion beneath and around the structure, leading to its eventual collapse. Traditional terraces and check dams tend to be permeable, being constructed of earth or unmortared stone walls, and are less susceptible but not immune to these problems.

Changes in ploughing techniques so the furrows follow the contour, as well as the use of bunds and the blocking of furrows have helped reduce erosion in the Highlands of Ethiopia. Here, soils are clayey and tend to get waterlogged, promoting overland flow. Therefore, improvements to soil texture and permeability are important (Whiteman, 1988). In many mountain regions, soils tend to be stony and farmers clear the fields, piling the stones into heaps. These can act as boundary markers between plots or territories. Development initiatives in Bolivia to exchange these piles for bunds or terraces in order to reduce erosion have not been taken up, however, because bunds occupy more crop area than piles (Clark *et al.*, 1999).

## Cultivation

The cultivation of terraces is usually accomplished by hand as the area is often too small and the slopes too steep to permit the use of mechanized vehicles. This is still the case even in developed countries, where the expense and limited effectiveness of adapting mechanized technology remains an issue. On larger, lower terraces a traditional plough, drawn by donkey or bullock, or a mechanized cultivator, such as that used for paddy fields, may be employed. In addition, commercial forestry has developed machinery capable of working on steep slopes and of selecting individual trees to fell. The use of machinery of any size, however, is not only limited by topography but is also associated with problems of gullying of tracks cleared for access and destabilization of topsoil due to churning.

The soil on terraces often needs considerable improvement and decades – or, indeed, millennia – of tilling and manuring produce excellent, fertile soils, but on new terraces the soils tend to be low in organic matter, stony and poorly structured. The need for manure to maintain fertility and to improve these poor soils remains an important factor in the agro-pastoral economy of many mountain regions (Plate 7.4). Manuring is an important annual task

**Plate 7.4**  Manuring of plots in Hunza. A basket of carefully collected and preserved manure is tipped onto each subdivision of this terrace. Fruit trees are grown on the same terraces as subsistence and fodder crops, making maximum use of limited space.

and the farmers go to considerable lengths to collect manure – even bringing it down from the high summer pastures if necessary. Even the type of fodder fed to livestock is recognized as affecting the kind of manure produced and its possible uses. Cattle manure is prized as it breaks down well, that of poultry and pigs much less so. Because of the shortage of cultivable land, and the investment in time and effort to construct the terraces, they are intensively used and the efforts well rewarded.

Where the growing season permits, two crops are grown in succession (see Chapter 9), known as double cropping. This practice is very labour-intensive, however, and may be constrained by household economics rather than environmental factors. In addition, arboriculture – the growing of trees for crops – is practised. The trees are often fruit (olive, apple, apricot, cherry, peach, etc.) or nut (walnut, almond) and represent not only an important part of the subsistence diet but, increasingly, a profitable cash crop.

The advantage of trees is that other crops, such as grain or lucerne for fodder, can be grown underneath them. A considerable part of the growing season is in spring, when the shading of trees is not a significant problem. In the summer, though, it is advantageous to have harvested the ground crop as shading by the trees in full leaf causes significant delays in ripening. This is why fodder crops are preferred as they are best cut when young and green. Only valuable crop trees tend to be grown on terraces, because of the scarcity of flat land. Trees for fodder and construction timber are located either in forests or plantations beyond the cultivated area, or grown by individuals along irrigation channels where they do not interfere with crops. This is such a critical issue that in the Hunza valley specific rules apply to the planting of trees (Whiteman, 1985, 1988): they may not be planted where they are likely to cast shade on a neighbour's fields (see Plate 7.2 on page 151).

Other factors affecting cultivation on terraces concern the micro-environmental conditions for growth. Each terrace supports a range of micro-environments, and some farmers exploit these by growing several different crops on one terrace. The ground at the foot of an adjacent terrace wall tends to be damper, being partly shaded and also benefiting from seepage of water from the terrace above. Crops here may benefit from extra water but may be retarded by shading effects. On the outer terrace edge, the soil is drier and more exposed, so more drought-resistant or sun-loving crops are grown. In some areas, different strains of wheat, rice or barley are grown on the same terrace, exploiting these different niches. In Nepal, 17 local varieties of rice are grown by hill farmers, with 36 per cent of farmers growing four or more varieties (Bhuktan et al., 1999). In other areas, completely different crops are grown, such as lucerne and a grain crop and possibly a vegetable such as beans as well. The shading effect is enhanced on steep slopes where terraces are narrower and the walls higher. It is modified by slope aspect and other topographic effects and produces the highly diverse mosaic of growing environments characteristic of mountain regions.

Conservation of soil and reduction of erosion are enhanced by the construction of terraces, which not only stabilizes the soil on the surface but also lessens the force of runoff. However, this does not make terraces immune to collapse, as discussed above. The soil itself is subject to several conservation measures, both directly and indirectly. In the Andes, ancient systems of raised bed cultivation are being resurrected in Peru and Bolivia for this purpose (Erickson, 1992). Prehistoric land systems have been examined, dating back 3000 years. They comprise raised elongated beds with intervening canals. The raised areas are cultivated and benefit from the drainage of water, and also, critically, cold air as these are areas of aseasonal climate with potential year-round frost at high altitude. Thus fertility and nutrient cycling could be enhanced, and irrigation or drainage achieved. Periods of abandonment are associated with higher erosion rates as these systems fell into disuse. An estimated population of 37.5 persons per hectare could be supported by these systems. Their reintroduction could promote higher sustained yields in the present day.

Similar systems of raised beds are also found in Papua New Guinea where they occur in the form of mounds. These are constructed from green manure (crop residues) heaped up and covered with soil. Potato runners are placed in the mounds, which protect them from surface frost and cold air drainage, whilst the green manure enhances nutrient recycling (Allen, 1988). This mechanism allows permanent settlement to be sustained up to 2300 m based on cultivation of the sweet potato, introduced in the sixteenth century. Harvesting and planting is continuous in this aseasonal climate. The strategy was originally designed by lower-altitude farmers to overcome problems of friable, nutrient-poor volcanic soils, and carried with them as they migrated to higher altitudes. Green mounds can, in fact, maintain soil fertility without communities needing to relocate, creating a 'stationary shifting cultivation' system (Sillitoe, 1998a). Other forms of green manure include leaves, as in upland Japan (Knight, 1997) and in the Philippines the planting of trees for timber or fruit crops on cultivated lands before setting aside for fallow is a potential strategy to promote both a long-term return from the land whilst in fallow, as well as allowing natural fertility regeneration (Poudel et al., 1999). Certainly in many areas, the depletion of fertility is a much more immediate problem than erosion (Renaud et al., 1998).

Cultivation in aseasonal climates such as the northern Andes and Papua New Guinea means that strict fallow systems are needed to maintain fertility. However, leaving ground bare increases its vulnerability to erosion, and at high altitudes the rate of decomposition of organic matter is very slow, so nutrients may remain latent in the soil for months or even years. In the Venezuelan Andean paramo, adaptations to overcome these linked problems involve a rotation of one to four years of potatoes followed by one year under cereals, after which the field is left fallow for seven to twenty years. The vegetation biomass accumulated over the restoration phase from natural regeneration is sufficient to fuel the next cultivation phase. When in operation the system occurs in a spatially staggered pattern so that not too many adjacent fields are fallow at any one time. Thus any erosion is limited as the soil is captured by the vegetation in an adjacent field, and the distance over which wind gains momentum to accumulate suspended particles is too small and subject to too much turbulence to have a serious effect. At any one time, 10 per cent of the area may be cropped, with 65 per cent under natural vegetation. Of the cropped area, 24 per cent is under cereals and the remainder under potatoes, which are the staple crop (Sarmiento *et al.*, 1993). The fallow period also serves to control horticultural diseases; potato nematodes are a major problem but the fallow period exceeds the survival of the nematodes in the soil. This system is highly effective when operating in this way, even though subject to severe problems, and the need for additional fertilizers or pest control of the fallow period is reduced.

## Pastures

Grazing lands are often located at higher altitudes than cultivated land, where natural meadows above the treeline are traditionally exploited by pastoralists (Plate 7.5). In the mid and high latitudes, such pastures are only accessible

**Plate 7.5**  Pastures in the Spanish Pyrenees. The fields adjacent to the village are manured and sometimes irrigated, producing better-quality pasture for milk cows and young beasts. The rougher pasture higher up is seasonally snow-covered and of poorer quality. The ragged edge of the forest in the top of the frame shows where recent lighter use of these pastures has resulted in encroachment of the forest. The bands of trees are to reduce the risk of avalanche, provide shelter and sometimes represent land boundaries.

during the summer months when the snow has melted. These pastures are therefore effectively 'rested' for several months of the year. However, the slow recovery of pastures from overuse, the effect of marginal climatic changes upon productivity, and the fact that productivity is lower and concentrated in time mean that these pastures are sensitive to overgrazing. Systems of exclusion with respect to times of access and those entitled to use such pastures have evolved in order to protect both the resource itself, and the claims of communities on it.

The regulation of pasture access is a communal affair, and takes a variety of forms, from the exclusive granting of rights to a given population, to the highly restrictive nature of reserved grazing lands such as the Moroccan *agdal* (see Box 7.3). In tropical mountain regions, high-altitude pastures are accessible all year, requiring communal control of access over time in order to allow grazing to recover. It is the breakdown of these systems of control over timing or numbers which leads to the idea of the 'Tragedy of the Commons' in the Andes where continual overexploitation results in irrecoverable damage (see Chapter 11). In many instances these pastures are being reduced as land is

---

### Box 7.3 The High Atlas *agdal*

In the High Atlas of Morocco, the occurrence of the *agdal* or reserved grazing land is a fundamental part of the maintenance of productive grazing lands at high altitude. The Oukaimeden *agdal* is located between 2600 m and the summit of Jebel Toubkal 3260 m on the north side of the High Atlas (Figure 7.3). Within the *agdal* are several types of land – stream banks of rich grass growth, higher slopes for sheep and areas for the cultivation of subsistence crops. The pattern of use is constrained by time – the *agdal* is closed between 15 March and 10 August. These dates were established by a saint in the seventeenth century and are rigorously upheld. Once it is open, grazing continues until the first snows come. During the open season, sheep may graze freely on designated upper slopes, but cattle must be kept in a paddock and fed hay cut from the meadows. It is not permitted to take hay away from the meadow, or to let cattle graze freely. This avoids spoiling the grass, thus maximizing the yield.

The rights to graze are established over centuries by the Rereya and Ourika tribes. Households of the tribes have a right to the pasture, but in reality access and use are constrained further by other factors. First, in order to have the right to graze, the household must own a corral in which to keep their cattle. Without this, they cannot bring cattle onto the *agdal*. These corrals may be lent or rented out to other families if the owners have no use for them. If a corral is unused for any length of time, it may be gradually dismantled as other users use the stones to repair theirs. When it ceases to exist as a recognizable structure, the owner has effectively lost their right to use the *agdal*.

*(continued)*

*(continued)*

   The second constraint is in the form of labour. As a result of the cut and
feed policy, it is necessary for several members of the household to remain
permanently up in the *agdal* during the open season. Many households are not
big enough to cope with the demands of their irrigated terraces and *agdal*
grazing, or have too few cattle to make it viable. Such households may rent
out their stock for the summer, or amalgamate herds with other households in
order to share the labour costs, and to maintain a presence on the *agdal*. A
third constraint to use is distance. The time taken to travel up valley to the
*agdal* may be considerable, and finding fodder en route may be difficult. Thus
households with few animals which live some distance (three to five days'
journey) away are less likely to use the *agdal*. Although the herders maintain
the right to establish new camps, the government has forbidden this.

*Sources*: Gilles *et al.*, 1986; Miller, 1984.

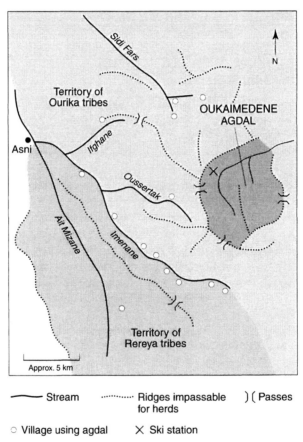

**Figure 7.3**  Location of the Oukaimedene *agdal* in the High Atlas, Morocco. Marrakech is
about 50 km to the north of Asni. The territories of the Rereya and Ourika tribes are shown
(note the former is also shown in Figure 5.1). The cross marks the location of the ski resort
(after Gilles *et al.*, 1986).

allocated on a private basis to individuals who claimed grazing rights in the past. This land is cultivated in response to the growing market economy, the need for more land for a growing population, and the introduction of modern technology to boost yields and therefore income. Common pastures therefore are fragmented, and pushed further up the mountains into more fragile and marginal ecosystems.

In Nepal, the keeping of *zomo* (Chapter 3) is a dairy industry based on the use of individually owned grazing lands. The ownership of a *gode* is critical in claiming rights to pastures. During the summer months parts of families live up in the *gode*, tending their herds and producing cheese and ghee. This requires significant labour inputs, and is a hard life. It is gradually being replaced by yak breeding, and by dairying concentrated nearer to the villages (Bishop, 1989). This relieves pressure on the intensively used middle-altitude grazing lands, but increases pressures at higher altitudes where yak are grazed, and nearer the villages where families tend to remain in the village all year instead of moving up slope.

The maintenance of reserved grazing lands can be an effective way of sustainable pasture use. However, in the case of the *agdal* at Oukaimeden, other external uses have impinged upon it. First the construction of a ski resort by the French. This activity does not coincide in time, as access to the pasture is denied during the winter by the presence of snow, but the effect of ski runs upon grass regeneration is not known, and the obstruction of runs by the presence of the corrals is a source of conflict. A second intrusion is the construction of a dam to provide water for the lowland and for Marrakech. This entailed flooding of some of the meadow land, and restrictions being placed on the presence of animals near to the water for fear of contamination. This is a source of conflict between traditional users and modernization.

In other cases, such as in the Middle Atlas, Morocco, the development of state-supported rangelands, with commercial cattle or sheep production has resulted in the enclosure of tracts of previously common land (which may appear to the state to be 'unused'). This marginalizes the indigenous nomadic community and, more significantly, may deny them access to water (Artz *et al.*, 1986; Bencherifa and Johnston, 1991). Changes in the political economy of the region of Himachal Pradesh, northern India, such as the construction of dams and irrigation projects, plantation forests and canals have changed the nature of the relationship between highland and lowland. As a result of this and of increases in populations of humans and livestock, the traditional systems of rangeland management and the institutions that maintained them are eroded (Chakravarty-Kaul, 1998). The taking over by the state of commons (grazing and forests especially) has bypassed the traditional systems of reciprocity which existed between herders and the communities they passed through en route between highland and lowland pastures. Thus state intervention has affected both pastoralists and agriculturalists in this region.

A more positive and encouraging example comes from southern Bolivia (Preston, 1998), where graziers use livestock as a source of capital and increase their numbers in response to labour migration. This trend has evolved over the last century and has maintained a sustainable use of the pastures, with degradation possibly less than previously, although the nature of the vegetation cover has changed.

## Forests

Forest resources are an integral part of mountain livelihoods and of fundamental importance to their effective functioning. The management of forest resources in a sustainable manner is vital to the effective survival of mountain communities, who rely upon forests for fodder, firewood, construction timber and a variety of food, medicinal and other products (Plate 7.6). The use and management of forests is also an emotive issue on a global scale, arising from the concern over deforestation, global climate change, erosion and lowland flooding.

Perceptions of forest use and the value of forests as standing timber vary considerably from indigenous peoples to national governments and Western scientists. These differences in attitudes and values lie at the root of conflicting

**Plate 7.6**  Karen village huts in northern Thailand made of wood from the local forest. The small clearings and secondary growth surrounding the huts reflect periods of use and abandonment over some years. Banana and other tree crops continue to provide important sources of food in the forest, even when the natural growth returns.

management strategies and stimulate protest groups such as the Chipko movement. For example, the cultivators of the Himalayas and Karakoram view forests as essentially a convertible resource, i.e. under increasing population pressure and growing demands for cultivable land, the conversion of forest into cultivated terraces means a much higher productivity can be extracted from the same area (Gilmour, 1995; Dove, 1995). Compensation in the form of planting on terrace edges occurs to make up for the clearance (Gilmour and Nurse, 1991). This contrasts with the national view of the value of forests as a renewable resource, with the need or desire to retain a forest cover over the land for soil conservation, and with a global view of protection for biodiversity and climate change purposes, irrespective of the local people's needs.

There is a considerable literature on the subject of forest management institutions, including 'traditional', development-orientated and coordinated 'community forest' projects. There is a difference between traditional and indigenous systems; the former term implies some antiquity, whereas the latter implies a system derived from internal or local influences, rather than external, imposed influences. Such systems tend to evolve as a response to changing circumstances – such as environmental or population change – and should be considered dynamic rather than static in structure and in terms of who is involved (Fisher, 1989; Stevens 1993).

Rights of access to forest resources may be determined by belonging (through lineage or residence) to a particular community, such as a village or tribal group. Rights may relate to a variety of different products, which may be gathered within time or spatial constraints and which may differ between different user groups, such as pastoralists and settled agriculturalists. Villages in many mountain areas tend to identify a particular forest zone as belonging to their community. The boundaries may be unclear until pressure and accusations of infringement by adjacent communities occur, when the boundaries may be discussed and negotiated by tribal councils. Thus, the definition of the boundary between adjacent village forests may not be an issue until population growth and demand for timber increase, causing concern over the identification and protection of each village territory (Fisher, 1989).

Regulations may define in which seasons certain products may be harvested, and what quantities are permissible. Arnold and Campbell (1986) identified four categories of controls:

- harvesting only selected products and species;
- harvesting according to condition of product;
- limiting amount of product taken;
- using social means of protecting areas.

Some of these are culturally determined, with the expectation of bad omens if these taboos are breached (see Box 7.4). In the Nepal Himalaya, firewood and leaf fodder may only be gathered from trees near to the village during the dry season, and the cutting of grass is banned during the monsoon. This allows regeneration of the grass during the peak growing season. The period when labour is most hard-pressed and individuals do not have the time to gather

## Box 7.4   Karen forest harvesting regulations, Southeast Asia

The use of certain trees is prohibited by a number of community laws:

- banyan, pipal and golden keruing, which are all dwelling places of spirits
- *se-Koh-Du*, which may only be used for coffins
- trees interwoven with vines
- trees with dichotomous branching
- trees which make certain noises ('crying') when felled
- trees whose fall would be blocked by other trees
- trees occupied by ants and other animals
- trees used for the umbilical cord ceremony (where the cord of a new born is attached to a particular tree and it becomes the dwelling place of the child's spirit).

In collecting forest products, further rules apply:

- Only one rhino, bear or gaur per hunter per year is permitted.
- Some species of animals cannot be killed as their spirits are part of the balance of the forest: gibbons, pythons and certain birds.
- Rattan shoots must be eaten in the forest and not brought home. This applies to other species or shoots, roots and fruits and serves to prevent overharvesting.
- Only two bamboo shoots or stalks per clump can be harvested.
- Some plants and herbs may only be gathered in certain seasons, again preventing overharvesting and allowing regeneration.
- On cleared swiddens, certain trees are always left intact, which provides a source of seed for forest regrowth.

*Sources*: Anderson, 1993; Trakarnsuphakorn, 1997.

wood from afar coincides with access to near-village resources. Once the pressure is reduced, villagers are required to gather fodder from further afield (Dewees and Saxena, 1995).

Similar conditions apply to wildlife – certain species may be taboo, or limits may be placed upon the number the villager is permitted to hunt. In Japan, there is still a strong belief in the existence of spirits in animals (especially wild boar, monkeys and bears) and plants, particularly trees, which are attributed souls (*tamashii*) which need to be appeased before killing/felling. Elaborate ceremonies are enacted by lumberjacks and by hunters who stand in real fear of danger (*tatari*) because their actions anger the mountain spirit (Knight, 1997). It is interesting here that in Japan, forest is synonymous with mountain and there has been a shift from a general fear of a mountain spirit towards the spirits of individual trees and animals, with the culture of ceremonial appeasement developing since the 1940s. This has arisen from the growth of industrial timber plantations which are clear felled. This close association of plants

## Box 7.5  The various forest systems of Khumbu, Nepal

In Khumbu, 'village forest' management systems initially assumed to be simple have in fact been shown to be highly complex, with different goals, rules and institutional actors and arrangements. The main divide is between sacred and secular forests.

*Protected forest types*       *Administration system*
Lama's forests       Village, elected/rotated (*shinggi nawa*)
(Sacred forests established by individual lamas; sacred nature linked to religious beliefs and culturally ingrained in the local population.)

Temple forests       Temple management
(Surrounding temple grounds; logging may be permitted by agreement.)

Private forests       Councils of monks
(Belonging to monasteries for their use.)

Rani Ban       Village council or Pembu (ruler) or guard appointment
(Secular reserve, early twentieth century. Logging limited)

Bridge forests       Village management
(Retained to supply timber for the repair of bridges until the 1960s when cable construction arrived.)

Avalanche protection       Village management
(Not an original function but a guardian system established in the 1940s following avalanche destruction caused by clearance of slopes immediately above villages.)

Uncertain status       Lapsed village management

*Source*: Stevens, 1993.

and animals with spirits which may be angered is a common theme amongst Hindu and Buddhist cultures in Asia. It accounts for one way in which the perception of the forest as a living entity that demands respect differs from that of the Western world where trees are not objects of worship and their connection with a spirit world is much less widely accepted.

The policing of the established rules is again a complex matter. Forest guardians are commonly appointed. They may be full-time, elected for a given period such as a year. Their role is to enforce the restrictions and report violations, which are dealt with by the tribal council or other authority. These guardians are respected in their positions, and it is a matter of honour to uphold the rules.

State intervention in indigenous forest-use patterns or management strategies arises for three main reasons:

- The first concerns national use of timber as opposed to strictly local use. Examples of this are the development of mining and industrial activity in the European Alps, which required timber for construction from the fourteenth century onwards. Although local communes attempted to impose bans on certain uses of their timber, these were not always successfully administered as the demand for wood was so high that locals would risk flouting the bans for individual gain (Ferdmann, 1959; Tromp, 1980). Mining companies later leased the forests, which was a welcome source of income as agriculture was declining and this resulted in uncontrolled clearance. A second example is the extensive clearance of forests, and their appropriation by colonial authorities in British India in the nineteenth century (Negi *et al.*, 1997; Schickhoff, 1995; Tucker, 1988). This intervention had a number of facets; first, the clearance of forest for construction of roads and railways by the British regime throughout India; second, the exemption of cleared land from land tax encouraged felling; and third, the development of commercial plantations, such as tea in Assam and Kerala, replaced the vast range of forest products formerly available with a system of wage labour based on monoculture of a single crop.

- Plantations represent the second key intervention of the state – that of alternative uses of forest lands to subsistence agriculture. The planting of large tracts of fertile land with single species stands for the purposes of commercial markets is a major land use in some mountain areas. Crops such as tea and coffee in particular, and even, on a smaller scale, fruit crops such as apples in the High Atlas, have replaced subsistence cultivation of mountain lands. In addition, softwood plantations in Scandinavia, Europe and the USA are an important source of 'sustainable timber' supplies. One issue which arises here is the question of who has rights of access to the forest resource – the indigenous population who may be highly dependent on it as a source of livelihood, or the state with the interests of its whole national population to be met? This conflict of interest is one area of development which requires mutual understanding and compromise.

- The final area of state intervention is that of protection. Protection of forests for biodiversity and for the purposes of combating global warming has also become an international issue so the interest in protecting a forest can originate with institutions far removed from those living there. Again, there are conflicts of interest over use versus preservation, and over whether conservation should involve local people. Western concerns with biodiversity and carbon reservoir protection can seem very selfish when they interfere with local livelihoods (Chapter 11).

State intervention in many mountain communities has often resulted in denial of access to forest resources and hijacking the commons. Indigenous peoples are scapegoated as the cause of careless deforestation, although this often arises out of an incomplete understanding of how the indigenous systems of management work. Where state intervention has resulted in exclusion of the local population, as is the case in Thailand and Nepal, the result is often more severe degradation of the remaining areas open to them. This compounds the state view of the hill farmers as careless and feckless, where previously the pressure of use was dispersed over a wider area. This accounts for the differences in observation evidence as discussed in Chapter 10.

Access to forest resources may be denied to the local population, whilst at the same time the state may encourage commercial logging in the interests of the wider economy. This naturally causes a great deal of distrust and protest, as has been the case in the Himalayas, Southeast Asia and elsewhere. Even in the Swiss Alps, the privatization of forests led to extensive commercial exploitation and the need for further state action to control felling in order to retain some protective forests (Price and Thompson, 1997). Where afforestation projects have taken place, the species planted may be suitable for commercial reasons (such as pine for pulp), but single-species plantations do not deliver the variety of products found in natural, mixed-species forests. Even secondary forest regenerating in areas of shifting cultivation recovers much of its diversity and thus continues to be a more valuable resource than plantations.

When farmers undertake planting in order to make up the shortfall in available supplies, they carefully select the species most needed, or most valued for different purposes, and also tend to plant a variety of species, thus fulfilling a variety of needs. In addition, the use of forests changes with time: in the Colorado Rockies, forests have gone through periods of extraction up to the end of the nineteenth century, a 'custodial' era where the US Forest Service followed a sustained-yield programme of controlled logging and replanting (1900–1950), to the present phase where the demand for timber has fallen and the costs of extraction have risen, with the result that recreational and protective functions are currently the main goals of forest management.

## Mining

One important area of external intervention in land issues is that of the development of mineral resources in mountains. The example of displacement in the Appalachians has already been given (Box 7.2). Mining is one of the greatest destructive forces in the mountain landscape. Not only is there surface stripping and clearance but also leaching of heavy metals and acid deposition, causing pollution of soil and watercourses. The Summitville mine in Colorado, closed in 1992, has left a $100 million clean-up bill to be paid by the US Superfund (Fox, 1997). The geological formation of mountains has made them very rich in reserves and many of these can be quickly accessed by surface stripping. Prospecting is facilitated by glacially stripped surfaces. Once problems are overcome, including those of access and transport, as well as acclimatization of workers to high altitude (in Chile mines occur at up to 4000 m) the returns on investment can be very lucrative. The development of mines also forms an alternative source of employment and even potentially markets for food which local producers can exploit.

The most important non-ferrous minerals are gold, silver, zinc and copper. The Chilean Andes produces 30 per cent of the Western copper demand, which has been growing at 1 per cent per year for the last 20 years. The largest copper mine in the world, Chuquicamata, is located at around 3000 m near Antofagusta in northern Chile, near to a large underground mine. Farming in the region is limited to scattered oases up to 4000 m. Emigration from these oases to nearby mining towns has occurred on a large scale, creating squatter settlements around the town of Calama and a decline in agricultural activity. Copper mines employ relatively few people for their size, leading to large numbers of unemployed migrants who contrast with contracted workers who immigrated to the area with the mining company (Bähr, 1985).

The environmental cost of mountain mining is enormous. In Himachal Pradesh some 1000 small- to medium-sized slate mines have caused the clearance of 60 per cent of the forest and triggered many destructive landslides of both hillslopes and tailings. In Papua New Guinea's Star Mountains, the Ok Tedi open copper and gold mine is a key source of foreign income to the country, exporting 600 000 tons of copper to Japan in 1991 (Denniston, 1995). The 2330 m mountain, which will be effectively levelled by the end of extraction, is sacred to the Wopkaimin people. Failure of a tailings dam in 1984 caused severe pollution and since then the company has been permitted to tip 80 000 tons of toxic tailings into the river. As the mountain is the source of the river, the whole system is thus affected.

Despite the calls for and implementation of regulations at all stages of mine development, there is a problem of enforcement. The implementation of standards for Environmental Impact Assessment and Environmental Management Systems is not always effective and where a country or region is dependent on the revenue generated by the activity, enforcement is often weak (Hughey, 1998). In Peru, a deputation of farmers met the Minister for Energy and Mines in protest at the soliciting of their lands for mining. The San Gregorio mine at 4100 m produces 75 000 tons of zinc per day for Lima, but at the cost of local rights to the determination of land use and control over their land. The compensation offered was US$4 per hectare per year for 50 years' use. Some 3000 ha have been effectively requisitioned for the expansion of the mine and since 1992 some 20 million ha has been sought (Centeno, 1998).

Once mines have been exhausted – or bankrupted, as in the case of Summitville – there is usually the unconsidered cost of clean-up and rehabilitation, which is substantial. For many companies the remote mountain locations of some mines makes it tempting to abandon the site to natural processes. It is not only the environmental fallout which is important, however, but the communities which have grown up alongside the mine to supply and serve their workers. Occasionally, as in the Cerro Rico de Potosí in Bolivia, the site has some potential as a tourist attraction.

## Key points

- Land is a critical resource in mountain regions. It is varied in its nature and potential use. The most common categories which comprise the basis of traditional livelihoods are high-altitude grazing land; cultivated land, which is often terraced and privately owned; and forest land on steeper slopes and remoter locations.

- Control and ownership of land is variable and generally includes both collective management of common land and private ownership.
- Access to resources is often by membership of a lineage, clan or community. These rights of access are fiercely guarded on an individual and community basis.
- The physical management of land includes the construction of terraces to provide flat land for cultivation. These are labour-intensive to construct and maintain, but critical to the successful cultivation of slopes.
- Specific adaptations in cultivation techniques have developed to cope with the mountain environment. The careful use of manure, water and the considered location of individual crops are highly attuned to the environmental mosaic of mountain slopes. The system is designed to maximize the use of scarce resources and to minimize risk by having a variety of crops and other sources of subsistence.
- Strict and diverse controls over common grazing land and forests are an important strategy to safeguard access by the entitled community, and also to regulate the use of resources to prevent overexploitation.
- Management of commons may be embedded in the culture of a community, but also controlled by both sacred and secular institutions.
- State intervention has often conflicted with local uses, either by exploitation for national use, by commercial plantation activities or by protection for biodiversity and global climate purposes. These often exclude the indigenous population and are a continuing source of friction between local people and national governments as they face these critical land tenure issues.
- Mining is a powerful force on the landscape and causes widespread environmental devastation. Despite the wealth of mineral deposits in mountains and the employment opportunities mining presents, much of the revenue accrues to international companies. Implementation and enforcement of appropriate environmental protection standards is difficult, given the costs of exploitation and need for foreign exchange by poorer countries.

# Chapter 8

# Water resources of mountain regions

Irrigation systems are closely associated with social structures in mountain regions (Banskota, 1999b). Without water, there is no life, and in many semi-arid and arid mountain regions, such as the Karakoram in Pakistan, agriculture is entirely dependent upon irrigation. Apart from agricultural needs, supplies of water are also needed for drinking water for humans and animals and for hydro-power. Mountains are increasingly recognized as 'water towers' for the whole world as this is where the headwaters of the world's rivers lie and a large proportion of river flow is derived from mountain precipitation, snow and ice. With increasing demands for water, issues of access and control, of urban versus rural use and of international conflicts over rights to supplies, water has become a resource of increasing political importance in many parts of the world. It is estimated that some 35 per cent of the global population is likely to experience water shortages in the twenty-first century, increasing the need to manage global resources more carefully, but also with a consideration for the varied uses to which they are put.

The management of water in mountain regions centres on supplies for agriculture. In more recent years, this supply has been claimed by national governments for lowland and urban supplies for domestic and agricultural use, and for hydro-power and industrial activities. This chapter will first explore the management of water for subsistence agriculture, before looking at how this has been changed as a result of increased demand for water by external agencies.

Several aspects of water in mountains have a bearing on its potential uses and thereby the sustainability of livelihoods. These aspects include the source of water, which determines its quantity and quality; irrigation systems which control the distribution of water in time and space; and other constructions which enable water to be captured in very arid regions ('harvesting') and stored for later use. Finally, the socio-political aspects of allocation of water control its actual use.

## Sources

The source of water will determine its quantity, quality, when it is available, and thereby its potential use (see Chapter 2). In terms of timing, rainfall is immediately available, snow when it melts in the spring and glacier sources in mid to late summer. The higher peak in mid to late summer may be important in areas which experience a dry period at this time of year. However, it is in spring that crops are most sensitive to drought, and water is needed to enhance early growth of all crops. Springtime snow melt supplies are dependent on the winter snowfall accumulation. If this is limited, then shortfalls of water may occur at this critical time, and allocation systems are more strongly enforced (see below). In the Hunza valley, Pakistan, there is very little precipitation at the altitude of the villages – around 150 mm (Kreutzmann, 1991) and cropping is entirely dependent on snow and glacier melt. In this situation, a dry winter has significant effects on the potential for cropping the following year. However, the valley has the benefit of spring snow melt supplies and summer glacial supplies, which provide a double peak in supply availability. In other valleys such as the Rereya and Ourika in the High Atlas, there are no glaciers, and whilst there is a greater occurrence of rainfall, snow melt remains a dominant source of water. Rainfall may also be restricted to certain seasons, and may be erratic and unreliable, in which case harvesting techniques may be employed, as is the case on the southern flanks of the High Atlas, which are in the rainshadow of the north.

In terms of water quality, rainfall is warmer and relatively clean. Snow melt is cold but clear unless it coincides with spring floods, when it may carry high suspended sediment loads. Glacial melt, however, is not only very cold, but also contains a high proportion of very fine rock 'flour' derived from glacial erosion. This may form a fine coating on leaves of young plants, retarding their growth. In addition, the coldness of the water can also slow plant growth (Whiteman, 1985, 1988). This is significant in areas where the growing season is already short, and the delay may prevent ripening of the main crop, or the possibility of planting and harvesting a second crop. Various strategies have been employed to reduce this effect: glacial water may be held in ponds or reservoirs to warm up before being let onto the fields. This also allows the sediment to be deposited in the reservoirs, which can be cleaned out during low flow times. Second, temperature-sensitive crops such as beans may be planted further from the source of water, thereby giving it more time to warm up (Butz, 1994). Thus the supply of water itself is complex and variable in time and space, reinforcing the need for effective distribution and allocation systems.

## Traditional irrigation systems

Irrigation systems are a characteristic of most mountain regions, and testify to the effective innovation of mountain farmers in developing appropriate technologies. There are relatively few examples of traditional irrigation technology in the developed world, although there are some examples in the Swiss Alps (Netting, 1974; Crook and Jones, 1999). There are a number of types of irrigation system arising from specific needs and circumstances. Vincent (1995) classifies systems into eight different types: offtake, underground, spate, collection, storage, lift, combined and wetland. The first is the most common,

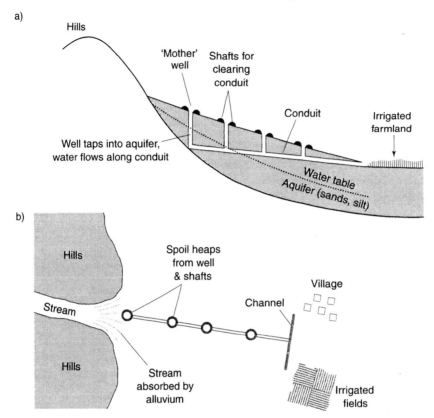

**Figure 8.1** A *qanat* or underground water conduit system designed to capture water at the foot of the hills. The conduit works by gravity, appearing on the surface some distance down slope. There systems have largely been replaced by wells and pumps tapping directly into groundwater supplies.

comprising a gravity-fed system of channels taking water off a river channel or from a glacial meltwater channel, and delivering it via a series of successively smaller channels to individual field plots. This may be combined with other systems, where other sources than channelled water are used. Simple diversion techniques where a stream is temporarily blocked and allowed to flood over meadow land also occur, as in the traditional hay meadows of the European Alps. The time allowed is that taken for the water to flow from top to bottom of the field, after which the stream is unblocked and the next meadow's turn comes (Netting, 1974).

Underground systems tend to prevail in flatter areas where subsurface channels carry groundwater from hillsides to lower land (Figure 8.1). They occur in many areas of the Middle East and North Africa, and are known as *qanat* (Middle East) (English, 1968; Honsari, 1989; Joffe, 1992) and *khettara* (North Africa). They are of considerable antiquity. But as they are expensive and difficult to maintain they have largely been replaced by pipes or by wells with

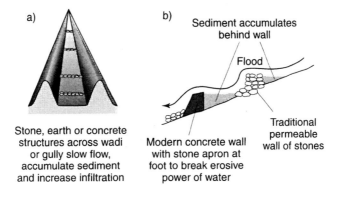

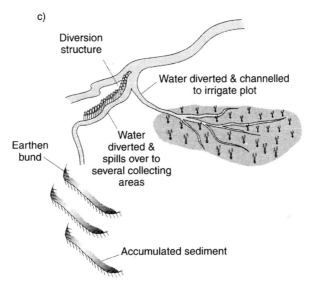

**Figure 8.2** Spate irrigation systems. In a) walls are constructed across wadi floors to capture ephemeral flow and encourage infiltration. In narrower, steeper valleys these resemble check dams. In cross section b) the walls are shown to entrain sediment and slow the flow of water. The upper wall is a typical indigenous construction of stones, which is semipermeable, allowing water to flow through it, reducing erosion damage. The lower wall is a modern concrete wall, which is more durable and has an apron of stones at its foot to reduce undercutting. In c) a diversion structure is shown where floodwater is diverted from the wadi and channelled onto irrigated plots (right) or allowed to spill over the surface onto separate impoundment structures (left).

pumped groundwater access. Spate irrigation utilizes floodwaters and thus is an intermittent and variable supply (Figure 8.2). When rivers rise in their beds, the water may be captured at higher levels and directed onto fields. Such systems occur in arid regions where floods are an additional but ephemeral supply, as in Yemen. Interestingly, different laws of access apply to this *sayl* water in the Muslim world (Vogel, 1987, 1988b): because the supply cannot be guaranteed, anyone who is able to exploit it may do so. This contrasts with standard supplies from ordinary river flow, springs and wells (*ghayl*) which are predictable and to which strict laws of access apply (Varisco, 1983). In addition to flood waters, surface runoff may be captured. In Yemeni this is called *sawagi*, but the techniques also occur elsewhere, e.g. in the Anti Atlas, Morocco (Kutsch, 1982). The water may be captured following major rainfall events by systems of bunds and channels, and stored underground, a form of harvesting discussed below. These are also known as collection systems and are related to storage systems where water is directed into underground cisterns and securely locked and stored for later times of low supply. Alternatively, temporary storage systems are in operation either to allow water to warm up and sediment to settle, or where the allocation for a particular sector of the landscape occurs at inconvenient times, such as the middle of the night. In this case, rather than missing out on their turn, the users may store the water in a pond and let it onto the fields during daylight.

The tapping of springs may be collective or individual. Spring water is prized as it is usually clear, of a good temperature and reliable. It is regarded as a preferential source of domestic supplies and strict rights of access may apply. Diverting flow through pipes to houses may be the responsibility of individuals or of neighbourhoods, which share in the expense of bringing it to their locality.

Finally, in specific regions of the world, the collection of cloud moisture and fog is a critical source of water. Cloud forests on the west coastal flanks of the Andes are supported entirely by fog. Developments in the harvesting of fog by systems of collection such as that in northern Chile (Schemenauer and Cereceda, 1992) can supply up to 22 litres of water to each of 330 residents of the village of Chungungo, whereas before they were obliged to collect a maximum of 10 litres per day by truck from a well 40 km away. The fog is caught on polypropylene mesh and directed to a 100 000 litre storage tank. The water has the advantage of being of relatively good quality. Similar projects have been established in other regions such as Oman, Peru, Namibia, Cape Verde and Yemen.

## Construction

The construction of offtake systems is a major task, requiring the cooperation of the whole community, as for terracing and other major constructional projects. The systems are gravity-led, not requiring any artificial means of propelling the water along the channel. This means that in order to water fields, the offtake (the point at which water is diverted from the main channel) must be higher than the highest field to be irrigated. Consequently, the offtake channel may need to be raised for long distances up to the point at which it joins the main channel. In the Hunza valley, some systems extend as much as 8 km (Kreutzmann, 1988) (Plate 8.1). Where there are problems of glacial advance and retreat, these precisely placed collection points may be overrun by ice, or left high and dry by retreating glaciers, as occurred in the European

**Plate 8.1**   In the lower left of this picture is a collection pond from a glacier offtake in Hunza. Water collects there, depositing sediment, before flowing along the side of the cliff in a channel constructed by blasting the rock face, building up an outer wall and lining with silt. The precision necessary to ensure gravitational flow with minimal, but constant hydraulic drop is successfully achieved despite the techniques. Here the channel is some 200 m above the river bed.

Alps during the Little Ice Age. The channel meanders down the side of the valley towards the cultivated land. The channels are often subdivided, with subsidiary channels coming off at intervals to direct water to specific sectors and plots. Each of these channels can be blocked, permitting absolute control over who gets water and when.

Building the channels may require digging into the hillside or even blasting with dynamite (Plate 8.2). It makes it all the more astonishing that this amount of labour can be conducted with precision without the aid of modern surveying techniques. The community has a detailed knowledge of factors such as slope stability, common avalanche tracks and landslide failure zones, as well as the gradient and character of the slopes (Netting, 1981; Kreutzmann, 1988). The materials used tend to be those locally available, although in recent years concrete and galvanized metals have also been used (see Box 8.1)

The construction of the main and secondary feeder channels is usually a community responsibility (see below) but the furrows that distribute the water on individual terraces are the responsibility of the terrace owner or user. Traditionally, all users contributed labour, or materials or even food to support the

**Plate 8.2**  An irrigation channel directed through a tunnel. Tunnels may carry water beneath roads, footbridges, across active debris slides, and even under houses.

labourers' families whilst they were working on the channels, which required a similar investment of time as terraces (for example, in the *pabchu* system in Ladakh meat is the form of payment, Banskota, 1999a). They might contribute agricultural labour to help on the fields of those constructing the channels in order that they might not miss out on a year's cultivation. It was necessary to have effective leadership to mobilize and organize the community, to decide on beneficiaries, and to ensure completion of the works. Now it is more common for labourers to be paid, and in some cases employment of migrant labour has replaced community involvement in construction in the physical sense. However, any families who hope to benefit from new supplies are involved in some way in order to protect their interests and to bear their share of the burden of expense. This emphasizes the common need for water, which fuels effective participation in maintenance work throughout the year.

Water harvesting techniques are employed in very arid regions. These usually comprise the collection of runoff and channelling into storage systems, and the construction of low bunds on gentle slopes which slow the flow and increase infiltration of water into the soil (Figure 8.3). These bunds may be 30–50 cm high, are discontinuous and systematically scattered across the hillside (Kutsch, 1982) (Plate 8.3). Over time they accumulate soil behind them and form low, gentle terrace-type features. Water tends to concentrate in the vicinity of the walls, and in the finer sediment accumulating behind them and

---

### Box 8.1    Valais *bisses* old and new

In the Canton Valais in the Swiss Alps, traditional irrigation systems have been updated to meet changes in crops as well as to make use of new materials, although the types of construction remain similar. Where channels are dug into the surface, the original earth lining has been replaced with concrete. Whilst this can increase the rate of flow and reduce losses by infiltration, the fine sediment which was deposited in earth channels and helped to seal them only blocks cemented channels so sediment traps become necessary. Artificial channels are built where slopes are prone to avalanche; these are often re-placed by rock tunnels which serve the dual purpose of piped drinking water and open irrigation conduit. Some of the wooden artificial channels have been replaced by galvanized aluminium, but wood is again preferred as it is more aesthetically pleasing, despite losses of 50–75 per cent from the channels by leakage and evaporation. Pipes and tunnels do reduce evaporation and theft, however. Sluice gates of wood and stone are largely replaced by metal, but their function remains unchanged.

Traditional irrigation of pastures was by blocking a stream and allowing it to overflow onto a meadow (*ruissellement*). This gave uneven supplies which enhanced the floral diversity. The increase in fruit and vegetable production in the lower valley and intensification of vineyards has meant that *ruissellement* is largely replaced by spray irrigation. The size of pipes and roses is controlled by community irrigation institutions in the way that *bisses* (channels) have always been. As spray irrigation requires a head of pressure, distribution is increasingly to reservoirs and tanks rather than to channels leading directly to fields, but following a similar timed rotation system of allocation. Spray irrigation is perceived to save expense, time in maintenance and water.

*Source*: Crook and Jones, 1999.

---

it is here that crops grow best. Thus the fields during the growing season have a striped appearance, with crops growing much better around the capture structures than on the remainder of the relatively barren and stony ground. Trees are planted at the foot of the little walls and bunds, also benefiting from damper conditions (Plate 8.4).

Apart from bunds, ditches and other obstacles may be employed to capture or direct runoff. Sometimes a large catchment area on a hillside serves to irrigate a small plot or collects in one cistern. Elsewhere, the angle of dip of the rocks below the surface may help; if impermeable strata dip parallel to the slope surface they can channel water to the base of the slope, where it may emerge as a spring or be tapped by *qanat* or similar systems. Much steeper slopes are less suited for harvesting; the steeper the slope, the greater the force of runoff and the greater the potential instability of the bunds and other constructions (Kutsch, 1982; Barrow, 1999).

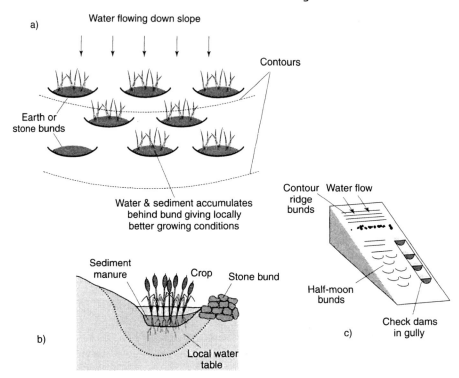

**Figure 8.3**   Runoff harvesting structures. In a) a series of half-moon-shaped low bunds are scattered across the surface. Cross section b) shows the area where soil is improved and water collects behind the bund. In c) a block diagram shows the distribution of contour-ridges, which cross the hillslope, and long low bunds, small curved bunds and gully check dams are all shown (adapted from Kutsch, 1982).

Other harvesting techniques may be employed to conserve supplies, apart from storage in reservoirs and cisterns. In the Highlands of Yemen and Ethiopia where there are two wet seasons, insufficient water may fall in one season to support a crop (Whiteman, 1988). Thus, the fields are left uncultivated, allowing the water to remain in the soil in the hope that the cumulative quantity from that and the following wet season will be enough to make cultivation possible. Rainfall may be collected over several seasons, so the land is only cultivated intermittently. The stony surface serves to armour the soil against erosion and to reduce desiccation by sunshine. Other strategies to cope with drought include the variation of planting densities according to an assessment of the rain or snow season yield (Whiteman, 1988). This is based on intimate knowledge and intuition concerning the weather and crop yields. In the Anti Atlas of Morocco, where the precipitation is relatively high (90–250 mm/yr), crop densities for wheat and barley are about 300–250 plants per square metre, and where precipitation is low (70–200 mm/yr), densities are 70–120 plants per square metre (Kutsch, 1982).

**Plate 8.3** Low, discontinuous stone bunds crossing a hill slope in the Moroccan Anti Atlas. These intercept runoff, increase infiltration and catch fine sediment, becoming 'pseudo-terraces'. During the growing season these bunds bear much denser and taller vegetation.

**Plate 8.4** Detail of a runoff harvesting bund. Notice the trees growing through the wall where water is more abundant. In the middle-right of the picture a penknife gives an idea of the scale of these structures.

## Allocation

The institutions which determine the allocation of water through these delivery systems are a fundamental component in effective agricultural strategies. As mentioned previously, the distribution of water in time and space may be highly variable, and it is the assurance of an equitable share to all entitled households which is critical to survival. Disputes over water, and violations of rights to it are one of the most common sources of conflict dealt with by village councils, as in Hunza.

In many mountain regions, rights of access to water are associated with the possession of land – until recently in the Andes, for example, land and water rights were combined so water rights were transferred with the land. However, with the intervention of state land reforms, and the development of additional sources of supply, water rights have become separated from land, and may be sold at auction. This therefore favours those who can pay, rather than ensuring access to all in the community. The situation arises, therefore, where individuals may hold land with no water. In Peru, the government grants water to communities separately from land, and it is up to the community to allocate it. Likewise in Yemen, the traditional allocation was one measure (*tasah*) of water to one hundred '*libnah*' of land. This applied only to traditional irrigation supplies – from runoff, wells, etc., where a supply was guaranteed (Varisco, 1983). Water derived from the capture of flash floods is exempt from Islamic law as this constitutes an uncertain source, and is therefore free for those able to harness it. However, this traditional allocation per unit of land has been bypassed, partly by the development of new supplies, but also by the increase in *Qat* cultivation (see Chapter 10). This is a crop of substantial economic importance, harvested throughout the year, and thus requiring more than its fair share of water. The economic status of the crop ensures that water rights can be bought or bribed away from subsistence crops.

In most irrigation systems, the tribal council or village community appoints a guardian, as in the case of forests. He may be full-time for a given period – in the Swiss Alps he was appointed for a year (Netting, 1974) – and was traditionally paid in crops and livestock. His duties were to oversee the allocation of water according to the timing rules of the community, and to police its use. He was also responsible for repairing minor damage, for supervising major damage repair and clearing of the channels at the beginning of the season. Many other mountain societies used similar systems, which survive to this day in areas where irrigation remains of critical importance, as in Hunza (Box 8.2).

In most mountain irrigation systems, a strict rotation occurs between areas of land fed by each irrigation channel. Specific channels serve defined areas of the valley, and water may not be taken out by people not entitled to it, even if it flows past their land. In the Canton Valais, however, reference is made to instances where channels have been 'diverted' to direct water to the land of some people rather than others – for example, to the labourers' girlfriends in return for sexual favours (Crook and Jones, 1999).

## Box 8.2   Hunza irrigation management systems

In the Hunza valley, a long-established system of irrigation management has evolved. Water is captured from the glaciers, or the meltwater streams and directed along channels to the terraces. At a number of locations, particularly where the channels diverge, a system of sluice gates enables water to be directed into a different part of the system (Plates 8.5 and 8.6). It is the duty of the *chowkidar* to operate these sluices, thus allowing water to flow along a given pathway towards a given sector of land. As there are relatively few points of offtake in the system, a number of villages may share major channel systems, and in this case, a *chowkidar* may be appointed by each village to ensure none is cheated.

With the construction of new irrigation channels to feed new crops and new terraces, issues over water rights have arisen. The AKRSP, which has overseen many development projects in the valley, has been at pains to establish the rights of access to the 'new' water, as well as any changes in 'old' water rights which may arise, before the construction proceeded. This occurred in the case of the new Aliabad channel, which was enlarged soon after its construction in order to supply sufficient water for the growing community.

*Sources*: Kreutzmann, 1988; Sidky, 1993; Parish, 1999.

**Plate 8.5**   At the point where this meltwater stream flows into the open cultivated Hunza valley, this wooden structure represents the equal divisions of the water supply to three villages and their associated land. When the water level rises to permit water to flow outside the structure, additional villages are then entitled to a share, but this ceases when the water level falls again.

*(continued)*

*(continued)*

**Plate 8.6**  A wooden sluice gate in a dry irrigation channel. These simple, robust structures are lifted to allow water to flow into specific channels for controlled periods of time. The allocation of water throughout the whole system is largely controlled by sluices.

Water may be available for a set time every few days; in the Zangskar valley of the Himalayas, households are entitled to 24 hours of water within a strict rotation cycle. This may not be distributed all at once, which is an advantage as it means water is available 'little and often' rather than in a deluge in which much may be wasted, with a long time gap until the next delivery (Crook and Osmaston, 1994). 'Tailender' issues arising from dwindling supplies at the end of the irrigation network are one of the commonest complaints, along with breaches in the rules relating to allocation. During times of scarcity allocations may be less frequent, but equally distributed to all the communities served by the system (Vander Velde, 1989). In the Indian Himalayas, tailenders may not ask for water during times of scarcity, as they know there is not enough to reach them, but they are also exempted from contributions towards the upkeep of channels during this time (Banskota, 1999a).

Superimposed on this is a more detailed system of timing, by which the amount of water delivered to individual plots is measured. In the Swiss Alps, Netting (1981) described the system of irrigating meadows as measured by the time it takes for water spilling over a temporarily dammed stream, to flow down and reach the bottom of the meadow. At this point the supply is cut off and the next sector of meadow watered in the same way. In Morocco, Mahdi (1986) describes the use of a number of techniques, such as the filling of containers which have small holes pierced in them, or the time taken for a shadow to move between two markers. Elsewhere, it may be the filling of a cistern or temporary storage reservoir which represents that sector's share.

In such a system it is essential that effective monitoring of usage exists, particularly in times of scarcity, where violations can affect many other users. Peer pressure and intimate knowledge of the system by all the farmers operate as an informal means of policing, but it is often the guardian or watchman who is held responsible for ensuring regulations are adhered to. In Hunza, for example, guardianship responsibilities are shared between benefiting user groups. The valley is divided into sectors, each with its own channel or a major share in a channel. The network of sluices is operated by the guardians to let water through the channels for two to three days at a time. This water is subdivided by the villages within the sector, and may be directed towards specific areas in rotation. The rotation may allow for all areas to receive some water every 24–48 hours, and it is the responsibility of individual farmers to ensure they open the sluice for the permitted time and let the water onto their own fields.

A priority system is enforced, particularly in times of scarcity. This is done to ensure all have enough for the essentials. Domestic uses take priority, then the watering of vegetables, grain, fruit trees and finally fodder. Thus no family is prevented from growing subsistence crops and no one can waste water by using it for fodder, which tends to be the least frequently watered crop.

Disputes over water are the commonest source of discord which the village council must deal with in many mountain communities. In Hunza, a portion of the water was allocated to the Mir and his family and lands. After his deposition in 1974, the water was used by those who acquired his lands. Many of the terraces which are supplied by this water have in recent years been converted to domestic use by the construction of hotels. This effectively represents a 'change of use' as domestic usage is higher in the pecking order of priority. However, one of the oldest villages, Ganesh, disputes their entitlement to this water, saying it was 'loaned' to the Mir, but should not continue to be so to those who are not his descendants, and should revert to the village. However, as Ganesh would use it for agriculture rather than domestic purposes, this is a 'lesser' though still important use, and the hotel users argue that their need is greater (Parish, 1999). They also argue that changes to the old system are inevitable and necessary in order to adapt to the changes in population and economy in the modern times. More formal systems developed in recent times tend to be dominated by local elites (Banskota, 1999a) and politicians.

In Nepal, where new systems have been introduced in some areas, conflicts have arisen concerning responsibilities for these systems. In Chakhola, upland farmers argued that they had sufficient supplies and did not need a new system, and resisted contributing to it. Lowland farmers, however, needed the new supplies. In order to resolve this conflict, an Executive Committee set a fixed charge of Rs 600/ha for *bari* and Rs 1200/ha for *khet* terraces, and arranged for the channel to be cleaned twice each year, before the rice and wheat seasons. This appears to have resolved the conflict (Banskota, 1999a).

## State intervention and water development

Two areas of recent change relating to water and irrigation systems involve extension, modification or other change to existing systems. This may arise from indigenous responses to the demand for new land, or from targeted development initiatives instigated by external development agencies. State intervention may result from the need to secure larger supplies for lowland urban populations, or the construction of dams to control flooding and to provide hydro-electric power.

---

### Box 8.3    Cape Verde irrigation developments

Efficient traditional systems were well adapted to the local environment of steep slopes and limited flow, and to agricultural needs. Management is locally based and the interrelationship between different systems within the same basin is important. Farmers may possess plots of land in different irrigation systems, which spreads the risk of failure of any one element throughout the community so it is less likely to affect individual households so much. The rights are well known historically and a series of tanks relating to separate systems is in operation.

Since independence, upgrading of the systems has been a priority, particularly in response to drought in the 1940s, when the need for water and assessment of resources was highlighted. However, development practitioners emphasize individual system performance, rather than the interactions between them. An official view of wastage and inefficiency of the traditional systems prompted reforms, but little evidence of increased agricultural production was forthcoming. The focus of effort was on the infrastructure – the physical distribution system, rather than the social one, but changes to the former overrode the understanding and operation of the latter. This has caused problems in traditional systems, and has required a revised approach based on river basin planning, farmer participation and a less direct government presence. Changes which have taken place include the use of spray application techniques and the use of PVC pipes which help to save water, but competition is so fierce that users threaten the system. Leakage continues to be a problem, with the system losing about 50 per cent, and also rapid overland flow reduces capture and creates damage.

*Source*: Haagsma, 1995.

---

The extension of existing systems can be a major undertaking. Irrigation systems may need to be modified to increase the water flow through the existing channels, or new channels may be constructed at the tailend, extending the network in space. However, where extensive new tracts of land are brought under cultivation, a whole new network might need to be constructed. Many developments taken up by indigenous users are small-scale extensions, or changes in allocation systems and methods of application of the water (see Boxes 8.1 and 8.3, and Chapter 10).

Other development initiatives involving water have centred on hydrogenerated electricity production (HEP). This is perhaps the best energy option for mountain regions, and potentially the least environmentally damaging (ICIMOD, 1997). However, despite efforts to develop HEP, less than 5 per cent of the rural population of Nepal has access to electricity.

Many valleys in mountain areas are well suited to such schemes, being steep-sided with a guaranteed supply of water. These developments take the form of small micro-hydel projects instigated by local entrepreneurs or development agencies for the purpose of local income-generation and improvements to the standard of living (Plate 8.7). The location of the consumers is a

**Plate 8.7** A micro-hydel station in the Swat valley of the North West Frontier Province of Pakistan. The stream is running into a shaft in front of the man and flows by pipeline to the generating station, the small square-roofed hut just beyond the shaft. The electricity is carried in overhead power lines leading off to the left of the station to serve the local village and guesthouses.

## Box 8.4   HEP in Bhutan

The kingdom of Bhutan is a tiny Himalayan state with a population of 90 per cent subsistence farmers and very low GNP. Whilst it has a very small demand for HEP it has about 40 billion KWh untapped supply. India and Bangladesh, however, have projected shortfalls which would easily consume this amount. Bhutan therefore has an interest in securing these export markets, but it is also important that it is a focus for broader economic development in the country. Bhutan has four major and several smaller rivers. The Middle Hills have no permanent snow or ice but are served by the Indian monsoon. This provides a variable supply of water. Development options include a 'run of the river' system which uses the headwaters directly and is subject to seasonal variations. The alternative is major dam and reservoir construction. The latter option is hampered by the probability of earthquake damage, as this region frequently experiences tremors of 5 or more on the Richter scale. Sedimentation is also likely to be a problem. There are few local uses for electricity at the projected sites of the dams and little use for irrigation supplies. In addition, there are few other terraced areas for displaced populations to move to, so there are considerable social issues to be addressed in addition to the lack of physical knowledge of the region.

*Source*: Dhakal, 1990.

pertinent issue; if local villages are to be supplied, then a series of micro-hydel schemes have proved successful in a number of areas. However, the provision of electricity alone is not enough to stimulate development – markets for products are also required, as well as extensive and continued support for users in the form of technical maintenance training and availability of parts.

The second type of project is that of major dam construction undertaken by the state for national (and often lowland urban) markets. Dam construction in mountain areas is troubled by hazards of flash floods, landslides and glacier fluctuations. Many of these areas are located in earthquake zones with potentially catastrophic consequences of dam failure; in the Garhwal Himalayas, the 260 m-high Tehri dam, which will displace about 200 000 people, is being constructed very near to the site of an earthquake in 1991 (Pearce, 1991). Others include the Tarbela and proposed Bashi dams in Pakistan. However, large dams are needed if water is to be transported some distance, maybe hundreds of kilometres to an urban centre.

Such projects are notoriously associated with the displacement of people and loss of farmland, wildlife habitats and livelihoods. In addition, poor impact assessments have meant that in many areas, reservoirs have rapidly been infilled with sediment, with the result that the dam loses its effectiveness rapidly, offering poor returns on the investment. The Sriram Sagar dam in Andra Pradesh lost one-third of its storage capacity within two years of its completion in 1990, and the Nizamsagar Dam in the same region has lost two-thirds of its storage capacity (Pearce, 1991). In China, the Laoying project was abandoned before completion due to siltation.

This exploitation for the benefit of distant urban dwellers or lowland intensive mechanized agriculture bypasses the local population and creates serious conflict and marginalization issues. The construction of dams does regulate the supply more effectively, as it reduces the risk of flooding. Farmers in the High Atlas have taken advantage of this by beginning to use the traditional floodplain for crops, only to suffer heavy losses during years of exceptional rain and floods.

There are a number of high-profile examples of large dam developments, most of which are highly political issues as they are funded by the international community. In the developed world the potential for hydro-power is largely already realized – as in the Swiss Alps, for example, and in the USA the

---

**Box 8.5    The failure of Morocco's irrigation development strategy**

National schemes for irrigating the lowlands have been a focus of development in Morocco during the colonial era and since independence in 1954. This vision was inspired by the concept of North Africa as the 'granary of Rome' during classical times. The French protectorate intended to make Morocco into a wheat-producing nation again. This inspired an irrigation policy (begun in the 1930s) and was spurred on by the vision of California's irrigated fruit and vegetable production.

The redistribution of land at the end of the protectorate was mainly to elite Moroccan political figures. Around 20 new dams have been constructed since 1960, ignoring the traditional cereal cultivation sectors where 90 per cent of the peasants were employed. The rapid rates of sedimentation of the reservoirs has largely been blamed on peasant land uses, but in fact these have not changed for centuries and the blame rests at the doors of poor geomorphological assessments of the rates of erosion prior to dam construction. The investment has favoured the elite and lowland systems, and has been geared towards providing food and water both for growing urban populations and for export. However, the strategy was a failure partly because of the shrinking capacity of the dams but also because the agricultural sector has declined as industrial sectors have grown. The export market also declined in agricultural products, reducing interest in investment. Morocco is unable to sell the product of its 650 000 irrigated hectares and suffers from substantial trade deficits. In addition, the whole vision was poorly conceived, inappropriate and badly implemented. It has still not addressed the issue of reform and integration of highland and lowland which is critical to future success.

*Sources*: Swearingen, 1988; Claassen and Salin, 1991; Mriouah, 1992.

major rivers are converted into linked reservoirs to meet the existing high demand for water. In the developing world, the controversial projects include the Three Gorges dam in the upper reaches of the Yangtze, China. This is the world's largest hydro-electric project (18 000 megawatts) and fierce opposition has arisen from the 1.25 million local residents who will be displaced, and from the international community concerned with loss of biodiversity in the region. Malaysia's 2400 megawatt, 210 m-high Bakun dam in Sarawak will displace some 8000 local people, inundate nearly 70 000 ha, and require a further 80 000 ha of intact rainforest to be cleared for power lines. Other projects are located in Lesotho, where the Senqu river is to be diverted into the South African industrial and mining region, and Chile's Pangue dam on the Bio Bio river. Many of these projects are funded by the World Bank, which has become a target for groups opposed to the destruction of indigenous peoples' homelands as well as that of biodiversity.

State intervention has a number of consequences for indigenous water management. The case of the separation of water from land in the Andes arising from central government land reform initiatives has already been mentioned. Many of these state initiatives have not been successful, partly through lack of understanding of local needs and institutions, as well as through failure to provide technical assistance for smaller projects.

## Conflicts and legalities

Conflicts over water access on an international scale are increasingly visible. This is due to the increase in population and demand for water and the recognition that mountains are perhaps the most important source – the 'water towers' of the world (Mountain Agenda, 1997). Thus there is a growing international interest in the conservation of mountain areas in the interests of water availability to the wider world. There is a conflict of interest between local users of water and of land and needs at national and international scales. International boundaries often follow natural watersheds, as in the Andes and Himalayas, for instance. In these cases, rights of ownership to the water are undisputed internationally, even if they are contentious regionally. Downstream, however, rival claims to water require international law to secure agreement on use and protection, as for major rivers in the USA, where the disputes are between states, and in Asia, where they are between countries. It is here that federal or international treaty law is brought into play in order to regulate access and to defuse or address conflict.

In India and its neighbouring countries, the use of international treaty law has secured agreements on the use of the Indus, Ganges and Brahmaputra. The objective is to ensure that the rights and needs of the tailender (e.g. Bangladesh) are respected, but it also influences the management of water and erosion in the mountains by creating obligations of environmental protection and soil conservation. The necessity for such international agreements is highlighted by the weakness of national environmental laws in Nepal, India and Bangladesh. Precedents lie in the existing Ganges-sharing agreement between

India and Bangladesh and between India and Nepal, and a model for such an agreement lies in the 1985 ASEAN environmental agreement (Robinson, 1987). These agreements override the disparities between national environmental law regimes and foster a sense of shared responsibility over water resources and an obligation to protect them.

There are, however, limits to the application of international environmental law with regard to water rights. These stem from the method by which rights are assigned – by marketing, which ignores the wealth distribution in the community and effectively marginalizes those who cannot afford to buy such rights (as in Peru and Bolivia, where auctioning has replaced land-associated rights). Assigning water rights on the basis of the various functions water performs cannot accommodate social and economic change, and allocation by volume does not take account of the efficiency of use achieved by different modes of distribution or use. In Hunza, for example, the portion of water to which the now deposed Mir was entitled has been taken over by those who now cultivate land previously owned by him. A neighbouring village disputes this as it claims the original 'gift' of water was to the Mir and that it should revert to the village. As it is further downstream it has a vested interest in protecting its supplies. However, the new uses to which the water is being put include a growing use for hotels and hostels, and since such domestic uses have always taken precedence over agricultural uses there is a stalemate in the conflict (Parish, 1999). Thus, changes of use and of entitlements are a highly complicated issue on a regional level, which is greatly magnified at international level.

Nevertheless, there are areas in which environmental law can and should be used to protect the environment (du Bois, 1994). International law is commonly considered to be 'soft' in that it is not enforceable in the way that national laws are. The legislation regarding international watercourses is dominated by issues of pollution, but there are cases where it has been brought to bear upon large dam projects (see Box 8.5). Other cases include Lac Lannoux (1929) between France and Spain and the Gut dam arbitration between the USA and Canada settled in 1968 (Birnie and Boyle, 1992; Sands, 1995). Both these issues concerned the downstream user rights, where one party considered their rights to be curtailed by the construction of a dam on the other's territory. Disputes of this nature occur frequently in the USA, and international agreements between users of large rivers (e.g. the Rhine, Zambezi, etc.) are increasingly common (Sands, 1995). However, these are concerned with the lowland downstream end of the rivers, rather than control over the headwaters themselves.

In the headwater region, degradation of the Himalayas and its assumed link with sedimentation in the Indian lowlands have been the focus of concern over the land use and environmental protection of the uplands (see Chapter 11). Another area where international standards of regulation are increasingly applied is that of Environmental Impact Assessment in relation to the construction of large projects such as dams. The World Bank, as a major source of funds for such projects, has come under pressure to demonstrate due diligence in assessing not only the physical but also the economic and social impact of

## Box 8.6  The Gabcikovo-Nagymaros (GN) Project and international law

The GN Project concerned the construction of the Gabcikovo dam and the diversion of the Danube river. The case was brought to the international law commission in 1992 by Hungary against the Czech Republic and Slovakia. There was an earlier treaty in 1977 agreeing the construction of a dam and two HEP plants involving the diversion of the Danube. The problem arose when the Hungarian government decided that environmental concerns should have precedence over economic and objected to the Czech unilateral continuation of construction. This is a key case where the precedence of environment over economy is at issue and raises questions of the relationship between equitable utilization and obligations to protect the environment.

*Source*: Sands, 1995.

its projects. However, it is debatable whether the real concerns of the indigenous population are fully integrated into such assessments. Despite the sources of funding being international, because of the problems of enforcing international law, it is difficult to force such assessments to be undertaken to a required standard. All too often in such cases big investment and political pressure win over environmental protection concerns. Schemes for assessing the impacts of dams on mountain environments such as that of Aegerter and Messerli (1983) attempt to integrate natural, social, cultural and economic issues and their knock-on effects, but remain complicated to administer, although nonetheless necessary. The 'hydro-debate' centres on the need for energy and the source of that energy; in the longterm, renewable sources are preferable but they do have transboundary issues and substantial local environmental and social consequences which may be incompatible with distant economic development at the user sites (Petts, 1990). This issue is increasingly visible as the media and various interested groups have a louder voice. There is also incompatibility between development and conservation, particularly of biodiversity which is so rich in the mountain regions (see Part IV).

## Key points

- The management of water resources is critical to sustaining life in mountain regions. Domestic use and irrigation supplies are the most important traditional uses of water.
- The supply of water varies in time and space according to its source; the water has different qualities and thus effects on crop production.
- Many traditional systems survive today. There are many types, according to source and topography; the offtake system with surface channels is one of the most common.

- Construction of irrigation systems is time-consuming and requires the cooperation of the whole community, together with effective management and adequate resources.
- Most offtake systems operate under gravity by capturing headwaters and guiding them along channels to be distributed upon fields.
- In arid regions, harvesting of water occurs by the capture of runoff or by conserving it in the soil or cisterns over a period of time until there is sufficient to support a crop.
- Allocation of water supplies is often a communal decision as it is such a critical resource. Allocation may be by space (watering sectors of land in rotation) or by time (period during which water is allowed onto a plot).
- Guardians are often appointed to oversee the condition of the system and the allocation arrangements.
- Extension and modernization of existing water distribution systems often follow traditional regulations and patterns of use. However, some development initiatives conflict with traditional systems by introducing new obligations or restrictions on access or use.
- State intervention in water resources often takes the form of large dam construction. This is often for the benefit of lowland, extensive, export-orientated agriculture or for urban populations.
- Such developments may bypass local users or cause them to be displaced. In addition, considerable environmental conservation issues are raised by the construction of large dams.
- As water is such a critical resource, there is an increase in the application of international treaties and agreements concerning the use of water and the obligations to conserve and protect it.
- International environmental law cannot solve all issues as there is no body to enforce it, but it can instil a sense of responsibility over shared resources.

# Chapter 9

# Mountain agro-economies and livelihood strategies

The livelihood strategies practised by mountain people are an assemblage of adaptations to local environmental conditions, and operate within the social–political structures which determine rights of access, etc. The preceding chapters have outlined these operational frameworks and examined various adaptations to the local conditions, such as terracing, soil management, irrigation and the patterns of control of resource use and access relating to grazing, water and forest. This chapter seeks to give an overview of how the livelihood strategies operate and how the different parts fit together to make an effective whole.

## Livelihood strategies

Traditional mountain economies are based largely on subsistence agriculture incorporating a diverse range of cultivated crops, livestock and use of forest resources. This is supplemented by craft work (production of clothing, household items and tools, etc.) and trade to obtain items not available from their own resources. The diversity of activities and ecological niches exploited is a deliberate strategy of risk avoidance, allowing for the failure of some parts whilst maintaining sufficient yield from other sources. Generalists tend to maintain both cultivation and livestock, and use other resources, notably forest, within a broad altitudinal range, forming compact or extended spatial distributions of settlements (see Figure 4.2, page 102 and Table Pt 3,1, page 144). Specialists, on the other hand, concentrate on either cultivation or animal husbandry, relying on trade links developed through reciprocal and other arrangements to make up the shortfall in subsistence. Thus, both strategies make use of the wide diversity of environmental conditions, using each in the most appropriate manner, and often involve movement of all or part of the population during the year (Figures 9.1 and 9.2).

Whether an individual group or village is tied to one main activity depends on its location and the distances involved; in the High Atlas families can exploit all zones, whereas in the Himalayas the much greater scale of the mountains means that some groups are obliged to specialize, either lacking grazing land, or being above effective cultivation altitudes. It is the cooperation and interaction between these different groups which makes up the effective use of the whole system and ensures that the whole system can support all its populations. Uhlig (1969, 1978, 1995) describes the chain of interaction from

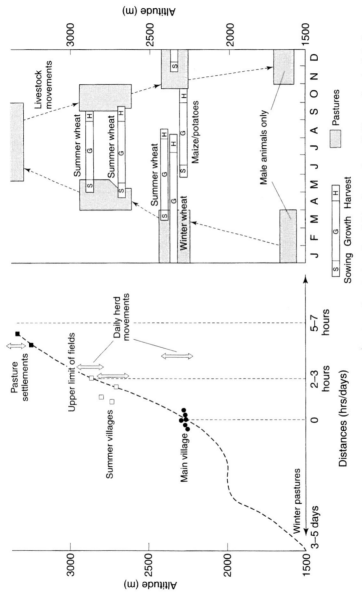

**Figure 9.1** Diagram showing the time–space distribution of agricultural activities in northern Pakistan. On the left the location of different settlements is shown, together with the altitudinal range of daily grazing movements at each site. On the horizontal axis the length of time it takes to move between different settlement locations is given. On the right-hand side the pattern of cultivation through the year is shown at each location. In the higher settlements, a crop may be planted as herds move up valley, visited during the growing season, and harvested on the return journey in the autumn when grazing of stubble and manuring of the fields can follow the harvest (after Ehlers, 1995).

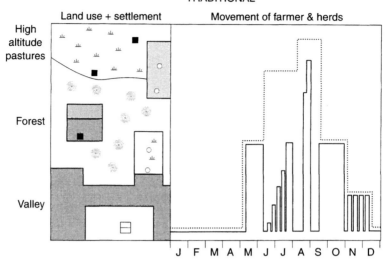

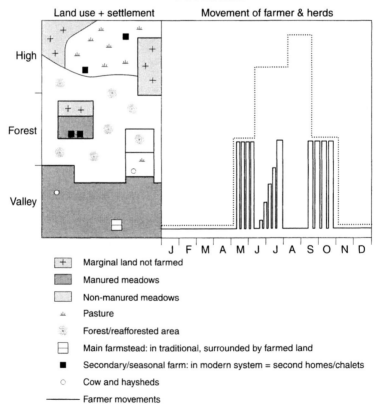

Kashmiri rice farmers in the lower Himalayas and valley floors, right up to pastoral nomads in Ladakh who only cultivate limited quantities of cereals. Extensive literature exists to describe the pattern of resource use and the interactions between production zones and communities, from which a few examples are given in boxes in this chapter.

Mention has already been made in the Introduction of the vertical organization of mountain agriculture, where different production zones occur with altitude. This is known as the *Staffelsystem* (Grötzbach, 1984; Uhlig, 1988, 1995). Lower slopes tend to be terraced for cultivation of grains and vegetables, with tubers higher up, whilst valley floors are used for grazing and hay due to their vulnerability to flood and frost risks. Higher slopes are either given over to forest or grazing, whilst summer high pastures often occur above the seasonal snowline but below permanent snowlines. The number of livestock is constrained by fodder supplies, as they need to be fed over winter. Forests are an important source of leaf fodder, necessary to supplement seasonal or sparse grazing for cattle kept near to the village. In addition, the forest is a valuable source of timber, food and medicinal plants. The animals provide wool, meat, dairy products, cash from the sale of lambs and calves, and also dung, which is of vital importance to the maintenance of fertility of the relatively poor mountain soils. Without all facets of the economy, subsistence becomes untenable; in the case of specialists, as mentioned above, great care is taken to maintain trade, tribe or other links in order to provide for the missing parts of their own economies.

In the tropics, pastures are located in the zone where frost risk occurs for all or most of the year. Stevens (1993) describes in detail the zonal use of land in Nepal and combines two concepts – those of verticality and of production zones, which can be combined to describe the mosaic of interactive livelihood activities of traditional farmers. In many areas these strategies are much modified in modern times, although they have always been dynamic. Recent changes are particularly noticeable where cash crop production has replaced subsistence agriculture, and where remittances from migrant labour have supported larger populations without increasing productivity.

Several detailed studies have been made of the energy production and consumption, and the degree of self-sufficiency of mountain communities. For example, Imlil in the Rereya valley of the High Atlas, Morocco has been studied by Miller (1984) and Dougherty (1994). In the former case, estimates of the productivity and the financial economics have been made, whilst the latter deals with a detailed analysis of the energy balance of selected households. It was estimated that 64 per cent of fuel and 66 per cent of fodder comes from communal lands. Of the fuel, 84 per cent was firewood and the

**Figure 9.2** (*Opposite*) Pattern of animal and farmer movement through the traditional and modern Alpine agricultural year. The uses of each altitudinal zone are shown on the right. Note that in the modern version, the range of movement of the farmer is reduced, as hay may be replaced by fodder and dairying by meat production, or alternative uses of the pastures, such as tourism, may have developed (after Ehlers, 1995).

remainder bottled gas. Food was derived from the environment – i.e. cultivated, with about 35 per cent of the grain needs being met. Fodder comprised a mixture of crop residues, hay and autumn barley, cut when green. In addition, access to reserved grasslands such as the Oukaimeden *agdal* gave further grazing. Shortfalls in fodder are common in late winter and early spring, before the first grass has grown. Cash crops in the form of walnuts and, increasingly, fruit have become more important parts of the economy, together with migrant incomes and tourism, providing cash to buy extra foodstuffs and fodder. What emerges from these patterns of use is a community which operates largely by subsistence but is not self-sufficient, and requires networks of exchange from household up to regional scales in order to ensure survival.

The critical issue is that of maintaining food security, an issue which has increased in importance with growing populations. In Nepal, for example, fewer than 50 per cent of households are self-sufficient in food for six months of the year and a diversity of strategies – of which external linkages and markets are a growing element – emerges as farmers cope with making up this shortfall. Operating these strategies is constrained by caste, ethnicity, gender,

## Box 9.1   Traditional mixed agriculture in s'Tongde, Ladakh

The integration of the different agricultural activities is illustrated on the agricultural calendar. The seasons are determined by climate and different activities dominate at different times of the year (Figure 9.3):

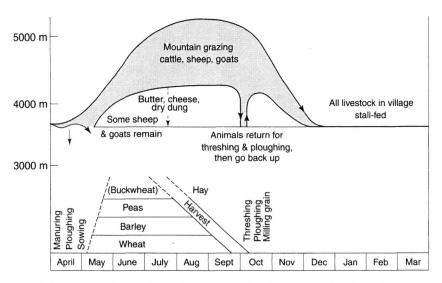

**Figure 9.3**   The agricultural calendar for s'Tongde, Zangskar valley, Ladakh (after Osmaston *et al.*, 1994). Some animals return briefly in late summer to plough and thresh and then return to the pastures until the snows come.

*(continued)*

*(continued)*

*Spring:* Clearing stables of manure and spreading (100 kg/ha). Livestock move to lower camps if snow has cleared. Ploughing and repair of irrigation systems. Ploughing has a ceremonial start and the date is set by an astrologer. The field is ploughed in arcs following the field edges rather than straight furrows. Wheat and barley are sown first.

*Summer:* Less arduous time; weeding of cereals – twice at about four-week intervals. The weeds are dried for fodder. Herds move to higher pastures. Irrigation of crops (time-consuming but not arduous). Weaving, dyeing.
Harvest in late summer is very busy; in July and August peas are gathered and dried communally but grain harvests are conducted on a household basis. Plants are pulled up by hand which clears the fields completely and threshed on each farm's threshing floor. Yak and *dzo* (a hybrid) are collected from the pastures and driven over the grain to thresh it. The grain is dried and bagged and the animals return to the pastures.

*Autumn:* Livestock return again to the village after the first snows. Some fields may be ploughed after harvest.

*Winter:* By the end of December all livestock are stabled, but are let out during fine days. They are fed and cleaned throughout the winter. The family occupy a ground-floor inner room surrounded by stables and fodder for warmth. It is a period of relative relaxation. As the rooms are dark, limited weaving and other activities are possible.

*Source*: Osmaston *et al.*, 1994.

household structures and health especially (Bohle and Adhikari, 1998). However, undernourishment and malnourishment still affect a significant proportion of the population and survival has become a permanent crisis.

A clear zonation of activity emerges in some mountain regions, reflecting the suitability of different crops at different altitudes. In the Peruvian Andes, for example, below 1500 m sugar cane, coca, fruit and rice are grown on haciendas or large holdings, at 1500–3000 m subsistence cereals are grown on smaller holdings; at 3000–4000 m tubers predominate and above 4000 m pastures. Similar zonations occur in other mountain regions. Second cropping is constrained by altitude – in Hunza, double cropping is possible only below 2500 m, despite adequate sun and water elsewhere (Whiteman, 1988). Aspect is also important, as only single cropping is possible on north-facing slopes. Thus again, it is more advantageous than not to have a range of plots scattered throughout the valley, so that each household has access to a range of conditions.

Other characteristics of traditional agricultural economies include the detailed knowledge and experimentation practised by the farmers, who try different varieties of plants in different locations, and address issues such as storage conditions, preservation of foodstuffs and new crops. The wealth of indigenous knowledge of varieties, growing conditions and cropping regimes has been more widely recognized in more recent years as development initiatives have begun to seek to meet the farmers on their own terms. Agricultural practices have, therefore, not been static, but have developed over the centuries to incorporate new crops, innovations and technology, and to meet new demands such as growing populations or climatic change. The introduction of new species, such as the potato, to mountains (see Chapter 6) has revolutionized nutritional yields, and therefore health, and trade opportunities.

## Crop production strategies

In Chapter 2 the climatic conditions of mountains were examined, and shown to be highly variable in time and space and to operate on a variety of scales. This section is concerned with micro-scale (terrace- or plot-level) environments. Chapter 7 mentioned how the environment varies within the confines of a single terrace, and this fact has inspired a number of innovations in cultivation techniques. Chapter 8 noted how the quality of water – glacial or snow melt – might affect crop growth and how settling tanks and distance from source were employed to modify water characteristics to suit the crops.

The mountain farmer seeks to eliminate or reduce the most limiting factor affecting crop growth. This will vary from species to species and thus the location of different crops will be determined by subtle differences in insolation and temperature. The use of local varieties of crops and livestock is advantageous as these are already adapted to local conditions, and can often give higher yields with minimal inputs than newer varieties. The farmer therefore tries to maximize production in time and space by careful cropping strategies (Butz, 1994; see Box 9.2).

Frost is a critical issue for farmers. It is seasonally restricted from about 15° latitude. In the aseasonal tropics, there is an increasing risk of frost with altitude – it may be year-round above 3500 m on Mount Kenya, in the paramo grassland belt of the Ecuadorian Andes (Lauer, 1981) above 4600 m, and above 4500 m in the seasonally arid Hindu Kush. Frost retards growth, kills young buds and can devastate fruit tree crops, for example. Frost avoidance strategies include the planting of fruit trees above the valley floor, thereby avoiding cold air ponding (see Chapter 2), and mounding earth and crop residues around potato crops as in Papua New Guinea and the Andes.

It is the unexpected frosts which cause the most potential damage, such as that caused by El Niño. This climatic oscillation reduces rainfall and under clearer skies there is greater frost risk in Papua New Guinea – up to three or four successive nights, two or three times a month (Allen, 1988). This causes substantial lowering of soil temperatures and kills the runners. The population then migrates down slope with their pigs until a crop is ready (see Chapter 6). Frost risk arises when plants, which respond to day-length stimuli to break dormancy, awaken before late frosts. Growing seasons can also be extended by protecting seedlings in seedbeds or covering the surface with vegetation cut and laid flat. In Peru and Bolivia maize and potatoes are restricted to the wet season, when frost risk is reduced. Above 3800 m in the Lake Titicaca region, *Chenopodium quinoa*, a local cereal, is the staple crop as it is more frost-resistant. Freeze-drying techniques have been developed over centuries by high-altitude farmers, enabling them to preserve potatoes by using the diurnal frost cycle. Potatoes were the main source of sustenance for the Inca empire. Now barley is grown up to 4100 m and camelids pastured above this. Cooling under clear skies in the arid Hunza, however, increases the sugar content of apricots to 20 per cent, making them particularly succulent. In this region the productivity of the soil is very high, with up to 90 t/ha potatoes being recorded.

Solar radiation is significantly affected by cloud cover. The thinner air at higher altitude is less able to retain heat so changes in incident solar radiation cause rapid temperature fluctuations. In regions of rainfed agriculture, the presence of rainfall marks the growing season but at the same time the clouds reduce solar radiation. Thus in tropical Indonesia the altitude of rice cultivation is limited to 1500 m, despite land extending to 3800 m. In arid subtropical Kashmir, and even in monsoon areas like Darjeeling and Simla, the altitude of rice cultivation is 2200 m, and in western Nepal and northern Pakistan 2700 m where the aridity results in unimpeded solar radiation and there are abundant supplies of irrigation water from ice and snow (Uhlig, 1978). Shading by trees and walls also retards growth by limiting sunshine. Even the loss of one hour of sunshine a day can retard the maturation of a crop by several days, which can make the difference between a safe harvest and possible damage from early frosts. Hence the rule that trees may only be planted at the southern end of terraces so as not to shade adjacent farmers' land (Whiteman, 1988). Topographic shading creates significant differences in growing environments: in Hunza, the Hunza villages occupy the south-facing slopes and are richly productive land, but north-facing Nagar opposite is much damper and cooler, having more trees and less productive cultivated land (Parish, 1999). Solar radiation is also important in warming glacial meltwater. By flooding paddies with water which has been allowed to stand and warm in the sun, paddy fields can be kept above 15°C for an extra five hours at night in the Karakoram. This is about 3.8°C warmer than those flooded with cold water and has the effect of shortening the growing season by several days (Whiteman, 1988).

The close interaction between temperature and precipitation creates variations in soil moisture conditions within short distances. In areas where most crops are at their ecological limits, and there is scarcity of land, differences in productivity are more critical. Avoidance of frost hollows, and strategies to warm the water by standing, or the soil by covering with dark organic matter to increase its absorption of heat are common and the location of crops is assessed in relation to their needs – manure, water, sunshine (Box 9.2). Where one element is a particular limiting factor, a farmer will try to reduce it in order to maximize yields, and minimize wastage (Butz, 1994).

Where water is the critical factor in rainfed cultivation, farmers must respond to precipitation in a number of ways. Planting takes place in response to actual or anticipated rainfall, and there is no control over the quantity as for irrigation systems. Rainfall may be local to a particular valley, or insufficient in one season to plant. Planting too soon results in loss of seeds due to surface erosion. Planting too late may result in too short a growing season. Planting density may vary in accordance with water availability (see Chapter 8).

## Box 9.2    Farmers' cropping strategies in Hunza

The distribution of potatoes, wheat, beans and alfalfa in Hopar, Hunza, is determined by their individual needs. Cultivation occurs between 2500 and 3000 m. Distribution is related to water, nutrient status and soil temperature and moisture conditions. The wheat is a local variety adapted to low temperatures and poor soils, and, although less productive than potatoes, is still grown on 70 per cent of the terraces. Its limiting factor is water so it is irrigated at 6.5-day intervals, but fertilized less as it does not show sufficient response to warrant manuring. Potatoes are a high-income cash crop and given the best land and manured, but are less sensitive to water so they are irrigated at 10–11-day intervals. Beans are closest to their ecological tolerances so are grown on the sunniest plots and irrigated most frequently and for a longer period per application. Only water which has stood to settle the sediment and warm up is used, so these plots tend to be furthest from the irrigation source. Alfalfa is a fodder crop and infrequently irrigated. It tends to be undersown beneath orchards where the trees might be irrigated by watering circular enclosures at the base of the trunk. Farmers, therefore, do not favour one crop above another but work to use their scattered plots to the best advantage, and to distribute their resources of manure and water where they will have the greatest effect.

*Source*: Butz, 1994.

## Cultivation patterns

Staple grains form the main crops, together with various tubers, vegetables and fruit. Wheat can only be grown at lower altitudes due to its sensitivity to frost, and the longer maturation. Higher up, barley remains important, and is either used for humans or cut green for fodder in the autumn or spring. Rye was more common but has largely been replaced by maize in Europe. Maize represents a higher calorific yield per unit area than some other grains, and has a relatively short growing season. Thus it is often planted after a first grain crop. However, it is more thirsty than wheat or barley. Potatoes are an alternative staple to grains, especially in the High Andes, and their introduction to many other mountain regions has been an important development. Miller (1984) estimated that about two-thirds of the farmers' plots were planted with grain, but that this still failed to meet the total demand.

In Southeast Asia, rice is the staple, and forms the central part of the culture. There are many different kinds of rice, but at higher altitudes constraints limiting insolation occur: at higher altitudes the frequent monsoon clouds obscure the sun and make it too cool and damp for rice. Rice is grown on wet irrigated paddy, where it is very high-yielding (Plate 9.1). Such land is highly prized by southeast Asian mountain farmers. Higher up it may be

**Plate 9.1** Wet rice paddy terraces in northern Thailand. Water needs to be carefully managed to maintain the right depth. Drainage of water needs to be controlled to ensure no adverse erosion of terraces occurs.

grown as dry upland rice, although the crop is much more risky and prone to failure. It is therefore to the advantage of farmers to have access to land where they can grow a range of crops. In addition to staple cereals, vegetables are grown on irrigated plots near to the homestead. Vegetables commonly include potatoes, beans, peas, onions, salad vegetables and other species. These are grown for domestic consumption.

In order to make the best use of constrained space and growing season, careful organization is applied to the crop planting regimes. Lucerne and fodder barley are planted underneath fruit trees to ensure no space is wasted. Terraces are planted in rows, even on very narrow strips. This makes use of the variable micro-environment (see Chapter 5). Developments such as new varieties have not always been successful. In India, new varieties of wheat have not been taken up as expected, partly because they require artificial fertilizers, but also because their short stalks are not suitable for all the traditional uses to which straw was put – particularly fodder and thatching (Negi, 1994). In Venezuela the traditional and modernized potato–cereal cycles are shown in Figure 9.4. The effect of the application of artificial fertilizers is to reduce the fallow time, which increases the potential for disease and replaces organic nutrients entirely with artificial ones. The reduction in soil organic matter weakens its structure and the reduction of fallow increases the area of soil at risk of erosion. This increases the dependence of farmers on such capital investment (Sarmiento *et al.*, 1993). The slow breakdown of organic matter

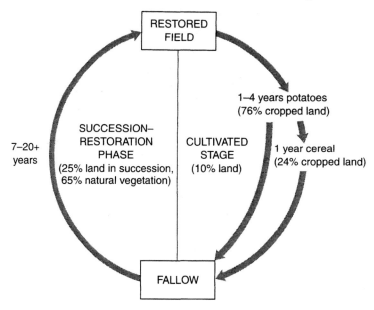

**Figure 9.4** Sequence of the status of fields in the Venezuelan Andes. The long period of succession of natural vegetation allows nutrients to accumulate and pests and diseases to be controlled (adapted from Sarmiento *et al.*, 1990).

means that long restoration times are needed for continually cropped soils. In Hunza, the substantial expansion of seed potato cultivation appears to be highly dependent on markets in the Punjab. In addition, new systems of exchange of varieties and seed stock have had to develop in order to prevent disease on terraces where monoculture of potatoes is practised year after year.

Technology is relatively unchanged – hoes, digging sticks, simple ards (single-share ploughs) drawn by donkeys, oxen or even camels in some lower arid areas. Ploughing by single ard is necessary as access to terraces is often by a steep, narrow path, and the field size too small to permit mechanization. This has often meant that agricultural development has lacked the investment of lowland areas because the returns on investment are much lower, as cultivation remains very labour-intensive. Manuring follows ploughing – heaps of manure are distributed on the terraces, carried up in baskets from the cattle stalls and paddocks. This is dug in, and the terrace marked out with small channels to distribute irrigation water from the main channels across the plot. Weeding, hoeing and thinning of crops to weed and thin out is also done by hand, taking perhaps two hours per terrace of *c.*100–400 m² (Plate 9.2).

Grain mills are driven by water, olive presses unchanged since Roman times. Threshing is done by hand or with sledges or animals driven across the cut crop. Most mountain villages have circular threshing floors with a central post. These may be earthen-floored or paved and some even occupy a terrace, but some areas have mechanized this process (Plate 9.3). The grain is cut just before it ripens, so less grain is lost through being shaken loose from the ears.

## Box 9.3 Thai hill tribe shifting cultivation

The hill tribes of Southeast Asia have traditionally followed a system of land clearance, cultivation for a few years and then abandonment. Originally, communities could clear new areas of forest each time (pioneer swiddening), but elsewhere land constraints have meant that each tribe occupies a defined territory and moves around it in rotation in a cyclical swidden system. Land is cleared of most trees, although some large ones are left with branches lopped. These provide sources of shoots for regrowth without interfering with cultivation. Once trees are cleared, the litter is left to dry and then burnt off in the spring.

Pioneer clearance could support cultivation for longer than secondary clearances, but took longer to recover. The Karen tribe used to use land for three or four years, followed by 20 years' fallow; the Mien allowed 10 years' fallow but this has been reduced to three. The Lahu can only afford to let 20 per cent of its land lie in fallow in any year, whilst the Akha have a five years' use/five years' fallow cycle. The Hmong are located at higher latitudes and used to have one year's use followed by five years' fallow.

Abandoned land has been subject to ownership issues, as these communities do not hold legal title to the land. Shifting cultivation has also been perceived as a wasteful technique, but in fact it allows new growth in forests and limits erosion. The use of chemical fertilizers, herbicides and pesticides is beginning to create problems of contamination of the soil and water, and also affect the regeneration of secondary forest on fallow land. Where cash crops such as coffee and terraces of ginger (which is hard on the soil) are grown, the investment in the land means that it is no longer fallowed but becomes continuously cropped. This is particularly true of tree crops, which need several years of growth before they can be first harvested. In these situations, security of tenure is needed to encourage investment in alternative agricultural strategies and sustainable land use in the increasing system of 'stationary shifting cultivation'.

*Sources*: Grandstaff, 1988; Anderson, 1993; Roder, 1997; Sillitoe, 1998a.

It is dried in stooks, threshed and winnowed and stored in granaries – often on stilts – like the staddle granaries in the European Alps and Picos de Europa. Scattered fields mean that harvesting is staggered, and labour can be shared within the community. Granaries may be communal, or each household may have its own – such as the rice bins in many hill tribe houses of northern Thailand.

In many mountain regions the cultivation of narcotics (cocaine, cannabis, heroin and opium) is an important, if illegal, source of income. Mountains offer the right soil and climate conditions and the relative inaccessibility is still an added protection to farmers where the trade is illegal (despite the use of helicopters for surveillance). There are three main areas – the 'Golden Triangle' of Thailand, Myanmar and Laos, now extending into southwest China; the

**Plate 9.2**   A Hunza farmer preparing his fields for planting potatoes. The small terrace area means that there is limited scope for mechanized techniques, so the bulk of farming is still labour-intensive. The extent to which potatoes have come to dominate the cropping patterns of Hunza is shown in Plate 7.2.

'Golden Crescent' of Afghanistan, northern Pakistan and Iran, which both produce opium; and the 'White Triangle' of Bolivia, Colombia and Peru for cocaine and crack. Between 1979 and 1987 in Peru around 180 000 ha were cleared for cocaine production (Denniston, 1995), whilst Myanmar accounts for one-third of the 6000 tonnes of opium produced each year (Renard, 1997).

Kif (cannabis) is grown in the Rif mountains of northern Morocco despite half-hearted attempts to quell its production. The crop provides a good income and adds a degree of economic stability to an otherwise impoverished area (McNeill, 1992). Originally produced for domestic consumption and for fibre and dye, it is now the most important export crop with 1500 tonnes going to Western Europe each year (USDS, 1996). An estimated 200 000 people rely on it as their source of income, despite the poor soils and dense population. Cultivation has been extended beyond good irrigated terraces to more marginal land, with resultant erosion (Laouina, 1994). It is not surprising that the crop is so important, despite the border controls and US$2.2 million drug control efforts of the King; the profit per hectare for cannabis resin reaches nearly Dh 50 000 (Moore *et al.*, 1998) – tomatoes at Dh20 come a rather poor second!

Opium has been a similarly lucrative crop in Asia (Figure 9.5). The rise in use and development of a market is attributed to Western colonial

**Plate 9.3** A threshing machine in Hunza. The straw is threshed, bagged and used for fodder or, in some cases, incorporated into mudbricks. Before the advent of such machines – albeit this is an antiquated example – threshing was carried out by animals driven over the straw on beaten earth or stone threshing floors.

expansion. In the seventeenth and eighteenth centuries the use of opium was not considered a vice and the expansion of its use in Asia and the Europe created a huge market (Renard, 1997). Once its less positive effects were noted, efforts were made to stop its production. Controls on marketing failed, partly as a result of the opening up of the Chinese market in 1858, which became the biggest importer by 1879. Incomplete suppression, ineffective enforcement, bribery and corruption of officials continue to weaken eradication efforts. The criminalization of opium production and Western development efforts to introduce alternative sources of income have also failed. In the first case, concentrating on opium has deflected interest away from heroin, which is much more dangerous, and at 10 per cent of the weight of opium, much easier to smuggle. In the latter case, projects such as the Thai–German Highland Development Programme have faced various cultural issues (Dirksen, 1997): opium was traditionally bartered to make up rice deficits, and so an alternative was needed. The growing of cash crops often fails due to lack of markets, and, simply, drugs are much more lucrative, making the risks worthwhile. Poor enforcement and insecure land tenure contribute to a reluctance to invest in alternatives such as trees, which take time to mature.

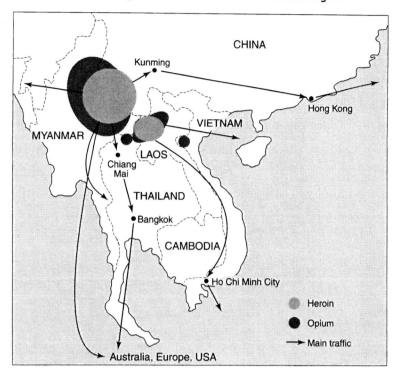

**Figure 9.5**  Map of opium and heroin production in the 'Golden Triangle'. The main trafficking routes are also shown (adapted from Renaud, 1997).

Qat is a mild narcotic related to privet. It is a social institution in Yemen, Ethiopia and Kenya, but it is banned in the more strict areas of the Muslim world such as Saudi Arabia, Iran and Iraq. In Yemen it is by far the most lucrative cash crop, and most of the agricultural income of households is centred on those whose lands are suitable for good qat. The quality is important; it must be consumed within hours of harvesting, which is almost the only incentive for improved communications in such areas!

## Livestock

The most common species are sheep and goats, due to their hardiness and ability to survive on the poor mountain pastures. In addition, cattle are kept. This may amount to only a few beasts to provide milk and cheese for the household. In other cases, the dairy production may be much more substantial, as for the European Alps and in the case of the *zomo* herders of Nepal (see Chapter 3 and Box 9.4). Other indigenous species include the Andean camelids – alpaca and llama – which are well adapted to the diurnal fluctuations in temperatures which occur throughout the year in the high puna of Peru and Bolivia (see Box 9.5), and yak in the High Himalayas.

## Box 9.4   Yak–cattle hybrids in Nepal

At Melemchi (2600 m) north of Kathmandu in Nepal, the use of hybrids for dairying has been a traditional adaptation to the use of the middle-altitude pastures. There is a ready market for butter for consumption and for religious festivals. The animals are privately owned and kept on private pastures (*kharkes*) unlike the communal herds kept on common land in Tibet and elsewhere in Nepal. The *gode* (homestead) on these pastures was occupied permanently by young families who were given the land on their marriage. The life is arduous, with two milkings a day, the production of cheese and butter, and the moving of the *gode* six to ten times a year. Elderly couples, or surviving spouses live in the village as they cannot cope alone and the land and herd are sold or handed on to the next generation.

Since the mid-1980s, however, there has been a shift from dairy production to the breeding of hybrids. Female *zomo* are sold to herders and male *zobo* to traders as beasts of burden. This change arises from the reluctance of young couples to take on the life of the *gode*, or a lack of young people due to outmigration to Kathmandu, India and Bhutan. Breeding hybrids can be done nearer the village and is less hard, but increases the number of animals and the pressure on grazing near to the village. The numbers increase because the hybrids cannot be sold until they are three years old, and are unproductive until then. Pressure on village grazing conflicts with cropping there. However, despite the new techniques which need to be learnt, this new initiative is seen as an economic move, but to date returns have been limited to a few successful herders and many have suffered losses of calves and unsuccessful breeding. There is also a shortage of butter both in the diet and for festival use, despite the growing market for it in Kathmandu. There are few other alternatives to these options; some tourism development associated with the nearby national park is beginning to emerge. Winter forage continues to be the main constraint on numbers of animals kept.

*Sources*: Bishop, 1989, 1998.

Dairying has long been a part of the economy of the European Alps. In recent years efforts to maintain the use of meadows have included the production of accredited labelled produce, marketed as specialities of the mountains, and subsidies have been granted to support the industry. In the Beaufort valley, France, an area adjacent to land flooded by a dam construction in the 1950s has been revitalized by a 'tradition in modernity' initiative where trademarked cheeses have found a good market and milk production has increased from 600 to 3000 tonnes. This milk is accredited a certain quality; it attracts higher prices than ordinary milk and has ensured the survival of local breeds of cattle (Warsinsky, 1997). Technological developments have helped, such as mobile milking machines. Milk is now piped down the mountain,

## Box 9.5   Andean camelid herding strategies

Llama and alpaca occupy the high puna and paramo grasslands of the altiplano of Peru, Bolivia and Chile. Average herds are 150–250 animals, with larger herds in areas where supplementary fodder is available. Herding is complex. Traditionally, male and female llama and alpaca would be herded in four separate groups, which may be further subdivided between white and coloured animals. Thus some amalgamation of herds would be necessary to enable the different categories to be herded separately. This arises because alpaca are less hardy, prefer moister pastures and the females suffer stress when pregnant, but they are valued for their wool and meat. Llamas only give wool but are valuable caravan animals. They suffer in damper conditions, preferring high dry puna areas. Most herders would want to keep some of each sex and breed. The variations in coat colour reflect markets; commercial weaving markets prefer white, which brings higher prices, but local markets tend to prefer coloured wool. White wool is also shorter and curlier, which is more suited to commercial weaving, whereas coloured wool is longer and straighter and easier for hand weaving. The differences in coats also affect the hardiness of the animals; long straight hair is less protective in wet weather than shorter curlier hair, so different environmental niches are needed. Times of shearing also vary; most are sheared in November–December at the start of warmer, rainy weather. The animals need less protection and the fleece is less damaged. However, llama may be sheared at any time before going on a caravan to the lowlands.

Herders have complex systems of shared herding, reciprocity and contract exchange for seasonal labour in order to help them meet the differing needs of their varied herds. Land reforms tended to privatize agricultural land but pastoralists have tended to retain cooperative systems in order to make use of economies of scale, given the fragmented nature of production. The preference to maintain the whole spectrum of animals represents a risk aversion strategy by the herders. Networks of traders requiring pack animals are carefully guarded, as are links with the market for selling animal products.

*Sources*: Browman, 1983, 1990.

whereas before it used to be made into cheese in the mountains and carried down. The relatively short (horizontal) distances involved mean that the movement of fresh dairy produce is possible each day. Elsewhere, a household in the Atlas or Himalaya might have only one or two cows to supply the household. These are kept near to the village and grazed by children or women. Strict regulations are enforced against trampling or grazing growing crops. Hay meadows in the valley bottom may be owned privately but the second cut of hay is traditionally left for those without such meadows.

Sheep and goats are particularly important in arid mountains. In the Arab world, lambs are important as a source of cash, particularly when sold for

## Box 9.6    Pyrenean agricultural change

In the western Pyrenees there has been continuous change in the pattern of agriculture. From prehistoric times until the fourth or fifth century, the population changed from a hunter-gather to an agro-pastoral-based economy. Following the collapse of the Roman empire, there was an abandonment of upland agriculture and return to pastoralism. The summer pastures were first used for grazing in the thirteenth century; this declined during the fourteenth century due to the plagues, but expanded again from the fifteenth to the eighteenth century, when a stable agricultural tradition of mixed agro-pastoralism emerged. Population growth from 1850 until the 1960s was supported by the expansion of the potato and an increase in numbers of sheep and goats which followed transhumance cycles of movement. During the summer they grazed large areas of cleared forest and in winter the stubble of lowland cereal production in the Ebro depression. Fodder production increased and firewood was exported from the valleys. After the First World War, there was considerable abandonment of less favourable land and at the same time greater use of valley floor land where intensive cereal and potato production was located. Settlements became more diverse, with the construction of specialized buildings. Following the Spanish Civil War in the 1940s, the growth of the urban meat market encouraged cattle production but greater outmigration to the cities also occurred. Since 1965 the meat and milk markets have become the mainstay of production, with the development of dairy cooperatives. Cereals are grown for fodder rather than human consumption. Meadows are given over to grazing whilst the highest pastures are abandoned as they are unsuitable for cattle. However, conflicts have arisen in the use of valley floor land. The construction of dams in the Teno and Gallego valleys, for example, has submerged large areas of productive agricultural land. In addition, tourism developments including roads, ski resorts and hotels have raised land values and further marginalized agriculture. These conflicts remain contentious issues and relate to varying local, regional and national needs for different resources.

*Sources*: Garcia-Ruiz and Lasanta-Martinez, 1990, 1993; Chocarro *et al.*, 1990.

various Islamic feasts, where the ritual slaughter of lambs ensured a ready market. Goats are hardier, and sheep do not always adapt well to winter stall feeding, but do survive long-distance transhumance. In Europe, long drove routes can still be traced between high summer pastures and lowland winter pastures, where fodder is provided by the stubble of cereal crops, and the farmer benefits from the manure (Box 9.6). The availability of lowland pastures is a critical factor in the maintenance of transhumance, particularly for people who are (or were) 'professional' herders.

Miller (1984) outlined the pastoral cycle of villages in Imlil, in Morocco. Each family might have 60–70 sheep/goats, producing 15–20 lambs/kids per year.

The movement of these flocks to summer high pastures and lowland stubble in winter is important as sheep and goats do not do well in close confinement so cannot be stall-fed so successfully as cattle.

In some areas, such as Hunza, livestock have declined in the face of economic changes. Patterns of livestock husbandry have continually changed in the region. During the nineteenth century, when the Mir held sway, substantial revenues were generated by renting out pastures to adjacent nomadic tribes. In 1935, 20 per cent of the households of Hunza and 75 per cent of neighbouring Gojal upstream participated in herding, but 50 years later only 1 per cent in the former region did so. The issue of taxation for using pastures which operated under the Mir was abolished with him in 1974. The current decline is therefore attributed to lack of herding labour, although this can be effectively covered by a few people now that herds are so much smaller. More significant is the increased income generated from cash crops – potatoes, seeds, cherries and apples – and from tourism, as well as emigration of young males. Alternative sources of meat, dried milk and cooking oil brought by road and paid for by cash mean that animals are no larger quite so vital in the diet.

---

## Box 9.7    Nomadic strategies in Ladakh

The Kharnak community in eastern Ladakh follow a true nomadic existence at average altitudes of 4350 m. They all belong to 'middle' castes, so there are no great hierarchical differences. They identify themselves as Kharnak with a defined territory. The landscape is very sparsely vegetated, with a few patches of grass near to springs, but is otherwise dominated by dry woody shrubs. Winter temperatures might reach 35°C and summers are brief, with convectional storms. The nomads follow well-defined territories in strict rotation, moving sheep, goats and yak for considerable distances to avoid winter snow. This is a form of 'horizontal' rather than vertical transhumance as the differences in altitude are much less than the horizontal distance travelled to a zone of different climatic conditions. Herds are split up to make the most use of different grassland areas and to take account of livestock needs. In May, sheep and goats move to satellite locations in adjacent valleys to the yaks grazing at higher altitudes. Only the female animals providing milk remain near to the main seasonal settlement. The two encampments are about 1.5 km apart and interactions occur daily between them. As each family revisits the same site each year they 'improve' the site by constructing windbreak walls, pits for tents and pens for livestock. Some have mudbrick or stone dwellings at the site of spring pastures where they may remain for four or five months. In 1993 there were around 67 families (about 363 people) but since then 25 of these have migrated to settle in areas with medical and educational facilities, so this is very much a declining way of life.

*Source*: Dollfus 1999.

They still remain an important, and in some aspects irreplaceable source of manure. The animals which are still kept are grazed on more accessible pastures rather than distant ones, and imported fodder is a valuable supplementary source.

Pigs are much more important that sheep and goats in parts of Southeast Asia. They cope well in the hotter humid conditions of the tropics and are an important part of the culture, taking the place of lambs in rituals. They play an important part in the wealth of these people, and act as a back-up against crop failure (see Chapter 6).

Livestock improvement schemes by development practitioners have concentrated on the improvement of breeds, making them higher yielding, However, they are often less well adapted to mountain environmental conditions and low-quality pastures than local breeds, which continue to be favoured for their resilience.

## Sylviculture

The importance of forest and tree resources is a critical element in the subsistence economy. Forests provide a range of products, including fuel, fodder, construction timber, foods and medicinal plants. They offer grazing for animals such as pigs, and wildlife for hunting, as well as entering into the imaginations, myths and beliefs of the mountain cultures they support. These products are essential to the maintenance of subsistence livelihoods, and the careful management of forest resources discussed previously reflects this. The importance of sustainable use of forest resources has long been a hotly debated and topical issue, arising from concerns about deforestation and its association with flooding, avalanches and siltation of reservoirs (see Chapter 10). However, it is increasingly recognized that in fact the knowledge of local farmers and their management strategies have been effective over centuries of occupation and exploitation, even though they are coming under greater pressures in the modern world.

In addition to forests, other trees in the landscape are important parts of the economy. Fruit and nut trees have long been cultivated in many areas, although many are relatively recent introductions – such as apples, which were introduced into the High Atlas by the French colonists in the nineteenth century. The comparative advantages of mountain climates in many areas – especially North Africa and the Himalayas bordering the more arid parts of the Indian subcontinent – mean that the cultivation of fruit has been an important economic development (see Box 9.8). In Hunza, productivity of fruit is very high, and there are some 60 local varieties of apricots, many of which are still grown by many families. It is only very recently, however, that efforts have been successful to dry, process, preserve and market this fruit to a sufficient quality and standard to make it acceptable to more than local markets. One of the problems is that the apricots are sun-dried, and unseasonable rain can therefore reduce a large part of the harvest to a rotting mass on the roofs of the houses. In addition, the standards required by many Western markets involve the grading of the fruit, and as cultivation is still a rather *ad hoc*

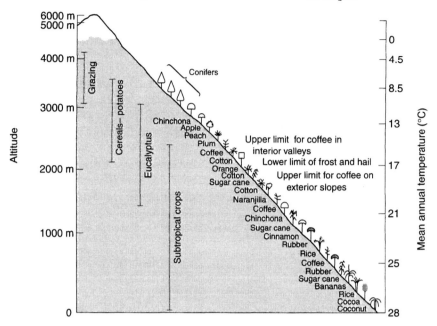

**Figure 9.6** Diagram to show the range of tree crops and their altitudinal locations in Ecuador (adapted from Stadel, 1992).

arrangement, with each household having only a few trees and little technology, the fruit may be delicious, but is often poor in appearance, and therefore rejected by the markets.

The growing of fruit such as apples, cherries and apricots and nuts such as walnuts and almonds has increased in recent years. Mountain climates fulfil the growing requirements of fruit trees in having a frost to break the winter dormancy. The apricots of Hunza are renowned throughout the world for their high sugar content, which is enhanced by the hot days and cold nights. Commercial development of fruit and nut production has enabled some farmers to develop a marketable product, although this has also led to changes in the balance of production for many households. An example of the altitudinal range of tree crops cultivated in Ecuador is shown in Figure 9.6.

Other parts of the Himalayas have successfully developed apples. As in other mountains, such as in Morocco, these often involve imported varieties such as the French Golden Delicious. In Hunza, though, surveys (Parish, 1999) showed that farmers were beginning to graft traditional stock onto imported rootstock because they preferred the taste, texture and keeping properties of their older varieties. These latter also had better resistance to local diseases and climatic vagaries. However, again market forces demand standardized products, which the new varieties meet better than the old.

Not all communities and cultures have taken advantage of the potential fruit production. In the Hindu regions of the Himalayas, fruit is seen as a luxury, worthy of offering to the gods. It is therefore grown incidentally, and most of the fruit is given as offerings. In addition, commercial production is constrained by the concept that as fruit is a luxury rather than a basic necessity, anyone is entitled to help themselves. Thus there is generally little fruit left for consumption, so it rarely forms a significant part of the diet. In this cultural context, the investment in growing fruit trees is not repaid, as it simply represents more offerings and more opportunities for the community to benefit, with no reward for the individual. Such innovations therefore require a significant shift in cultural values to be successful.

In response to a shortfall in fodder trees and timber for construction as a result of increased pressure on forest resources, many farmers in the Himalayas have systematically planted trees selected for specific purposes. Gilmour (1995) and Dove (1995) describe this phenomenon as 'rearranging' the trees in the landscape. Where large tracts of land have been cleared or degraded, this replanting represents a spontaneous response to specific shortages, and is equivalent to a significant proportion of estimated forest clearance. The species planted tended to be high-value fodder, particularly fodder available at certain times of the year, such as winter and spring, when other sources were scarce.

---

## Box 9.8 Moroccan fruit production

There has been a tradition of tree crops throughout the Mediterranean at least since Roman times; olives were a staple element of the food production and are still grown in the lower reaches of the valleys of the northern High Atlas. At higher altitudes, walnuts and almonds have been produced for local consumption but are increasingly grown for the expanding market in Marrakech. Brokers buy up the crop before it is ready, and it is then delivered to the market. Whilst this means the price is lower, it secures a sale for the crop in advance. In recent years the demand for apples, cherries, plums and other tree fruit has resulted in increased planting. Initially, planting was concentrated on the edges of terraces but the market appears so lucrative that whole slopes are given over to trees, together with new terraces and, more and more frequently, the lower valley floors. This pushes out the traditional subsistence crop production and the valley floor grazing. Tree planting represents a substantial investment for farmers as it is at least three years before a crop is harvested. In addition, the new expertise must be acquired and possibly new terraces in more marginal areas. The costs of planting and development of new varieties attest to market-orientation of the economy. Despite the risk of flooding on lower lands, which results in the need to replace all the trees, they are still planted there. With a number of significant floods in recent years and the potential for greater flooding in the future, this represents a strong economic force overriding climatic change concerns.

*Sources*: Parish and Funnell, 1999.

**Plate 9.4** A sawmill in the Swat region of Pakistan. Sawmills using simple technology provide additional income and employment. In this region the cedar *Cedrus deodara* is a valuable source of timber. This sawmill is powered by electricity provided by a scheme similar to that shown in Plate 8.7.

Fodder species are also categorized by quality and the effect they have on the quality of dung (critical for the fields), milk yield and animal health. Other species planted were for construction – in Hunza, for example, poplar trees are often grown along stream banks and water courses as there is very little standing timber in the region. These trees are highly prized.

## Non-agricultural activities

A variety of subsistence activities is undertaken by households. These range from collecting water, firewood and food and medicinal plants, to hunting and gathering, to the crafts which produce clothing, tools, implements and household items. They may be roughly classified as processing activities, such as making dairy products and drying fruit and nuts; domestic activities concerned with the running of the household, such as collecting fuel and water, and cooking; and crafts such as weaving, carving and embroidery. In the last case the items produced were traditionally for home consumption, but increasingly there are outlets for selling such artefacts to tourists, either directly or through craft trading enterprises and middlemen.

A few individuals, or in some societies, a particular caste may be responsible full-time for tasks such as blacksmithing and wood carving. For example, in the Swat valley of the North West Frontier Province, wood carvers are well known for their skill, and some became itinerant workers, supporting themselves by the carving work they did on buildings. Some artisan activities have been supported by the provision of electricity, increasing their productivity (Plate 9.4). In Thailand, some tribes had a blacksmith for each village, who was a respected artisan and provided the tools for agriculture, etc. Such individuals have generally disappeared with the advent of mass-produced tools available in markets, or in the wake of modern-style housing (Lewis and Lewis, 1984).

Crafts such as clothing production and weaving have also changed – the example of changes in the type of camelid hair produced to suit mechanized production was given above. Other crafts have been adapted – almost exclusively in some cases – to the tourist market. Thai hill tribes tend to weave cloth and embroider bags, clothing, etc. to designs and patterns demanded by the tourist industry, rather than using their traditional motifs. The impact of tourism has had the dual effect of both reviving traditional crafts, but also adulterating them.

Other economic activities which are alternative or supplementary to agricultural subsistence include emigrant labour to mines and construction sites (Chapter 6) or involvement in tourist activities (Chapter 12).

## Key points

- Traditional livelihood systems in mountains integrate cultivation, livestock rearing and forest use. These use different altitudinal zones and environmental niches and spread the risk of failure across a wide range of resources.
- In areas where the physical distance was too great to permit one community to use all niches, or where existing communities already occupied certain areas, some peoples specialized in cultivation or pastoralism. They developed exchange networks which enabled both communities to survive.
- Creative production strategies are followed by mountain farmers to make the best use of scarce resources and to minimize the risk of crop failure. These include scattered plots planted with different crops at different times. This also distributes the labour requirements throughout the year more evenly.
- Many of these traditional systems have adapted and developed continuously. Many others have undergone great recent change as a result of development initiatives or in response to population pressure.
- The main change is an increasing orientation towards cash cropping rather than subsistence. Many farmers are able to combine the two strategies but others have concentrated on income generation.
- Changes in herding and cultivation patterns have also occurred as a result of emigration of some or all of the population. Elsewhere, revitalization of local economies has stemmed the outmigration.

Part 4

Managing mountains: development,
conservation and degradation

# Introduction

This final section concentrates on the issues of degradation and conservation which are particular foci of international and national concern for mountain environments. They are closely connected with the wider issue of sustainable development and economic change within mountain communities which has introduced different sources of income, shifts in balances of power, population growth and social change.

There is a perception of a crisis in mountain environments, of deforestation, catastrophic erosion and collapse of societies. However, others argue that communities are resilient and adaptable, as they have been over millennia. Whilst institutions may need to change and adapt in order to cope with the rapid pace of economic change in particular, it is increasingly recognized that the capacity and the incentive to cope lie within mountain communities themselves, as much as in outside initiatives.

The location of many mountain regions at the periphery of nations has raised security issues concerning illegal immigration, the environmental impacts of refugees and the ability to cope with natural disasters, the intimate details of which are relayed almost immediately to the rest of the world. Another security issue is the political struggle for autonomy of mountain peoples, such as the Basques of the Pyrenees, who frequently hit the headlines with acts of violence. Finally, the mountain regions are one of the most important sources of drugs which reach the world markets – cocaine from South America, 'kif' from the Rif and elsewhere, and opium and heroin from Southeast Asia; international concern over drug trafficking brings mountains back into the forefront of the international agenda.

There is a dichotomy – in part artificial and a product of the development process – between traditional management and development and externally generated management and development. It is the real, or imagined, existence of this dichotomy which perpetuates the myths of collapse and crisis and also the difficulties of implementing and encouraging effective development initiatives. The perception is changing, however, with a growing recognition of the resourcefulness of mountain people in fully using their environment. There is an increasing acceptance of the fact that, whilst many indigenous strategies and institutions struggle to cope with the rapidity and enormity of economic and environmental change, many are quite capable of adapting, although these coping strategies might be misunderstood or misinterpreted by external agencies, or stifled by lack of autonomy, self-determination or resources for implementation.

Traditional, or indigenous, management of mountains has already been examined at some length throughout this book. Colonialism paved the way for the 'right to interfere' of foreigners, by enforcing or introducing a variety of changes – spiritual, as a result of missionary zeal, especially in South America; in health, through the introduction of new diseases such as measles, the eradication of others such as malaria, and improvements in health care; and through the introduction of different education systems, languages, and the integration of these regions into the national economy and political control. The colonial legacy has meant that in the current world of globalization, international concerns for specific environments and issues, such as biodiversity, still clash with national traditions of state supremacy and autonomy in international law, and also in a local context of survival of marginal livelihoods. There is also a conflict over meanings of concepts such as development and sustainability, and difficulting in reconciling differing priorities for local, national and international agendas (see the biodiversity conservation example in Chapter 12).

The following three chapters examine some of these issues: first Chapter 10 discusses approaches to development, what it means, its successes and failures, along with a consideration of the role of law in managing mountains and also considers whether we should be aiming at mountain-specific policies and agendas, or whether their meaning and appropriate management lies in their integration with adjacent lowlands, in which case the traditional sectoral approaches to development are most appropriate. Chapter 11 looks at conservation of biodiversity, the establishment of national parks and the evolving concept of the role of local populations in both conserving the environment and generating revenue from conservation. Chapter 12 examines the rapid growth of tourism, which has become the main 'crop' in many mountain regions, with its inevitable impact on social and cultural conditions.

# Chapter 10

# Catastrophe or capacity?

## The nature of the crisis

The idea that the Himalayan mountains were in a state of crisis with regard to soil erosion and deforestation emerged in the 1960s and 1970s. Confusing observations from travellers gave the impression of wholesale destruction. Greater media coverage of world affairs highlighted problems of siltation and flooding in India, which was attributed to degradation upstream in the mountain regions – a process termed the 'Theory of Himalayan Degradation'. The fact that erosion of highlands and deposition and flooding on lowlands is, literally, as old as the hills became submerged in this new era of catastrophic thinking. The causes of this heightened destruction were placed firmly at the feet of the 'reckless and selfish, unthinking peasants' who farmed the hills. Population growth caused by improvements in health care reduced mortality, particularly of infants, and the growing pressure on resources apparently reduced the peasant population to a horde of desperate tree-fellers and soil scrabblers. In Southeast Asia, misunderstandings between hill tribes and governments led to the consideration of shifting cultivation practices as nothing but wasteful.

Since the 1980s there has been a welcome reevaluation of this vision of destruction. Substantial research into the nature of the assumed linkages in the 'Theory of Himalayan Degradation' (THD) have led to a debunking of this myth, and to a more measured and balanced assessment of the nature, or even existence, of any such crisis (Ives, 1987; Ives and Messerli, 1989). Confusion has arisen with regard to a number of aspects, for example, from conflicting field observations. Fürer-Haimendorf (1984) commented on the serious depletion of forests in Khumbu, but Houston (1987) was of the opinion that forest cover was, if anything, more complete. Later work by Byers (1987a,b,c) questioned the idea of wholesale devastation, submitting that there was devastation but that it was very localized and in some cases species-specific.

In fact, these differences could be attributed either to the location of observation, or to national policy as to forest protection, rather than to the extravagances of the hill farmer. In Nepal, the privatization of forests not only protected large areas from any use (and may unwittingly have reduced their use potential once they were left untended) but meant that other areas open to use were much smaller and correspondingly under greater pressure. Similarly, the constraints placed upon shifting cultivation practices by the establishment of

221

## Box 10.1    The impact of Afghan refugees on Pakistan's mountain environment

Some 3.5 million refugees have fled to Pakistan, especially to the North West Frontier Province. They come from all areas of Afghanistan. Some have left the refugee camps and are thereby unable to claim any assistance from the state. The refugee camps were located by the government in areas of pristine forest rather than in the more degraded and unvegetated, but less accessible regions. Thus large areas of oak and deodar cedar forest have been cleared for construction and more for fuel and resin extraction for sale. Urban Afghanis have tended to settle in Peshawar. Some former nomads have settled in the Punjab. Others are limited to the high pastures around Chitral. The Chitralis have prevented their settling on the lower land but the already sparsely vegetated, arid upper regions are under increased pressure of grazing. Traditionally, the Chitralis had common property rights to these regions and the influx of around three head of sheep or goats per refugee has distorted this arrangement. Other pastoralists have lost their animals and some even continue to import animals for consumption from kin over the border in Afghanistan. The increased pressure of grazing and deforestation has implications for the sedimentation of the reservoir behind the Tarbela dam. It has also provided Pakistani nationals with opportunities to carry out illegal logging operations.

*Sources*: Allan, 1987.

political boundaries confined tribes to ever smaller areas, requiring them to reuse their fallow lands more often. The effect of this, combined with natural population growth, and also with the impact of refugee immigration, such as the 120 000 who left Tibet in 1959–1960 and moved into Khumbu and Mustang (see also Box 10.1), means that naturally there are areas which have suffered substantial clearance and increased pressures of use.

However, evidence suggests that degradation of these areas is localized. In addition to this, the farmers, far from being helpless and desperate, have more often been innovative and adaptive in coping with the constraints imposed by national governments. In Nepal, for example, the farmers have begun to plant trees on private land to make up for degraded resources, whilst in Southeast Asia, adaptations in cultivation, such as the switch to high-investment market crops like coffee, have helped shifting cultivators continue to survive in a new, stationary cultivation regime. From this picture there emerge two elements for discussion before the idea of development should be considered: first, the nature of the assumptions which led to the perception of crisis in the mountains in the first place, and second, the nature of farmer responses and adaptations. From these two issues, the role of development and its objectives can be explored, and how development initiatives have managed, or failed, to address changes and needs in mountain communities. Finally, there is a rethinking of what the nature of the crisis in mountains actually is, and whether the term 'crisis' is really appropriate.

## Debunking the myth

The Theory of Himalayan Degradation is based on a number of largely un-substantiated assumptions linking population, land use and degradation in the mountains with flooding and siltation in the lowlands of India and Bangladesh (Ives and Messerli, 1989). This highland–lowland connection has, geomorphologically, an element of truth, as the Indian plains are constructed of over 4000 m of eroded material derived from the uplifting Himalayan mountain range. As we saw in Chapter 1, there is a link between uplift and erosion in the mountains and deposition in the lowlands. However, the linkage in the THD (Figure 10.1) suggests that there is an increased rate of erosion due to overuse of mountain lands, and this is related to the perceived increase in frequency and magnitude of lowland flooding. Thus at its simplest, it suggests that recent deforestation and overgrazing have caused increased lowland flooding and siltation; in other words, a great enhancement of the natural process due to pressures of land or overexploitation of upland environments.

In the Mediterranean basin, the 'younger fill', a sedimentary unit found throughout the region, is attributed to the collapse of abandoned terraces at

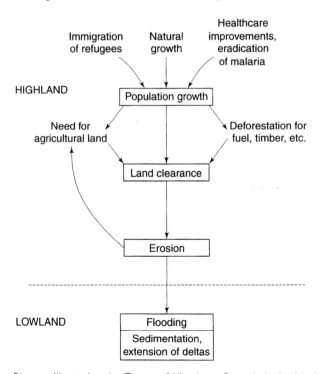

**Figure 10.1** Diagram illustrating the 'Theory of Himalayan Degradation', with the cause placed in the highlands and the consequences in the lowlands.

the end of the Roman Empire. However, it also coincided with a period of wetter climates, which would also have enhanced erosion (Vita-Finzi, 1969). Likewise, in the Swiss Alps at Davos, erosion in the eighteenth century is ascribed to deforestation rather than natural causes, as for earlier phases of enhanced erosion (Price and Thompson, 1997).

It is, therefore, necessary to assume that there is an increased rate of erosion and flooding, and that these are directly linked; also, that deforestation is the cause of the increased erosion, and that there is, in fact, an increase in the frequency and magnitude of flooding. Whilst the underlying linkages exist on long timescales (Chapter 1) and the effects of climatic change may enhance these rates (Chapter 2), the idea of recently enhanced rates and the directness of the link in the lowlands, as well as the reality of flood frequency and magnitude can all be fairly comprehensively 'debunked' or at least placed in perspective by more recent research.

The first key linkage is that between the rate of erosion and deforestation. Much of the erosion of mountain lands arises from mass movement and reworking of this debris. Whilst trees can reduce erosion, this is primarily limited to surface erosion and shallow mass movements. The tree canopy and leaf litter break the fall of raindrops which compact the soil. In many forests where leaf litter is regularly gathered for fodder, however, this effect is reduced. Tree roots can anchor the soil, but although this can reduce landslides, this effect is limited to shallow-seated slides only; deep slides are controlled by geological structure. The presence of a continuous turf mat is also a more effective soil stabilizer than trees – i.e. it is vegetation cover which is important, not trees *per se*. Cutting down trees does not always increase the rate of erosion from landsliding, but extensive disturbance as a result of road construction and mining has been far more destructive to the environment than deforestation. Enhanced landsliding does occur along road cuttings and logging routes but this tends to have only localized effects on erosion rates (Schelling, 1988; Haigh *et al.*, 1995).

The second key linkage is that between lowland flooding and upland degradation, which is an important geopolitical issue. The erosion of the Himalayas is blamed for the formation of islands in the Bay of Bengal. There is a natural link, but the timescales required to move this amount of sediment far exceed the few recent decades of perceived increased pressures in the mountains. Some 2 billion tons of sediment per year are deposited in the Bay of Bengal (Rana, 1993) and whilst these sediments do come from the Himalayas, they take between 1000 and 10 000 years to arrive, so recent erosion cannot be directly responsible for the flooding (Grosjean *et al.*, 1995). There is a substantial time delay between the two events.

In addition to this time delay, the rivers of the Indus basin meander and change their courses across the plains as they have done since they were formed. However, with the expansion of urban areas and the increase in density of population in all areas, this movement is constrained by artificial infrastructure. Just as the recent floods and avalanches have had an increased effect in the European Alpine valley floors, the hazard is increased due to the growth of human activity in this area, in the path of the hazard (Chapter 1). There is no consensus of evidence that there are more frequent floods, earthquakes or any other mountain hazard when these are considered in the context of centuries rather than decades (Hofer, 1993), but their effects may be felt by more people, because there are more people 'in the way'.

The debate about the increase in flood frequency and magnitude is now intensified, and the greater risk because of the increase in population and economic activity in the hazard zone compounded by the effects of global warming. Again, this is an enhancement of a natural process, but sufficient consensus exists to predict changes in temperature and precipitation. However, the actual effects on erosion in the mountains and flooding in the lowlands are hard to predict with any certainty.

It is this uncertainty which has prevailed in all aspects of the research into the Himalayan 'crisis' (Thompson and Warburton, 1988). Uncertainty exists in the data itself – in defining degradation and deforestation (Jaroz, 1996). Does deforestation comprise any interference in natural forest or full-scale clearance? (Plate 10.1) What does lopping of branches but retaining of tree

**Plate 10.1** The issue of deforestation is a contested one. The lopping of branches from trees for fodder, as shown here, may not clear landscapes of trees, but thin the crown cover. Likewise, the collection of dead wood for fuel reduces the risk of fire in arid regions, and does not contribute to clearance. In most mountain regions, people need to search further afield and higher up for such resources, and in some areas have taken to planting selected species near to their farms to ensure a supply of fodder or fuel – for example the tall poplars in Plates 1.1 and E1.

crowns count as? What about coppicing or replanting in different areas? If retaining the vegetation cover is the critical issue, what does the conversion of forest to shrub or grassland count as? For example, the estimates of deforestation rates in the Nepal Himalaya alone vary by a factor of 67 (Ives and Messerli, 1989), and the worst-case scenario predicts complete clearance of forest by 2000. This has clearly not happened, in fact quite the reverse is happening, both as a result of government reforestation initiatives and local individual farmers' decisions to plant to ensure their own supplies. Whilst short-term, small-scale erosion cannot be the cause of lowland flooding, this does not absolve the farmers of their responsibility to use their land sustainably, and there are many cases where nothing is further from the farmers' intentions!

The perception, therefore, of the whole Himalaya as disintegrating into a state of collapse is also patently ridiculous. The region is highly diverse in environmental and socio-economic conditions. Areas of the eastern Himalayas which are much more arid have always supported sparse vegetation and have always experienced high rates of erosion. In the western Himalayas, vegetation has been substantially altered by human activity, but well-maintained terraces can be argued to be more effective soil conservation than poorly maintained forest. High-density populations and intensive use of these lands are actually associated with high levels of maintenance, as it is in the farmers' interest to care for the carefully and painstakingly constructed terraces, which are the most precious resource, namely cultivable land (Gerrard and Gardner, 2000).

Does this leave us with any real catastrophe in the mountains? Can the farmers justifiably be blamed for the misfortunes of the lowlanders? We now need to consider the mitigating effects of mountain peoples' ability and need to cope with the highly sensitive nature of mountain environments in the face of growing population pressure. The concept of 'more people, less erosion' is becoming a dominant focus in mountain research. The term derives from the title of a book (Tiffin et al., 1994) which explores the nature of intensification of agriculture in Kenya without increasing erosion. It is this intensification which will allow greater populations to be supported and is becoming a major theme in development. However, it does not hold true everywhere, as other research, in Bolivia, has shown that fewer people means less erosion, maintenance of the long-term resource base and diversification of livelihood strategies (Preston et al., 1997).

However, there has to be a reason for maintaining populations in the mountains; do we want to keep them populated to prevent collapse of abandoned terraces such as that at the end of the Roman Empire? If so, we could afforest such areas and conserve them. Keeping people in the mountains encompasses their own self-determination of the future, giving them an equitable partnership and a voice in national and international affairs. It means their having meaningful and fulfilling lives in the mountains, and not just being cultural capital for a tourist attraction. It is these questions and issues which are considered in the nature of development and its outcomes.

## Indigenous capacity: the farmers' response _____

Throughout this book, the adaptive flexibility of mountain communities has been demonstrated and it is this resilience in the face of changing environmental and political conditions which is their trademark. This is not to overromanticize their struggles and efforts at winning through; neither is it to propose that traditional institutions and strategies should be left untouched, as there are many instances where they cannot cope well with the exceptionally rapid and global nature of the economic and environmental changes which many communities now face. However, there is much to be said for using the farmer as a starting point in development – that he or she is neither the subject nor the object of development but a generator of it. In the last decade or so, development practitioners have increasingly emphasized the capabilities of indigenous people to cope with change and sought them out as part of the solution. This is a change from externally imposed development, through putting the farmer first, towards the stage where the farmer calls the tune, not the development agency (Chambers, 1994). This is discussed below, but first it is important to explore the nature of indigenous capabilities before we consider how these interact with development approaches.

Indigenous knowledge is a broad and elusive concept. It is essentially a personal, practical knowledge related to the meeting of individual and community needs. It is immediate and corporate, but represents a different world view from Western science in that it is not constrained by rules and hypotheses (Agrawal, 1995). Indigenous knowledge is not static, any more than the cultures and institutions of the communities who hold it (Roling and Brouwers, 1999). Thus knowledge relates to navigation; territorial definition; plants, animals and other resources; the capacity of different soils for different crops; planting densities according to rain- or snowfall, etc. It also extends into a wealth of experimental knowledge, particularly relating to cultivation practices; soil fertility and erosion management; varieties of crops; storage and the understanding of the properties of plants, particularly their medical uses (Box 10.2). In other words, the knowledge specifically addresses the particular conditions of the mountain environment (mountain specificities), as shown in Table 10.1.

This wealth of knowledge and attitude towards experimentation within conservatism – i.e. trying new ideas and crops but without risking all on an untried novelty – has been 'rediscovered' by Western development practitioners and in a sense 'created' by the process of discovery and communication. Whilst this can facilitate releasing it into the wider world, it runs the risk of being commodified, simplified, removed from its context, classified and thus fossilized. In reality, because it is often qualitative and intuitive, it is difficult to incorporate into quantitative, scientific milieux (Funnell and Parish, 2000). However, it remains crucial to the contemporary 'inverted' development approach of participatory learning on the part of the agencies (Sillitoe, 1998b). The problem remains how to utilize it in a meaningful way.

Some indigenous knowledge is clearly exploited; many cases of this exist in relation to biodiversity and peoples' knowledge of plants which gives Western

**Table 10.1** Indigenous strategies in response to mountain specificities.
(The columns are labelled with aspects of mountain specificities, whilst the rows represent indigenous strategies. In the boxes are aspects of each strategy which address each specificity.)

| | Inaccessibility (remoteness in space and from external linkages) | Fragility (vulnerability to irreversible damage, low carrying capacity, high costs) | Marginality (economically isolated, high dependency, limited production options) | Diversity (complex constraints and opportunities, interdependence of production bases/activities) | Niche availability (small, numerous specific activities with comparative advantages) |
|---|---|---|---|---|---|
| Diversification and self-provisioning | Spatially and temporally linked activities. Scattered settlement pattern. | Spatially and temporally linked activities. Local resource-focused recycling. Self-sufficiency. | Scattered settlement pattern. | Spatially and temporally linked activities. Local resource-focused recycling. Self-sufficiency. | Spatially and temporally linked activities. Local resource-focused recycling. Self-sufficiency. |
| Folk agronomy | Cultivars with different attributes. | Cultivars with different attributes. Cultivation–livestock–forestry linkages. Intercropping, fallowing, rotations. | Cultivars with different attributes. Cultivation–livestock–forestry linkages. Intercropping, fallowing, rotations. | Cultivars with different attributes. Cultivation–livestock–forestry linkages. | Cultivars with different attributes. |
| Ethno-engineering | Small-scale transport and access logistics (rope bridges, etc.). | Terracing, slope management. Protective vegetation, contour farming. | Terracing, slope management. Protective vegetation, contour farming. Traditional irrigation. Drainage management. | Traditional irrigation. Drainage management. | Traditional irrigation. Drainage management. |
| Collective arrangements | Crisis period sharing systems. | CPR. Social regulation of fragile resources. Crisis period sharing. | CPR. Community irrigation, etc. Crisis period sharing. | CPR. Community irrigation, etc. Crisis period sharing. | Community irrigation, etc. Crisis period sharing. |
| Upland–lowland linkages | Trading in specialized mountain goods (high value, low weight). Periodic migration. Transhumance. | Periodic migration. Transhumance. | Trading in specialized mountain goods (high value, low weight). | Trading in specialized mountain goods (high value, low weight). | Trading in specialized mountain goods (high value, low weight). Externally planned exploitation of mountain niches. |

*Source*: After Jodha, 1990.

## Box 10.2   Some examples of indigenous knowledge and experimentation

- Storage of potatoes – farmers selected between different coloured eyed varieties according to storage conditions; this was later 'proven' by scientific research (Rhoades, 1985; Rhoades and Bebbington, 1995).
- The 'new green technology' of Asia – indigenous selection of rice varieties to maximize production on land holdings (Fujisaka, 1995, 1999). Farmers had on average 0.7 ha per family divided up into a mean of six parcels with different growing conditions (Bhuktan *et al.*, 1999).
- New crops adopted – flowers, ginger, spices, different kinds of bananas in the Andes to maintain diversity and find niches in expanding markets (Quiroz, 1999).
- Sweet potatoes in Irian Jaya are selected according to local micro-environments and societal preferences (Schneider, 1999).
- Agro-forestry developments by farmers in Nepal; they planted tree species selected for their different fodder properties, enabling farmers to control the milk yield of cattle and alter the consistency of dung for different uses in different seasons. Further selection was based on seasonal availability of different species in order to maintain a year-round supply, and on preferences according to ease of harvesting, which was based on gender (Rusten and Gold, 1995).
- Grazing strategies in highland Bolivia have been adapted, with the introduction of alpacas, to take account of labour availability and quality of grazing (Preston, 1998).

scientists clues as to their potential importance as cures for diseases. Once the plant enters the domain of Western science institutions, it is very rare that the original source of knowledge is even recognized, let alone given a share in the patent or profits. Steps have been taken to address this, but problems arise with the application and enforcement of international law in this field (see Chapter 11). What is important about indigenous knowledge in a mountain development context, however, is simply that local people know, to a large extent, a lot more about their environment and their needs and capabilities than many development agencies. It is with this in mind that greater sensitivity is shown in the approach to development options and the way they are presented to these communities.

A great variety of institutions exist and which have evolved and adapted over centuries to manage individual and common property. These have been considered in Parts 2 and 3 of this book. Whilst many are healthy and strong, others are weakened and are less able to cope with the nature of the changes faced by mountain communities today (Stevens, 1993). The greatest of these changes, exceeding climate change by far in rapidity, is the penetration of the capitalist market economy into remote regions. The impact of this on agriculture

## Box 10.3   Rural development in Hunza

In the Hunza valley, government initiatives using modern surveys have failed on a number of occasions to instigate effective new irrigation channels. This is because of their lack of local knowledge of the slopes and landscape, but also because the villagers do not feel it is their project, but one imposed on them. They therefore have little incentive to use, maintain or participate in the construction of these channels.

The AKRSP has filled the gap left by the deposed Mir in acting as a catalyst for development. Operating through Village Organizations which bid for funds, it has achieved extensive development in agriculture, health and education. Where new initiatives are proposed, the AKRSP has been assiduous in trying to ensure that issues such as user rights and allocations, as well as sources of contributions of labour and materials are assessed and secured before construction goes ahead.

Nevertheless, the difficulty lies in the fact that as any new land for cultivation tends to be at higher altitudes than existing land, whole new channels are needed, which breaks new ground, literally and metaphorically, and impinges on areas of hill slope known to be unstable. The new Aliabad channel in Hunza, sponsored by AKRSP, is 8 km long and is already having to be widened and deepened to increase its capacity. Disputes arose as to the allocation of rights to this water whilst the channel was under construction, which further delayed its completion. Especially contentious were the rights of the new 'capital', Aliabad, established after the deposition of the Mir, to tap the waters of the Ultar glacier which were traditionally reserved for villages to the east.

*Sources*: Saleem *et al.*, 1994; Nayab and Ibrahim, 1994; Parish, 1999.

has been demonstrated in some of the examples in Chapters 9 and 12. The most obvious change is the move to a cash economy and the change towards market production rather than subsistence. This has ramifications for social structures as reciprocal exchange networks are eroded. The balance of power shifts in the community towards those with new wealth and away from traditional hierarchical or lineage structures.

The outcome of these changes is to weaken some institutions, but also to introduce alternatives, such as the Aga Khan Rural Support Programme (AKRSP), an Ismaili organization in the Hunza valley (Box 10.3). This example illustrates the interactions between old and new institutions and also some of the reasons why what seem to be good ideas do not bear fruit in reality. What appears to constitute appropriate technology may be a non-starter if the institutions do not exist to implement it in the first place. These are issues relating to the nature of development.

Although communities have indigenous capacities, given the nature of the changes they face in the contemporary world, they need a helping hand –

technology, funding, ideas, etc. There is a fine line, however, between helping and 'overhelping' and overreliance on institutions can become a cause of failure if these are not appropriate, or strong enough, to bear the brunt of the forces of change. Institutional failure is one cause of development failure (Griffin, 1987; see below), but sometimes local communities can mobilize themselves to form new institutions – the AKRSP Village Organizations, for example, or PRATEC in the Peruvian Andes, which is a group set up by local writers and activists to revitalize Andean culture and re-imbue its people with their own, traditional world-view and livelihoods (Apffel-Marglin, 1997). This institution has been able to revive the people with the idea of their own ability to sustain themselves, and to reinvest in their environment and culture, bringing a new cohesion to local communities and a new pride in their cultural traditions.

## The nature of development

The evolution of development theory and practice is complex. Development itself is generally considered to be a 'good thing' (Dower, 1988), something which ought to happen. Lampe (1983) claims that it is 'every enlightened government's goal [to achieve] harmonious and comprehensive development of rural areas'. But what ought to happen and how, is much contested. Sustainability is also subject to multiple definitions; the most widely accepted definition of sustainable development is that which 'meets the needs of the present generation without compromising the ability of future generations to meet their own needs' (WCED, 1987). This therefore takes a long-term view, incorporating our obligations towards future generations. More specifically, sustainable agricultural development might be defined as the 'ability to maintain and enhance production performance without damaging its long term production potential' (Jodha, 1990).

Development is both a process – the process of change towards an envisaged outcome, and the outcome itself. It may be undertaken for a variety of motives (altruism, professional status of individual practitioners) and is often misapplied or generated as part of a big industry which can appear to serve itself more than those it seeks to develop (Chambers, 1993, 1994). Much of the current perception of development arises from the traditional approach to it as something brought to barbarians and savages by the enlightened West. It is seen as a scientific approach of rules and laws enforced and imposed on an unruly population who are irresponsible and ignorant – a view which originated in the colonial era that introduced the idea of a 'right to interfere'.

The dichotomy between the 'enlightened West' and the 'primitive barbarian' lies in their contrasting attitudes to life and the environment. Most indigenous peoples are closely attuned to their environment and can work with it to bring out its best yields whilst minimizing risk and loss. The juxtaposition of this view with a Western, capitalist, materialist approach causes a change in the aspirations of each. For the mountain communities it introduces quite alien ideas of economic prosperity. Western cultures are highly materialist and have a tradition of individual ownership backed up by force of law. Many traditional mountain communities have very different conceptions of owner- ship (Chapter 7) and greater communal responsibility. They do not have such materially based aspirations. Wealth lies in livestock and land; the state of the former is determined by the ability to feed it in winter, and the latter may be limited by the availability of labour to work it (it is only in the last century that the concept of wage labour has grown appreciably).

Although the 'enlightened West' is itself increasingly subject to a new re- naissance of development (the recognition of inherent knowledge and values) there is still a tendency to commodify knowledge and to control outcomes. However, a more flexible, sensitive and forgiving approach to the process is emerging in the attitudes to those 'being developed'. In the 1990s the growing emphasis on 'participatory' development has meant that local institutions, local knowledge and local politics have almost been hijacked by the develop- ment machinery in order to demonstrate the practice of this more sensitive approach. Power structures are often shifted – the investment of special author- ity and responsibility in individuals rather than in organizations means that an element of individual gain over community or collective gain is introduced. This trend is reinforced by the capitalist market economy into which mountain communities are increasingly drawn – both by their own efforts and by the development opportunities arising – cash cropping, for example.

An additional problem with development is the new and different aspirations it brings. These may not of themselves be a problem for individuals or even communities – for instance, construction of new houses, piped water, education and health care are all laudable improvements. However, the problem lies in the introduction of aspirations which cannot be met – for luxury goods, visas to foreign countries and cash. The kingdom of Bhutan has one of the lowest GNP per capita, but it argues that it is one of the happiest nations (*Independent*, 19 April 1997). In other words, this represents 'affluence in poverty' (Sahlins, 1997). The introduction of new aspirations into this arena emphasizes the poverty and breeds dissatisfaction. Bhutan has jealously guarded its integrity and only recently opened up to carefully controlled tourist activity.

This is itself a thorny ethical issue: are we trying to maintain these people as a traditional museum piece, or is the objective of development to improve the quality of life in some areas (schools, agricultural improvements, health, etc.) but to deny the people access to television and motor transport and manufac- tured goods because it 'spoils' them or because satellite dishes on mudbrick houses look odd to tourists? This is hardly allowing them self-determination and respecting their right to decide their own futures. Modern development approaches have begun to address this issue. The terminology which emerges is that of 'facilitation', 'capacity-building' and 'helping them to help them- selves'. Such development involves the provision of options and allowing a choice of additional or alternative sources of income or subsistence which are of a scale and nature appropriate to their needs and aspirations. This is, of course, costly and more difficult to implement and is much less impressive on the World Bank balance sheet than a shiny new dam!

The objectives of development therefore need to be clear. The emphasis is generally on poverty reduction and the fulfilment of basic needs. Although the exact nature of basic needs is itself contested, they are usually taken to involve basic water and sanitation, health care, sustenance and shelter. These are the areas which take up much of the aid which is sent to these regions in the event of a natural disaster. However, development needs to go further and to breed hope and vitality into areas which are remote, cut off, oppressed, or filled with refugees. It is about building lives, or rather making it possible for people to build their own lives. In many mountain regions it involves the creation of economic alternatives, and not just tourism. It may require the control of tourism and almost certainly involves the concept of 'Integrated Watershed Development' where the watershed becomes the unit of planning and management (Karan, 1989; Hamilton and Bruijnzeel, 1997), and a 'niche-based' approach to sustainable production systems whereby the microenvironment and small-scale community can be incorporated into the development agenda (Partap, 1999). Biodiversity and conservation are a major current focus, and the 'more people, less erosion' intensification-without-degradation idea is currently being applied to the sustainable development of mountain regions. Such development is as much about building futures as solving today's problems.

## Constraints and opportunities

With the best will in the world, not all development works, but whilst there are both successes and failures, it is very difficult to identify reasons for either. This results from the diversity of development initiatives, local conditions, the complexity of the human and physical environment and the apparently unpredictable response of individuals. However, several studies have attempted to identify the constraints (i.e. reasons for failure) and opportunities (reasons for successes) in development initiatives. These will be considered below, and it is interesting to refer back to earlier parts of this book to identify the context of some of the patterns of change described earlier. It is important to remember, though, that the lists of reasons for failure are not exhaustive and the reasons for success do not provide a fixed-recipe approach to 'good' development, but are illustrative of the elusive conditions which tip the balance between success and failure.

Lampe (1983) outlines a number of problem areas arising from development:

- *Natural constraints*: the nature of the mountain environment – steep slopes, high-energy processes and physical remoteness and accessibility problems (the mountain specificities of Table 10.1) – represents a barrier to successful construction projects. Examples include the sedimentation of reservoirs; earthquake hazards; landslides associated with road building and infrastructure arising from opening up remote areas; tourism and industrial and extractive activities. In the case of road construction, associated developments tend to be concentrated near to the road, thus having a limited impact further afield (Rawat and Sharma, 1997). Construction is costly and may not have substantial impacts on subsistence livelihoods (Banskota and Jodha, 1992) but in the case of the Karakoram Highway the increased access has changed the agricultural base to one of import–export and also opened up opportunities for tourism (Kreutzmann, 1991).

• *Economic constraints*: the nature of the physical environment means that any construction projects are more expensive to implement. This is compounded by the fact that most projects can usually only be on a small scale, given the limited land and other natural resources, and the fact that markets tend to be small and distant. Thus the returns on investment are much smaller than for comparative projects in the lowlands. This can be offputting to potential investors. The marginality of mountain climates for some forms of agriculture increases the risk of failure and thus the costs of investment. In some cases (e.g. the Moroccan fruit trees in Chapter 9) the economic incentives are sufficient to encourage an increase in the acceptable risk of failure. Farmers' assessments of economic gains are also important. Where high-yielding varieties (HYVs) of crops are introduced, they weigh up the potential returns against the capital outlay on new seeds each year, fertilizers, pesticides, etc. and in many cases decide against new developments (Renaud, 1997). This was the case for HYV wheat in the Indian Himalayas where the costs arising from the loss of the benefits of other uses of the crop (e.g. for fodder and thatching) outweighed any marginal improvements in yield (Negi, 1994). Improvements in yields and livestock with a view to producing a surplus require a market, and there is no advantage in increasing production in a vacuum. An additional factor in the higher costs of development projects in rural areas is the need to train individuals to attain the required skills to operate and maintain new technology and to learn new skills of husbandry. This is an ongoing, and often hidden, cost of development. Finally, quite simply, many countries cannot afford development on large scales.

• *Social constraints*: language difficulties arise from the fact that many mountain peoples speak local dialects and the communication of new concepts and ideas can be difficult, as can technical language. Religious and cultural conditions may also determine a reluctance to take on development; this is a particularly recurrent issue in relation to women and their role in societies – the cases of the bathhouse and of women's access to high pastures were mentioned in Chapter 4. There may be limitations on women's handling of money or employing other women – in parts of the Atlas women engaging in such activities are considered to be prostitutes (Steinmann, 1993) – and on their integration into development initiatives (Byers and Sainju, 1994). Likewise, there may be caste barriers to the take-up of particular options. Development projects may be orientated towards individual gain, which may destabilize community networks and power balances. Finally, many development initiatives are presented as 'technical packages' to farmers, who are then not able to pick and choose, and who may not see individual benefits, or understand how such benefits link to their own lives. This is the case for some soil-conservation initiatives introduced to hill tribes in northern Thailand (Renaud *et al.*, 1998). Despite the appropriate and entirely rational nature of the propositions (leguminous hedges, grass strips, intercropping, etc.) many farmers were reluctant to participate on the basis of not seeing the gains, the outlays involved, insecure tenure, which discourages investment, social marginalization of some tribes and people and labour constraints.

• *Local constraints*: each community reacts differently to development initiatives, which may be hard to reconcile with apparently marginal social factors and concepts. For example, the introduction of an apricot kernel-crushing machine in order to mechanize the production of nut oil was abandoned by Pakistani women in Hunza. The machine needed servicing and parts but, more importantly, they could not see the advantage in saving time; the job was an important social event for the women, who gathered together and 'put the world to rights' whilst working. The machine was noisy so they could not talk, and thus they were deprived of the social interaction. Also, they could see that a saving of time on one job only meant they would have to fill the time with extra tasks so as not to be seen to be idle. As it was, the manual task filled the time to the right degree at certain times of the year and provided a cohesive social function. On the other hand, they welcomed the opportunity to keep chickens to produce eggs for sale; chickens did not take up much extra time and provided them with egg money for their own use, so direct advantages could be gained. Other local constraints arising from local customs and laws which are often unwritten may prevent certain activities from being readily acceptable (see Box 10.4).

## Box 10.4   Local constraints on irrigation development

The village of Ganesh, (see Box 10.3), despite being Shiite, acquired funds from AKRSP to construct a channel for new irrigation supplies. Unfortunately, this was damaged beyond repair and instead of rebuilding, a diesel pump was installed to bring river water up to the fields (Plate 10.2). The fields to which this water was dedicated were located on a piece of land extending into the Hunza river. This was flooded and severely damaged in the 1960s. Although the fields have to some extent been restored, and irrigation supplies are now available, the land is not cultivated and the pump remains unused because there is a continuing dispute over who owns the land. In other areas, installation of too many diesel pumps, as in the Anti Atlas, has lowered the groundwater below the level of traditional wells, thus not only overusing supplies, but also denying those without pumps access to water. Some pumps have fallen into disuse through the unavailability or expense of parts and of diesel. The ownership of pumps in the Anti Atlas area has increased enormously as a result of the burgeoning remittance economy which makes cash available to families to buy them.

*Sources*: Parish, 1999 and unpublished observations.

*(continued)*

*(continued)*

**Plate 10.2** The infamous Ganesh diesel pump. Until the land dispute is settled, this pump, which replaced a damaged irrigation channel, is destined to decay where it stands.

- *Political constraints*: The power of a local leader may control changes affecting his area, preventing the take-up of apparently suitable options, whilst nationally there may be differences in attitude between rural and urban and local and national actors as to what constitutes acceptable development. In the Alps, it was shown that the government's encouragement of tourism development was welcome to the locals, who needed the revenue, but unwelcome to urban people, who saw the mountains as already overdeveloped and becoming unaesthetic (see Chapter 12). In addition, government intervention may be unwelcome as this represents a 'takeover bid' by external agencies, depriving the locals of their means of subsistence (see Box 10.3). The conflict between international pressures to preserve forests for biodiversity and climate purposes, national needs to exploit them for timber revenue, and local needs for a multiple-use resource typify this political tangle. In addition, short-term outlooks and quick returns sought from projects conflict with the long-term need for sustainable rural livelihoods. An additional factor in the politics of development is the issue of disputed regions such as Kashmir and Tibet, and separatist claims of peoples such as the Basques and the Kurds. These often have more political value as unresolved bargaining points than resolved issues. Continuing conflict slows the opportunities for development and thus its progress is reduced to emergency relief in the wake of natural disasters.

- *International aid and professional constraints*: not all the problems of development lie with particular countries or the local farmers. The tangle that the politics of international aid and development intervention represents is almost impenetrable and often seems far removed from the farmers' experience. In many cases, it seems that development agencies' agendas are orientated towards their own self-perpetuation rather than concentrating on the task in hand. Large proportions of some aid budgets are spent on administration of organizations rather than on development projects themselves, and in Britain it is the custom for development charities such as Oxfam, Red Cross, Christian Aid and the like to submit statements on the proportion of their budgets spent on administration in an effort to be accountable to donors and to encourage support of their operations. There is a distinct lack of joint projects, coordination across sectors and across organizations, resulting in conflicting projects and duplications in target regions (Lampe, 1983). In addition to this, the pressure on individuals and organizations to be accountable and to achieve targets of outcomes in terms of quantifiable gains (publications, funding, numbers of people reached or projects completed) lead to short-termism and a concentration of activity and effort on self-perpetuation. This is, of course, not true for all organizations or individuals, but it is a trend present in the development 'industry' at present. It is important that the main aim of development – that of benefiting peoples' lives – should not be lost in the machinery set up to accomplish this.

It is not, of course, all doom and gloom! Many development initiatives have worked and do work. Some of these have arisen from indigenous groups of farmers or individuals, whilst others have been nurtured by external agencies. It is difficult to determine what works and what does not – as Bebbington (1997) points out, so much is spontaneous and circumstantial and cannot be orchestrated that it appears to be a bit of a 'hit and miss' affair. However, there have been several attempts to identify what works in general and what does not. The Mountain Forum's electronic conference, 'Investing in Mountains' (Preston, 1997), provides a useful summary of the points marking out successful projects:

- sensitivity to mountain specificities;
- matching new developments to traditional systems, creating appropriate, adapted development options;
- a degree of choice for farmers to be selective in the direction of change;
- autonomy and control over local resources, particularly land tenure security;
- equity in access – between and within genders, households and regions;
- transparency and accountability of institutions;
- support to implement and maintain technological developments and comprehensible extension work;
- low-cost, effective, viable and appropriate options;
- self-perpetuating development with a future.

**Table 10.2**  Using traditional strategies as the starting point for development initiatives.

| Traditional strategies and their key elements | Appropriate strategies for new development options |
| --- | --- |
| Diversification and interlinkages of agricultural components | Systems approach with focus on multiple-use species, adaptability and complementarity with other crops and activities. |
| Balanced land-extensive and land-intensive agricultural activities | High-yielding, fast-maturing species to increase land-intensity, retaining their multiple-usage potential; alternative activities not competing for land (e.g. aquaculture, apiculture). |
| Local-resource orientation of folk agronomy, other technologies | High value, low weight, low perishability, recyclability of agricultural products with lower dependence on external inputs (pesticides, etc.), using comparative advantages of mountain environments to overcome inaccessibility and marginality constraints. |
| Land-extensive crop rotations, nutrient cycling through cropping systems | Plants selected to build and bind soil; legumes in rotation to accumulate nutrients and reduce need for fallowing; biological pest control. |
| Ethno-engineering for resource management and conservation | Bio-engineering and mechanical devices for slope stabilization, soil and drainage management; agro-forestry; use of indigenous knowledge in land management. |

*Source*: After Jodha, 1990.

In many cases, successful development initiatives have arisen where the macro-scale (in political, economic and other terms) meets the micro-scale (Lama, 1997), but the location of this meeting point is different in each case, for such is the effect of diversity on development. Thus, where farmers' practices are used as the starting point of development thinking, the outcome in terms of sustainable options is more likely to be acceptable to them, particularly if they have had the opportunity to develop them (Table 10.2).

Sometimes, a fresh perspective is needed on a problem. For example, whereas traditionally stones cleared from fields in Bolivia are piled high to act as boundary markers and as a way of minimizing lost land area from the heaps, this could be changed to bunds which would help reduce soil erosion, even if it meant losing a small area of cultivable land (Clark *et al.*, 1999). The farmers' reasons for not taking up the option of stone-walled terraces in preference to their traditional stone heaps are the labour needed to construct the terraces, the skills involved, and the resultant less suitable ploughing environments, even though the potential gains would be a more stable environment in the long term, the opportunity to grow trees as well as crops on the terraces, retention of top soil and of soil water. Adaptations to ploughing and slope engineering techniques would facilitate the transition from old to new. More creative and sensitive approaches to 'marketing' technologies by using farmer-to-farmer

## Box 10.5   Reintroducing commons law in northern Norway

The search for solutions to environmental problems which are focused on neither state nor individual goes against the modern Western capitalist ethos. This is reflected in the controversy over the reintroduction of commons law in mountain regions of Norway. One issue is the degree of individual freedom, which may be greater under state control, where all are treated equally, than under collective control, where local rules may constrain individual behaviour in one region as opposed to another. The degree of sustainability of resource use is also an issue; under state control, the land may become over-used and underprotected (or vice versa) but under commons management it could be sustainably used in the long term in such a way as to reflect local needs and use patterns. The land in question is the northern part of Finnmark, which has a long tradition of nomadic use and was not part of the original Norwegian state. Aboriginal rights to the area have been contested, and the reintroduction of commons law to this region represents a transfer of rights back to the original inhabitants.

*Source*: Sandberg, 1998.

dissemination rather than government official-to-farmer would be helpful, as would avoiding short-term artificial incentives such as subsidies, etc. Whilst start-up assistance is important, it is equally critical to ensure that this is not the reason for farmers adopting an option and that the longer-term economic and environmental benefits are meaningful to them (Bunch, 1999).

Where the opportunity arises, farmers as individuals or groups are often able to capitalize on the comparative advantages of the mountain environment in terms of production, but the critical link is access to markets and non-exploitation by external buyers. In Chile the local farmers adopted contract farming first for tobacco in the 1950s and then fruit export and grapes in the 1970s and 1980s (Korovkin, 1991). This has in fact resulted in the emergence of an individual-biased capitalist economy, but the benefits of corporate participation extend throughout the community. In southern Ecuador, farmers have been able to adapt to changing markets and environmental conditions by incorporating alpaca into their grazing regimes on the basis of a growing market for alpaca wool (White and Maldonado, 1991). This has brought economic gain at little environmental cost.

The role of institutions and different actors forming links in a chain between farmers and markets, for example, has enabled some communities to expand and diversify their economic base. In Ecuador and Bolivia, successful entrepreneurial developments have arisen from the activities of individual priests, professors or local networks to provide an opening for new developments such as horticulture. This 'social capital' is critical in the success of these projects, but lack of it has been a cause of failure elsewhere (Bebbington, 1997a). Social capital and institutions alone do not constitute a successful project; they need to be plugged into the wider system, and thus networks incorporating farmers and markets are critical. These networks can also go some way to ensure that farmers can enter the market on an equal footing with others, and that exploitation of their efforts is dictated by their choice rather than their need or obligation. Of course, suitable options and physical environments are also necessary, but mountain farmers have often appeared to make a living out of little!

This social capital approach to development has grown since the criticism of the Green Revolution of the 1970s where projects were considered to be impositional, inappropriate and degrading of local cultural, social and economic values (Bebbington, 1996, 1998). It goes some way towards meeting some of the factors characterizing the success stories listed above. The Mountain Forum's conference 'Investing in Mountains' provides an interesting and encouraging summary of the diversity of contemporary ideas, projects and initiatives emerging from mountain regions (Table 10.3). This provides a refreshing and invigorating reflection of the diversity and resilience of mountain communities. It also demonstrates that we are a little further along the road of understanding how mountain communities operate and what their priorities and aspiration are, and thus knowing a little more clearly 'where to hit it' (Sanwal, 1989; Thompson and Warburton, 1985).

## Key points

- The 1969s and 1970s were dominated by perceptions of eco-catastrophe in the mountains: deforestation, refugees, societal collapse and erosion causing lowland flooding.
- The last 20 years or so have questioned this perception and recent research indicates resilience of communities and mainly localized environmental problems.
- Research into environmental and economic change in mountains is shrouded in uncertainty, making it difficult to assess the real nature of the crisis, and, indeed, whether there is a crisis at all.
- The resilience of indigenous communities lies in their capacity to cope with change by adaptation and innovation. The range of knowledge encompasses all aspects of environmental management and resource use and is uniquely adapted to mountain specificities.
- This knowledge has become the focus for contemporary development as approaches have evolved from externally imposed big projects to locally generated, self-determinant, participatory approaches.
- Development itself is dogged by conflicting ideas of what it entails but it is generally accepted as a positive move. Sustainable development takes a long-term perspective, including future generations in its remit.
- Development is met by both constraints and opportunities of an economic, institutional, social, political and physical nature. There is no one reason for

either success or failure but there are particular elements which contribute to one or the other.

- Sensitivity to community perspectives and the nature of the mountain environment is a critical component of potentially successful development projects, but many successes have arisen from local, spontaneous activity which cannot be externally orchestrated though it can be externally facilitated.

Table 10.3  Innovative mechanisms for investing in mountains.

| Mechanism | How it works | Examples |
|---|---|---|
| **TENURE RIGHTS** | | |
| Property rights | Legal rights to use, manage or own land, property or resource | Community forest user groups, Malaku-Baran conservation area, Nepal |
| Transferable development rights | Legal rights to develop property which can be traded | Mountain protection plan, Virginia, USA |
| Conservation easements | Legal agreement of sale or donation of right to develop property | Conservation easements in Vermont |
| Tradable water rights | Legal rights to use water resources which can be traded. Restrictions apply to the use of water | Tradable water rights in Chile |
| **USER FEES** | | |
| Royalty fees | Fees charged by government for use of a national resource | Mountaineering royalties, Sagarmatha National Park |
| Entrance fees | Fees for entry into protected areas | Annapurna conservation area; gorilla viewing fees, Rwanda |
| Tour operator fees | Fees charged to tour operator rather than tourists | Contributions to conservation, Nepal and India |
| Hunting and fishing fees | Fees charged for right to hunt or fish | Control of species, New Zealand; Akagrea Domaine de Chasse, Rwanda |
| Environmental taxes | Fees attached to goods and services | Lodge taxes, Langtang National Park, Nepal |
| Redirection of user fees | Fees channelled back to protect the resource being used | New York City Watershed Agricultural Program, USA |
| **MARKET STRATEGIES** | | |
| Regional trademarks | Exclusive legal rights to production and sale of high quality, locally produced foodstuffs 'Apellation Controlée' | Cheese production in the Beaufort valley, France |

(*continued*)

**Table 10.3** (cont'd)

| Mechanism | How it works | Examples |
|---|---|---|
| Green marketing tools | Tools which capitalize on value addition from environmentally benign products | Ecotourism marketing, Sikkim |
| Micro-enterprise development | Training and support for developing new small businesses | Hindelang Nature and Culture Program; Micro-enterprise development, Nepal |
| Cooperatives | Entrepreneurial, self-managed systems of associations with roots in the local region | Trentino region, Italy |
| Micro-finance | Credit and savings for low-income people | AKRSP, Pakistan |
| Revenue from genetic resources | Strategies enabling communities to derive appropriate economic value for their biological diversity | Bioresources Development and Conservation Programme, Cameroon |
| EXTERNAL FUNDING SOURCES | | |
| Foreign aid | Bilateral and multilateral assistance for countries needing financial support | Global environmental facility |
| National trust funds | Money invested at national level for long-term funding | Mgahinga and Bwindi Impenetrable Forest Conservation Trust, Uganda; Bhutan Trust for Environmental Conservation |
| Debt-for-nature swaps | Hard currency debt of one country exchanged for conservation or preservation of globally significant natural resources | National Trust Fund for Protected Areas, Peru |
| Local trust funds | Money invested by local community or organization for long-term funding | Warm Springs Indian Reservation Trust, Oregon, USA; Wolf Compensation Fund, Rocky Mountains, USA; Snow Leopard Trust, Mongolia and Tibet |
| PRIVATE SECTOR FUNDS | | |
| Mobilization of private funds | Use of private-sector funds for conservation | ShoreTrust Bank, Washington, USA; Recreational Equipment Inc., USA. |

*Source*: From Preston, L. (ed.) (1997) 'Investing in Mountains: Innovative Mechanisms and Promising Examples for Financing Conservation and Sustainable Development.' *Synthesis of a Mountain Forum Electronic Conference in Support of the Mountain Agenda.* Mountain Forum, The Mountain Institute, and the Food and Agriculture Organization of the United Nations. Franklin, West Virginia. 48 pp. (Table 1).

# Chapter 11

# Conservation and biodiversity

Conservation and the issue of biodiversity is one of the most prominent areas of global interest in mountains. Conservation extends not only to the protection of flora and fauna in designated areas such as parks, but also to the soil and ecosystem. Increasingly, there is an integrated ecosystem approach to the conservation of mountain landscapes. There has been a significant change during the past century away from a protectionist, exclusive approach to landscape conservation, and towards this integrated approach, which includes the incorporation of local peoples' needs and sustainable uses as part of the conservation objective. This arises partly from the impractical and unacceptable exclusion of people for the purposes of protection of landscapes but also because traditional land uses do much to shape and maintain what external agencies then seek to conserve: without continued and sustainable levels of use, the 'capital of conservation' is eroded as much as it would be if it were overused.

Mountain regions are unusually high in endemic species, and the long history of use of plant resources by indigenous peoples has maintained a high diversity in many natural systems. Many others have been degraded, or converted to monocropping, but even in the agricultural sphere there is a vast wealth of native varieties and knowledge of their properties. Global interest in biodiversity arises from the increasing 'discovery' of plants which have properties useful to medicine and agriculture. The exploitation of plants with special properties, particularly by the pharmaceutical industry, has become a highly controversial issue as it involves a degree of exploitation of indigenous people.

Genetic modification of plants and the production of new varieties of crop with higher yields, pest resistance, etc. are an important part of the development objective of sustainable agriculture. However, these new crops often do not meet the multiple uses to which plants are put by local farmers, lacking properties which they alone deem to be useful and important. As a result, the uptake of new varieties in many areas is slow and disappointing for the development agencies. Where they are taken up on a significant scale, however, it is at the cost of the survival of the many local varieties which were used to fill micro-environmental niches in the landscape.

Conservation has increasingly become an international issue, with claims upon the 'global commons' by the international community which conflict with local claims on resources for subsistence livelihoods. In this respect the

precedence of rights of access between local, national and international institutions touches upon issues of state sovereignty and indigenous peoples' rights
to life and self-determination which are increasingly seeking resolution through
international law. This chapter will explore both these issues in the context of
mountain regions, with a consideration of the effects of conservation initiatives on local populations and their livelihoods.

## Mountain protected areas

The conservation of mountain landscapes has come a long way since the first
national parks appeared in the USA under the influence of John Muir in the
nineteenth century. The growth of protected areas began in the New World –
it might be argued, because it is sufficiently affluent and has land to spare.
However, the establishment of the Sagarmatha (Mount Everest) National Park
in 1976 in Nepal represents a key event in the move to protect mountains
(Box 11.1). Right from the start this initiative involved local people and
respected their subsistence needs in the landscape.

The current global coverage of mountain protected areas is shown in Table 11.1.
The dominance of Arctic and Antarctic coverage is very high due to the large
area of Greenland which is designated as a protected area.

There are several categories of protected areas: those designated as national
parks may include a variety of degrees of resource use or protection. Others
appear on the World Heritage Site list and are entitled to additional sources of
funds and stronger protection. The World Heritage Convention (1972) is
one of the earliest pieces of international environmental legislation. World
Heritage Sites include designated areas of environmental interest which are of
international importance. Designation not only adds power to enforcement

---

### Box 11.1   The protection of the Himalayan peaks

In one sense, it was the advent of mountaineering which initiated global
interest in the Himalayan mountain peaks, and now most of them, including
Everest, Annapurna and K2 are all encompassed by national parks and conservation areas. Everest has long been a sacred peak for the local people, and
the conquering of it by humans seemed sacrilege. The impact of climbers has
been to divert human resources from farming, and introduce new problems
for the region, including litter and demands for food and fuel. In 1976 the
Himalayan Trust and the New Zealand government assisted Nepal in establishing the Sagarmatha National Park, incorporating Mount Everest. It has
had positive effects on forest regeneration and economic opportunities for the
local population through controlled tourism.

In the mid-1980s a survey was carried out in Nepal and Tibet by the
Mountain Institute, to assess the wider ecosystem. At this time the park area

*(continued)*

*(continued)*

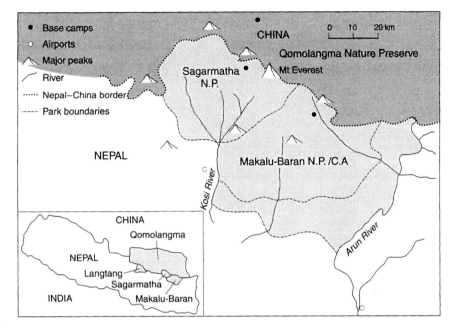

**Figure 11.1**   Sketch map showing the various parks and conservation areas in the vicinity of Mount Everest in Nepal and adjacent Autonomous Region of Tibet, China (adapted from Taylor-Ide *et al.*, 1992).

was expanded by the designation of Langtang (1985) and the Annapurna Conservation Area (1985). The area comprises over 8000 m relative relief and climatic variations from precipitation of 250 to 4500 mm/yr within the space of 10–20 km. It contains some 3200 flowering plants, 30 forest types, 440 bird species and several rare or endangered mammals, including the snow leopard, red panda and musk deer.

The protected area was expanded again by the designation of the contiguous Qomolangma Park in Tibet (1989) and the Makalu-Barun Park in Nepal (1991) to make an area equivalent to Switzerland subject to national park and conservation area status (Figure 11.1). In addition, the K2 area in Baltistan was designated in 1993. It has no permanent settlements but 15 000 local residents retain access to grazing lands and 14 000 tourists visit each year.

Management of the park includes partnerships with local government, local agencies and local residents. Cultural diversity is as respected as biological diversity and science are integrated with policy and 'muddy boots' expertise.

*Sources*: Taylor-Ide *et al.*, 1992; Fuller and Gemin, 1995; Campbell, 1997.

**Table 11.1**  Current distribution of mountain protected areas.

| Biogeographical region | Number of areas | % of total number | Area (ha) | % of total area |
|---|---|---|---|---|
| Afrotropical | 42 | 8.8 | 20 427 439 | 7.8 |
| Antarctic | 15 | 3 | 3 232 582 | 1.2 |
| Australian | 3 | 0.6 | 2 649 148 | 0.1 |
| Indomalayan | 42 | 8.8 | 7 204 043 | 2.7 |
| Nearctic | 96 | 20 | 153 804 175 | 58.1 |
| Neotropical | 103 | 22 | 34 454 473 | 13 |
| Oceanian | 8 | 1.8 | 3 643 048 | 1.4 |
| Palaearctic | 164 | 35 | 39 090 448 | 14.8 |
| Total | 473 | 100 | 264 505 356 | 100 |

*Source*: After Thorsell, 1997.

but also instils a sense of responsibility to care for the site and gives access to a World Heritage Fund. To date there are around 42 sites included on the list which are located in mountain regions (Table 11.2).

Another 67 protected areas are designated as biosphere reserves. These formed the focus of the UNESCO Man and the Biosphere programme, beginning in 1971 (Price, 1996) and have become a vehicle for sustainable development, as in the case of the Kluane/Wrangell–St Elias National Parks in Yukon and Alaska (Slocombe, 1992) and the Tatra mountains (Byrcyn, 1992). These areas recognize the importance of resident indigenous populations and resource potential as the reserves are meant to promote and demonstrate a balanced relationship between humans and the biosphere (Schaaf, 1997). They are based on the principle that an inner core area is better protected if surrounded by a buffer zone in which controlled activities might take place (Thorsell, 1997). The reserves are designed to fulfil conservation, development and 'logistic' (education, research, training etc.) aspects and lie within individual states' sovereign jurisdiction (Schaaf, 1997).

Other types of protected areas include border parks where the transfrontier region between two or more states may be designated a protected area. This may induce cooperation in protection, shared funding and prevent or deflect border disputes. Many such border parks are designated as 'Parks for Peace' and exist to nurture peaceful cooperation (Table 11.3).

This idea is an old one with a new name, as transfrontier conservation areas have long been proposed as a tool for fostering cooperation, but this role has been emphasized by the growing international insecurity and environmental conflict in the twentieth and into the twenty-first centuries. The rationale lies in the existence of a shared landscape, ecosystem or resource, which would benefit from shared management, and which would also benefit the participating parties (Hamilton, 1998). Joint training, funding, safety, security (e.g. in smuggling and drugs issues) and tourism initiatives would benefit from economies of scale. However, problems include the nationalistic tendencies of

Table 11.2 World Heritage Sites in the mountains.

| Name | Location |
| --- | --- |
| Los Glaciares National Park | Argentina |
| Greater Blue Mountains Area | Australia |
| Tasmanian Wilderness | Australia |
| Pirin National Park | Bulgaria |
| Rocky Mountains Parks (7) | Canada |
| St Elias Parks (7) | Canada/USA |
| Waterton/Glacier International Peace Park | Canada/USA |
| Huangshan | China |
| Huanglong | China |
| Jiuzhaigou | China |
| Talamanca/La Amistad | Panama/Costa Rica |
| Virunga National Park | Democratic Republic of Congo |
| Kahuzi-Biega National Park | Democratic Republic of Congo |
| Galapagos National Park | Ecuador |
| Sangay National Park | Ecuador |
| Simien National Park | Ethiopia |
| Pyrenees/Mount Perdu | France/Spain |
| Darjeeling Himalayan Railway | India |
| Nanda Devi National Park | India |
| Shirakami | Japan |
| Mount Kenya National Park/National Forest | Kenya |
| Gunung National Park | Malaysia |
| Kinabuhe National Park | Malaysia |
| Royal Chitwin National Park | Nepal |
| Sagarmatha National Park | Nepal |
| Te Wahi Pounamu/SW New Zealand | New Zealand |
| Tongariro National Park | New Zealand |
| Aïr Ténéré | Niger |
| Rio Abiseoé | Peru |
| Huascaran National Park | Peru |
| Manu National Park | Peru |
| Rice Terraces of the Philippine Cordilleras | Philippines |
| Virgin Komi Forests | Russia |
| uKhahlambe/Drakensberg Park | South Africa |
| Kilimanjaro National Park | Tanzania |
| Ruwenzori National Park | Uganda |
| Yosemite National Park | USA |
| Hawaii Volcanoes National Park | USA |
| Great Smoky Mountains | USA |
| Olympic National Park | USA |
| Yellowstone National Park | USA |
| Grand Canyon National Park | USA |
| Canaima National Park | Venezuela |

*Source*: http://www.unesco.org/whc/heritage.html

Table 11.3   Mountain transfrontier/border parks and reserves.

| NORTH AMERICA | |
|---|---|
| Wrangel-St Elias/Glacier Bay (USA) | Kluane/Tatshenshini (Canada) |
| Glacier (USA) | Warterton Lakes (Canada) |
| Cathedral/Manning/Skagit/Cascade (Canada) | Pasayten, N Cascade (USA) |
| Arctic Wildlife Refuge (USA) | N Yukon (Canada) |

| EUROPE | |
|---|---|
| Tatrzanski (Poland) | High Tatra (Slovakia) |
| Pyrenées Occidentales (France) | Ordessa (Spain) |
| Vanoise (France) | Gran Paradiso (Italy) |
| Swiss (Switzerland) | Stelvio (Italy) |
| Sarek, Padjelanta, Stora, Sjöfallet (Sweden) | Rago (Norway) |
| Berchtesgaden (Germany) | Various sites in Austria |

| ASIA | |
|---|---|
| Manas (India) | Manas (Bhutan) |
| Khunjerab (Pakistan | Taxkorgan (China) |
| Sagarmatha (Nepal) | Quomolangma (China) |

| AFRICA | |
|---|---|
| Volcanoes (Rwanda) | Gorilla (Uganda) |
| Virunga (Democratic Republic of Congo) | Queen Elizabeth (Uganda) |
| Nyika (Malawi) | Nyika (Zambia) |
| Gebel Elba (Egypt) | Proposed Gebel Elba (Sudan) |

| LATIN AMERICA | |
|---|---|
| La Amistad (Costa Rica) | La Amistad (Panama) |
| La Neblina (Venezuela) | Pico da Neblina (Brazil) |
| Puyhue and Vincente, Perez Rosales (Chile) | Lanin & Nahuel Huapi |
| Bernardo O'Higgins & Torres del Paine (Chile) | (Argentina) |
| Sajama (Bolivia) | Los Glaciares (Argentina) |
| Los Katios (Colombia) | Lauca (Chile) |
|  | Darien (Panama) |

| AUSTRALIA | |
|---|---|
| Australian Alps National Parks | NSW, SA, Victoria, ACT. |

*Source*: Thorsell, 1990.


many states which are in conflict with their neighbours, together with difficulties in access, and isolation, as in the case of La Amistad International Park between Costa Rica and Panama. Language and cultural barriers may create substantial gulfs between aspirations and attitudes and are a problem in Europe, where the problem of weak–dominant partners also occurs, for instance, between the Czech Umava and German Bavarian Forest Parks. Economic differences may mean that states do not share the same priorities and view the environment with different perspectives where resource use is concerned. Political differences, including the status of each state with regard to international conventions protecting the environment, present another obstacle to cooperation.

Internal civil strife can severely affect not only internal priorities towards the environment but also the environment itself, and international law cannot always be effective in protecting it (Shine, 1998). The contiguous Gorilla Parks of Uganda, Volcanoes in Rwanda and Vinigas in Democratic Republic of Congo are an example of this. In South Africa, the Drakensberg-Maloti Park between South Africa and Lesotho carries with it a long history of internal strife involving historical tribal conflicts as well as the impact of colonialism, apartheid and independence. Regional aggression between states resulted in the displacement of peoples and periodic redistributions of land. The San people of KwaZulu-Natal were eliminated and leave only the legacy of their rock art, which is the basis for the nomination of this area as a World Heritage Site (Sandwith, 1998).

In all parks and protected areas, there are emerging priorities. First there is the adequacy of coverage in mountain areas; second, individual priorities such as cloud forest protection (Box 11.2); and finally, coping with the effects of climatic change. The extent of protection of mountain areas is disproportionately high when compared with temperate grasslands and savanna and other ecosystems (Thorsell, 1997). However, the distribution of this coverage throughout the world is uneven. Very large areas of the polar regions and the New World, where population pressures are less, are much more extensively protected (Table 11.1). However, it is the areas under greatest pressure in the Old World which are in most need of protection and where the integration of conservation and development is the most promising route and sustainable use overrides total protection. Areas of concern are the total number of sites; cooperation in their management; science and research on the characteristics of protected areas; social conflicts arising from conservation; and the potentials

---

## Box 11.2   Conservation of cloud forests

Tropical montane cloud forests are one of the world's most threatened ecosystems. They are dependent on clouds as their source of moisture and thus potentially affected by changes in weather patterns. If areas of forest are cleared, the mechanism for collecting the water is also removed, which makes regeneration very difficult. Up to 170 per cent of local rainfall equivalent may be captured. Changes in air quality may also have an impact through the deposition of pollutants in fog.

Cloud forests are also characterized by extremely high diversity and endemism, much of which remains to be studied. The introduction of alien species is a serious threat and some 90 per cent of Andean cloud forest may be been lost to clearance for grazing land, cultivation and fuelwood. Priorities include making an inventory of locations and species as well as raising awareness and management.

*Sources*: Hamilton, 1995; Aldrich *et al.*, 1997.

and problems of nature or ecotourism (Hamilton, 1998). Sites such as Hohe Tauern in the Alps, Huangshan in China, Yosemite and Banff in the USA, and Nahud Hudpi and Machu Picchu in the Andes show particularly high concentrations of tourism and careful management is needed to manage the impacts whilst maintaining the revenues (see Chapter 12).

The impact of climate change on ecosystems is difficult to predict, but evidence of past changes suggests that vegetation assemblages would move up slope in response to global warming. The actual response is likely to be highly variable and modelling of the predicted change in different latitudes has been attempted to give a general idea of what may happen (Halpin, 1994; Chapter 3). This raises problems with respect to conservation in that the movement of species and the incorporation of buffer zones to allow for this movement needs to be considered in the design of parks (Figure 11.2).

In the Old World, protected areas tend to suffer from the 'doughnut' problem of being a protected area directly abutting against heavily used land (Mount Kenya, Cotopaxi in Ecuador and Vanoise in France). This isolates fauna and flora to a small area, prevents interaction and migration of species and thus limits the ability of ecosystems to cope with environmental change, and restricts the gene pool, reducing diversity. Protecting 'hot spots' of biodiversity is not sufficient (Miller, 1997). One way of overcoming this is to create linkages between protected areas in the form of networks or corridors. The existence of contiguous protected areas such as the Everest complex and border parks is an important start. For example, the ibex is free to follow its natural annual migration cycle, which takes it to summer pastures in the Vanoise National Park in France and to winter pastures in Gran Paradiso in Italy.

The Natura 2000 initiative in the EU is an interesting case. Its objective is to establish a Europe-wide network of protected areas which would benefit from the strong legal protection available in the EU (although the efficacy of enforcement is debatable). This effectively links mountain regions with lowlands in a continuum between all different representative ecosystems. This would serve to allow free movement (the sixth fundamental freedom of the EU!) of species in response to natural forces of change. However, the implementation of this network is slow because national governments fear the loss of sovereignty over designated land (Scott, 1998). A second important example is that of the Meso-America corridor. Here the 'big picture' is of a continuous protected area running 15 000 km from Cape Horn to the Bering Strait. In Central America the Pasco Pantera (Path of the Panther) already exists, stretching 1000 km from Colombia to Mexico and involving seven countries. This area has some 70 per cent endemic plants and faces population and economic growth of around 3 per cent per year. In the last 30 years some 461 parks have been established in Meso-America (Godoy, 1998). It incorporates native populations' interests and serves as a starting point for peaceful collaboration, respecting indigenous peoples' rights (Thorsell, 1997). The Andes–Rockies chain offers an excellent opportunity to establish such a corridor. Past environmental change has been shown to involve species migration north and south, so it would allow some room for adjustment to future climatic

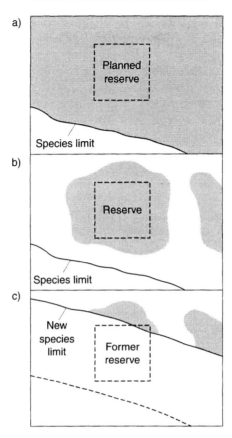

**Figure 11.2** Diagram showing the need for planning in the location of nature reserves. In a) the future reserve is located within the range of the species of interest. In b) the nibble effect resulting in the 'doughnut' problem is shown, where the reserve becomes an isolated island of original ecosystem conditions. In this case erosion of the gene pool of the species will occur, resulting in loss of vigour. In c) the species range has shifted as a result of climatic change and the former reserve is no longer within the species limit. If there is no original habitat for the species to migrate to in order to adapt to changing climatic conditions, it is highly liable to become extinct (after Peters and Darling, 1985).

conditions. There are problems, however – differing political will, funding, institutional fragmentation, and poverty and tenure issues in particular.

## Trends in mountain conservation

There have been two main trends in the development of mountain protected areas. The first is the switch from a single species orientation towards an integrated ecosystem approach. The second trend is away from pure protectionism towards shared management involving cooperation between local people, international funding sources and expertise, and national governments. This approach integrates conservation with development, recognizing the important roles that sustainable use plays in the maintenance of particular ecosystems. It also arises out of the somewhat obvious conclusion that in many mountain regions with a long tradition of relatively dense population and intensive use, it is not possible to exclude people from their livelihood for the sake of comparatively nebulous protection objectives.

The first trend, from single species to whole ecosystems reflects the increase in holistic approaches to nature and conservation. In the early decades of the twentieth century, public interest in the affluent west was aroused by individual species, particularly in those 'large furry mammals' such as pandas, leopards and gorillas. Strict nature reserves preserving undisturbed wilderness on state-owned land were the dominant formula. Early international laws reflected this approach – for example, CITES (Convention on the International Trade in Endangered Species, 1973). However, whilst individual species continue to be a symbolic focus for drawing attention to the need for conservation, there is an increasing acceptance that preservation of habitat is necessary for the successful preservation of species, and that laws restricting trapping and hunting are insufficient in themselves. For instance, in the Himalayas the snow leopard continues to arouse legendary affection in the minds of people who may know little else about the region, and thus encourages interest and support for conservation objectives. This has evolved into the more holistic ecosystem approach, incorporating the wider environment with its less appealing components, which we are more familiar with today. This holistic conception is extended to include human activities, although these may be modified to align with conservation rather than exploitation objectives. The emphasis at the present time is on the protection of the diversity of whole ecosystems rather than selected parts.

Wildlife, of course, remains an important focus of conservation activity (Fox et al., 1994; Jackson, 1996). In China, for example, strong legal protection exists for certain species such as the musk deer, white-lipped deer and blue sheep, but they are still exploited by poachers rather than by local populations (Harris, 1991). Species populations around Buddhist monasteries are afforded greater protection as the religious ethic against killing works in their favour. Despite legal protection for many species, there are problems of enforcement in remote areas (partly due to corrupt or negligent officials) and the fact that cultural practice and economic needs override respect of the law.

The listing of endangered species is an important aspect of legal protection; CITES operates by specific species being named on Threatened with Extinction and Endangered by Trade appendices to the convention. International organizations such as IUCN (International Union for Nature Conservation) also hold lists of species to be afforded different levels of protection. The IUCN is an important source of guidelines for many national governments seeking to apply conservation initiatives. These listings are generally reflected in national laws, and indeed this is required of signatories of the appropriate international conventions. This is particularly important in a mountain context in countries with significant areas of highlands, such as the Himalayan and Andean states.

## Box 11.3   Listed species in Nepal

Most of the species which are listed in the National Park and Wildlife Conservation Act 1973 (and amendments) are also listed by CITES and by the IUCN; many of these are in the endangered category. Endangered mammals include the wild yak, Tibetan antelope, nayan (Tibetan sheep), swamp and musk deer, snow and clouded leopards and tiger. In addition, there are several bird species, including three species of pheasant, and reptiles.

A number of species are harvested solely for scientific purposes, including the Himalayan black bear, wild boar, blue sheep (naur) and the spotted and barking deer, several small mammals, reptiles and birds of prey.

*Sources*: IUCN, 1990; Heinen and Yonzon, 1994.

However, the protection of wildlife can be contentious on a number of counts: it might be morally indefensible to those subscribing to a 'speciesist' ethic, whereby human needs and wants come before 'lesser' animals. This idea sees the environment purely in instrumental terms and contends that the protection of wildlife for the sake of it is an unaffordable luxury. In other societies, many species are considered pests – e.g. monkeys and carnivores such as tigers and wolves – or important sources of food to which people have long-established entitlements – e.g. bears. In these instances carefully conserving such creatures may be at the cost of livestock, crops or personal safety. Such is the case in the Langtang National Park in Nepal where the 35 000 local residents of the park are in conflict with park managers over this issue (Kharel, 1997).

Changing indigenous attitudes towards resources in order to foster a more conservative approach not only potentially interferes with their rights to self-determination and with their livelihood strategies, but is also potentially so alien to their culture that it is impractical. In western Yunnan, China, efforts have been made to instil a modified hunting ethic in the Lisu people, who traditionally have no taboos on hunting (Harris and Shilai, 1997). This is successful when species such as gibbons are involved as these pose no threat to their livelihoods, but not for wild boar, which are seen as the cause of crop damage. There is also a sense that by not killing animals themselves, they are leaving more for others to kill, reflecting the 'tragedy of the commons' attitudes which are difficult to overcome. In Pakistan, however, the promise of revenue generation and employment through hunting fees and training in conservation has proved to be successful in introducing new attitudes to the environment amongst the local population; they can see other ways of making it work for them which do not involve continuous exploitation (IUCN, 1996, 1997). In the USA, a Wolf Compensation Fund was established to allay the fears of loss of income due to predation by reintroduced wolves in Wyoming (Preston, 1997), and similar proposals have been made regarding snow leopard depredation in the Himalayas (Mountain Forum, 1998). Compensation schemes have been criticized, however, as reducing the incentive for herders to improve husbandry and revive guarding strategies. They are also costly to administer and insurance schemes are sometimes inappropriate for these herding communities.

There are arguments against unnecessary protection and conservation of wildlife, which are based on the idea of natural regulation without interference. This stems in part from the scientists' incomplete understanding of the interactions between species – for example, the predator–prey relationships between wolf and elk in the USA (Boyce, 1992) – and thus the effect of intervening in one species. One problem with this is the degree of human activity which might be accepted as part of the 'natural regulation' in areas where long-standing practices of hunting are followed. Once the exploitation enters the realms of economic gain – i.e. hunting is for market rather than subsistence – then it might be considered commercial and thus unacceptable. But this does not address the issue of communities who have traditionally traded products as part of their livelihood strategies.

Other natural protection functions of protected areas include landscape stabilization – maintaining a vegetation cover in order to reduce erosion. This is common around reservoirs, in an effort to extend their lifespans. In addition, the Andean altiplano protection rationale includes the need to protect the areas of groundwater recharge. In arid regions this is very slow and the control over use of water helps to speed up the process of recharging aquifers which have gone dry through overuse or due to natural aridity (Messerli et al., 1997).

The second trend, towards the integration of human activities as part of the conservation initiative rather than as a tolerated necessity, is in part an extension of the holistic ecosystem approach to conservation. It goes much further, though, in that it seeks to incorporate local management practices and local livelihoods directly into the management of conservation areas. Thus sustainable development becomes an integral part of the landscape and conservation effort. This is illustrated by the evolution of Biosphere Reserves (Price, 1996), which reflect the whole general trend away from genetic and landscape conservation outlined above. Biosphere Reserves are centred on the core–buffer concept where the buffer zone acts as an intermediate zone of use and protection between conserved and unconserved landscapes. In the USA, interaction was fostered between protected areas, forming clusters where information and expertise could be exchanged. Unfortunately, differences between state practices made it difficult to instil a single federal approach to conservation and so collaboration continued to be scant. However, changes in the philosophy behind conservation helped to overcome this reticence.

The World Conservation Strategy (1980) was an important step towards the realization of combined conservation and development. In 1986 the concept of the buffer zone was redefined as a zone of transition or cooperation to reflect this (UNESCO, 1986). But the 1990s strengthened the conception of cooperation and the interaction of development and conservation initiatives. The World Conservation Strategy also recommended active local community participation and that training and communication strategies should be developed for local participants rather than just for conservation officials (UNESCO 1995). It encouraged the establishment of networks for the exchange of information and funding initiatives to implement these ideas.

## Parks and people: the marriage of conservation and development

Conservation involves practical and scientific issues relating to the assessment of the threat to landscapes and ecosystems and of the reasons for their protection (Warren, 1987). But it also involves unavoidable moral and value-laden issues; choices must be made between use and protection.

Decisions concerning alternative land uses affect the livelihoods of people and become highly political issues. For example, in the USA, cases have been brought to the federal courts concerning the conflicting interests of conservation and economic exploitation. In one landmark case the state of Hawaii ordered the removal of feral sheep which damaged the habitat of the endangered *Pallila* bird. Under the US Endangered Species Act, states are required to act to preserve habitats. This decision, however, overrode the economic interests of the people who used hunting revenues as one of few possible sources of income. This was considered to be a violation of their human rights and was pitched against the 'rights of the environment' in the federal lawsuit. Although this case set a precedent, clearly enforcing the precedence of environmental over economic considerations, there has been little evidence to date that it has had much effect (Juvik and Juvik, 1984).

Similarly, there is conflict between livestock grazing in the White Mountains of California, which causes erosion which affects the habitat of endangered fish species where sediment is deposited. Livestock grazing continues as it is embedded in the local Land and Resource Management Plan as part of the maintenance of other terrestrial habitats. Thus there is a conflict between terrestrial and freshwater preservation which has cost US$260 000 to control in the period 1981–1991 (Kondolf, 1994). In the Tatshenshini mountain wilderness in British Colombia, plans to develop an opencast copper mine were overruled in 1993 in favour of protection. The area adjoins the St Elias World Heritage Site and is earthquake-prone, making mining a potentially catastrophic activity. The local First Nations people and the recreational revenues were an important consideration in the decision as the area would be devastated by the mine, settlement, roads and pipelines. The mining company were paid Can$26 million in compensation and the political 'fallout' continues (McNamee, 1994). In La Amistad park, Costa Rica, some 35 per cent of the land has been leased to mining companies, representing a major management problem. A consortium of NGOs, indigenous people, mining company and government representatives formed a foundation to address the issue of management within such a diversity of claims to use (Miller, 1997).

## Box 11.4   The northern spotted owl versus the US logging industry

Perhaps the most notorious conflict was that between the spotted owl and logging industries in the Pacific Northwest of the USA. Here, action was being sought to preserve the remaining stands of old-growth (in excess of 1000 years) forest from logging. The area comprised Douglas fir, white fir, ponderosa pine and hemlock. It is the habitat for many species, including the bear, peregrine falcon and bald eagle but it was the discovery in 1972 that it is home to populations of the northern spotted owl which sealed the fate of the lumberjacks. As a federally listed endangered species the northern spotted own was entitled to legal protection, which it received, but at the cost of the logging industry, which had been active since 1900. There is a history of non-compliance with federal environmental law on this issue and it was a prominent feature of the presidential campaign of 1992. The Clinton administration made efforts to foster integrated conservation and development of state forests which incorporates the needs of the native American populations, such as the Warm Springs Tribe, who have also been affected by logging. The Warm Springs Indian Reservation Trust was set up to fund claims for compensation from flooding of land by the Dalles dam (the settlement was US$4 million), to establish a forest products industry and to maintain employment and education for the local population.

*Sources*: Preston, 1997; Moore and Sehgal, 1999.

In the New World the conflict between 'unused wilderness' and its potential economic exploitation represents the dichotomy between intrinsic and instrumental values of nature. In the former, nature is valued for its own sake, supporting a protectionist policy where use is strictly controlled and the emphasis is on recreation and sustained use of natural resources. Many of the movements based on single-species protection arise from this ethic, where the right of sentient beings to life is applied, and pitted against the right of human beings to kill such animals.

Instrumental values of nature are concerned with the protection of nature because it is useful to humans. This may be demonstrated in the protection of natural renewable resources such as water, timber, plants and animals, but also in the protection of landscapes because although resources lie there, the technology does not yet exist to exploit them. This is particularly true of non-renewable resources such as minerals where advances in technology in the future would allow these to be developed, but until then any other uses may be controlled. The outcome of this strategy is short-term economic gain at the cost of environmental damage, albeit some time in the future. The symbolism of wilderness should not be disregarded either; it is a deeply embedded phenomenon in the psyche of many people in the New World, reflecting pioneering ancestors, and escapism from the modern materialist world. In this way the 'purity' of 'unaffected' wilderness is regarded with a degree of fervour which matches the religious zeal associated with sacred peaks in many Old World mountains.

In the less developed world more insidious choices exist between different sets of instrumental values of a single resource. We have already come across this in the context of forests, which may be viewed as convertible or renewable according to what role they play in a livelihood strategy. Forests in particular have been the subject of conflict between the protectionist policies of national governments, the economic power of the commercial logging industry, the interests of indigenous populations who rely on forests as part of their subsistence, and global concerns over climate change and biodiversity – the last for its potential future instrumental use for health and sustenance. It is often not possible to meet all these needs in any one area all the time, but in recent years collaborations between industries, indigenous people and national governments have begun to share the burdens and benefits of maintaining a forest resource.

The conservation and management of forests lies in the hands of a number of actors. The traditional indigenous management strategies involved the control on harvesting products according to season, volume and entitlement (Chapter 7). National governments' attempts to protect forests on exclusionist principles, as in Nepal in the 1950s, conflicted with the traditional patterns of use. Whilst state forests were largely protected, the pressure increased elsewhere, causing greater damage than would otherwise have occurred. Healthy forests require maintenance and the abandonment of forests causes build-up of litter, which harbours disease and represents a fire risk. Selective felling ensures a mixed-age stand which promotes stability and longevity. Controlled burning is an important strategy for rejuvenation of forest growth and for reducing the accumulation of dead material. In seasonally dry areas such as the Mediterranean region and parts of the USA this is important in preventing major forest fires.

Where forests have been replaced by plantations of commercial species, the serried ranks are often not only less aesthetically pleasing but are also a much more impoverished ecosystem than the natural older stands. There is also misunderstanding where indigenous people have been excluded in the interests of protection, only to watch 'their' forest being commercially exploited under a licence granted by the government. Such is the case in Southeast Asia, where the problem is compounded by the uncertain ethnic status of many hill tribe people. Without national identity they cannot claim tenure or rights of exploitation so their traditional, long-established entitlements to forest lands become strangled by national bureaucracy (Ganjanapan, 1996). Here, the politics of conservation is used by both the government and by the hill tribes in different ways to make their different claims legitimate. For example, the government uses ethnicity in a negative way, citing traditional forest use as a reason for exclusion of local people and instigating protection, whereas the hill tribes claim that their methods are sustainable and more protective than commercial logging. Thus ethnicity becomes a dividing line between different perspectives of forest use and conservation (Ganjanpan, 1998).

The civil disorder arising from the conflict between logging and farming communities, and the difficulties in enforcing laws of protection and exclusion in remote mountain areas, as well as the failure of government afforestation schemes have led to a general trend towards participatory development and conservation. This reflects the 'people in parks' idea and also the participatory development approaches discussed earlier (Chapter 10). Community forestry is an important concept in the management and protection of forests and a participatory approach to their management is a key focus of development initiatives and a critical factor in their sustainable use (Fisher, 1995). A variety of methods have been employed to balance local and external uses and needs, and to facilitate sustainable use and management (Table 11.4; Butt and Price, 2000).

There are other examples which reflect the impetus behind the collaboration of external and local actors in conservation. The role of local institutions is an important focus of control of local resource use. Community leaders can

**Table 11.4**  Mechanisms for sustainable management of mountain forests.

| Type | Name | Location | Aims/How it works |
|---|---|---|---|
| Legislation/ policies | Pan-European operational level guidelines | Europe | Promotes sustainable management and requires ecological, economic and social issues to be considered in national and local policies. Not legally binding, but a consultative approach. |
| | National biodiversity policy | Slovakia | Aims to develop databases for sustainable management, to increase proportion of natural regenerated forest and to monitor changes and impacts of development on forests. |
| | EU rural development regeneration | Europe | Negotiated between states; financial subsidies for ecological stability and sustainable management of forests in rural development context. |
| | NATURA 2000 | Europe | System of protected sites across Europe of biodiversity value. |
| | Forestry and mountain laws | Switzerland | Support local control, communal decisions and reinvestment in local communities. |
| | National park declaration | Nepal | Protects forests from exploitation, provides tourist attraction, boosts local economy. |
| | Forest master plan/Mining Act/afforestation strategy, etc. | Bhutan | Reforestation, conservation of biodiversity, buffer zone. Long-term strategy. |

(*continued*)

Table 11.4   (cont'd)

| Type | Name | Location | Aims/How it works |
|---|---|---|---|
| Multi-functional land-use strategies | Sloping agricultural land technology | Philippines | Replanting native trees, enhancing productivity. Integrating trees (mini-forests), food, cash crops in farm economy. |
| | Parma mushrooms | Italy | Local enterprise; protects forests of limited commercial value. |
| | On-farm native tree cultivation | Indonesia, Sri Lanka, Philippines | Protects natural regeneration of timber; enables soil conservation; provides fuel; gives income from fruit trees; enhances biodiversity. |
| Support/ extension | Hills leasehold and forage development project | Nepal | Enables subsistence farmers to lease long-term blocks of land for regeneration to diversify resource base. |
| | Annapurna Conservation Area | Nepal | Alternative energy sources promoted to replace fuelwood. |
| | Nepal Agroforestry Foundation | Nepal | Supplements existing agroforestry; trains farmers so they become self-sufficient in techniques. |
| | WWF/World Bank | Papua New Guinea | Funds community foresters in sustainable timber production; enhances viability of community forests and protects biodiversity and environment. |
| | Gia Lai agroforestry extension | Vietnam | Improvement of resource management of smallholders; responds to local needs, enhances food security. |
| | Food-for-work | Honduras | Planted trees managed in a similar was to regenerated forest; balances tree and crop production. |
| | FECOFUN | Nepal | Raises forest user group awareness of rights and responsibilities. |
| | Joint forest management | India | Government-led; promotes self-determination and local mobilization; integrated with autonomous community groups. |
| | Integrated mountain development | China | Diversifies tree species in afforestation within watershed management. |

(*continued*)

**Table 11.4** *(cont'd)*

| Type | Name | Location | Aims/How it works |
|---|---|---|---|
| Community organizations | Village forestry committees | India | Self-determined community planning. |
| | Community timber management | Yunnan, China | Community-managed forests with rights to timber market; revenue reinvested in schools, roads, water, electricity. |
| | | Nepal | Forest user groups – local rights to harvest timber. |
| | Mahila mangal dal | India | Protects forests from quarrying. Nursery and plantation work. |
| | Savings and credit cooperative | Nepal & elsewhere | Represents common interests of wider community. Self-determined economic power. Long-term investment. |
| Economic instruments | Joint implementation | Latin America & others | Markets in carbon credits to offset global warming. Nature conservancy runs projects funded by sale of credits. |
| | Compensation/ subsidies to mountain populations | Austria, Costa Rica, France, Switzerland | Transfers some economic benefit from upstream mountain forest conservation to upstream populations. |
| | Costa Rican territorial & financial consolidation of biological reserves | Costa Rica | Ministry of Environment & Energy acquires threatened land for protection; increases carbon storage. Funds from 'carbon bonds' sold by government, guaranteed for 20 years. |
| | Forest certification | Europe, Melanesia | Ensures products sustainably produced. Raises management standards. |
| | Sustainable practice tropical timber production – Futuro Forestal | Panama | Plantations of native species and teak to slow erosion, improve soil fertility, even out stream flow, enhance biodiversity. |
| Religious practices | Ridam | Bhutan | Ban on entering designated forests between mid-August and mid-October. Protects young animals in late monsoon. |
| Coordinated knowledge sharing | European Observatory of Mountain Forests | Europe | Shares experience, views, concerns, ideas on sustainable management and local development. Long-term view. |

*Source*: From Butt, N. and Price, M.F. (eds.) (2000) *Mountain People, Forests, and Trees: Strategies for Balancing Local Management and Outside Interests. Synthesis of an Electronic Conference, April 12–May 14, 1999.* Mountain Forum and The Mountain Institute. Franklin, West Virginia. 52 pp. (Table 2).

have a significant impact on the attitudes of their communities. However, cooperation is not all plain sailing and there are many instances where the complexity of actors and institutions involved and the politics of a situation have militated against effective conservation collaboration. In Ethiopia, for example, the management of the Bale Mountains National Park is split between several government departments which lack effective coordination and thus the management strategy is greatly weakened (Hillman, 1988).

The degree of conflict has increased in many countries as a result of the regulation of more private land alongside higher populations and greater economic activity. The public voice is important both in encouraging conservation but also in stating the varied interests, and as more private land is affected, especially by planning regulations, the conflict is much greater. In southern Norway, for instance, the expansion of the national park has emphasized the need for a more sensitive and creative approach to the designation of protected areas and their use (Kaltenborn *et al.*, 1999).

The potential for the generation of revenue and provision of alternative economic sources is an important way of encouraging cooperation, but not without negative effects. In Nepal, the residents of Melemchi reacted with caution to the designation of the Langtang National Park. Its regulations for firewood, timber and fodder, fees for grazing, and restrictions on access represent a curtailment of their activities, but the potential revenue generation from tourism is seen as a positive outcome. The park authorities control tourism, however, and development in Melemchi was limited as it does not lie on the main trekking routes and residents do not speak Nepali. The park brought national government to their doorstep, replacing the Chini Lama authority which they were used to. Many misunderstandings emerged, but these have been cleared up as more people have a better appreciation of the rationale of the park and become used to the changes in access and the regulations (Bishop, 1998).

Elsewhere, the revenues generated from visitors coming to view wildlife (the gorillas in Uganda), mountaineering fees, licences for hunting (New Zealand, Rwanda), etc. can prove useful sources of income, which can be reinvested locally. In addition to this, revenue promotes a conservationist approach to the landscape and wildlife which may not have been in existence before.

In Switzerland, the designation of the national park involved the promise of employment and tourism revenues, but in reality the distribution of benefits is uneven. Finance was directed primarily at the park administration rather than the wider community and the opportunities for income generation are limited by park regulations. There has been a negligible increase in local income – much of the income received by households involved in the park is spent outside the region, representing a drain on the economy. Very few jobs have been created but even the limited positive outcomes have been valuable in this region (Elsasser *et al.*, 1995). Most of the tourism revenue, however, is generated from accommodation, which is located outside the park, and much less comes from the park itself. The inaccessibility of the park in winter is a severe restriction on economic development.

# The thorny issue of biodiversity, biotechnology and biopiracy

Biodiversity is the diversity of life. It incorporates genetic diversity (the diversity of genes in organisms), species diversity (the range of different species) and ecosystem diversity (the variety of living things in ecosystems). By extension it also includes cultural diversity and human racial diversity. Biodiversity conservation may be seen as the preservation of individual species or of whole ecosystems, including their human components. Biodiversity is greatest in the tropics (which, ironically, have the smallest extent of protected areas), partly because these regions were not substantially ice-covered during the Quaternary. The ecosystems have migrated and evolved as a result of climate change, contracting to refugia and expanding as conditions dictated, but the landscape was not effectively cleared as in the polar and parts of temperate regions.

The interest in biodiversity lies in the fact that the whole world is dependent upon a healthy planetary ecosystem, either directly or indirectly. In many mountain regions people live very close to nature in that they derive their livelihoods directly from it. The diversity of wild cultivars in mountain regions is considerable. In Sierra Leone there are some 49 cultivated varieties of rice, each filling a different ecological niche, and in Nepal 42 traditional rice cultivars are used, 17 of which are upland varieties, with 93 per cent of upland farmers growing between one and five varieties (Bhuktan et al., 1999). In northeast Thailand, 50 per cent of foods consumed are from wild food gathered from rice paddy. They include snakes, fish, insects, fungi, fruit and vegetables (Bragg, 1992). Many more are collected from the forests (Anderson, 1993).

The loss of diversity includes agricultural crops and livestock, and arises from intensive farming of a limited range of crops. The 'Green Revolution' introduced new high-yielding varieties with disease resistance which were often not suitable for the multiple uses to which farmers put them (see Chapter 10). Local social movements in the Andes have promoted their own varieties and uses of crops, such as 'popping beans' (Zimmerer, 1992). In addition, such varieties required more irrigation water (which may be in short supply) or chemical fertilizers which are costly in mountain areas. Such varieties often did not stand up to the vagaries of mountain climates as well as the tried and tested versions. Farmers themselves experimented to select different varieties to suit even the variations of conditions on a single terrace. Other causes of loss of diversity are habitat fragmentation, alien species which may cause local species extinctions or overexploitation, as in the case of the cedars of Lebanon which were so widely availably in historical times and now exist in only a few remnants. Pollution and global climate change are additional environmental stresses which may cause extinctions. Mountains are particularly vulnerable to such extinctions because many species are at their ecological limits, and the gene pool is relatively small and thus more easily lost.

In mountains, as alternative economic options begin to separate the people from the land, greater risks to biodiversity ensue. Ecotourism can, however, link people back to their landscape as that itself is the object of the tourist visit (see Chapter 12). In the Western world, though, most people are separated from the natural world but still rely on it for food and medicines in particular. The discovery of new cures for (often Western) diseases in the tropics has made this a new hunting ground for pharmaceutical companies. The exploitation of both the flora and fauna and the indigenous knowledge which is so important in 'discovering' new potential drugs have become serious international issues. In the past, pharmaceutical companies were free to visit, take samples, acquire knowledge from local people and later in the laboratory to patent their discovery and reap vast profits. In this way, the original source receives no recognition, still less any return on what was 'their' resource. Whilst the pharmaceutical company might argue that such materials are part of the 'global commons' and free for all, international law respects the sovereignty of states over their own resources, and the developing world is increasingly taking a poor view of such exploitation, given the wealth of knowledge which is inherent in these countries (Hynes *et al.*, 1997; Olsen and Helles, 1997).

The Bioresources Development and Conservation Program, an NGO in Cameroon, has trained local people in gathering materials for genetic resources. The project runs in association with the DFID (UK) and Limbe Botanical Garden. The Program owns the samples and adopts a licensing approach to companies which wish to use the material. There is a benefit-sharing agreement between all parties (Preston, 1997). One problem in direct reimbursement of indigenous peoples is their different attitudes to ownership: they consider many resources as common and so they do not see themselves, as individuals, as being eligible for payment for others' use of these resources. In addition, such knowledge is not specific to any one individual but common to whole communities, so there are difficulties in justifiably rewarding one person for common knowledge; this is effectively privatization and could erode the unity of a community. Indigenous knowledge has become a commodity in the sense that it is necessary to define it in order to protect it by law. The 1994 UN Declaration on Indigenous Peoples' Rights goes some way to redressing this, but this document is not enforceable law.

## The international arena

The preservation of biodiversity and other conservation issues are therefore firmly in the international political arena, and very far from the mountain-specific agendas which were considered in Chapter 10. But it is precisely because these issues are of importance to mountain regions that it is crucial to put mountains onto the international agenda in a visible way, because they are part of the big wide world.

Funding and technology to set up gene banks and the like are needed in the south in order to preserve what cannot be maintained *in situ*. This has to come from the north, which is driven by capitalist market motives rather than altruistic ones, so everything has a price. International law, albeit 'soft' in that it is unenforceable, has a number of pieces of relevant legislation, including the Biodiversity Convention 1992 which emerged from the Rio Conference, the World Heritage Convention 1972, and the 1994 UN Declaration on Indigenous Peoples. There are negative powers too, of course; the GATT is a strongly market-based, sovereignty-orientated agreement which does not permit restrictions to trade for the purpose of nature protection (Scott, 1998). Other agreements and conservation strategies include debt-for-nature swaps, where funding from one country pays off part of the debt of another in exchange for the setting aside of tracts of land for the purposes of conservation. One example in Peru was set up in 1992, with the fund being managed by a non-profit organization, PROFONAPE. Funds are derived from the Global Environmental Facility (US$5 million) and donations of bilateral debt. The public–private structure of the organization is unique and has been able to maintain both government interest and private faith in the initiative (Preston, 1997).

Such initiatives have, however, been hampered by the perception on the part of the less developed countries of neo-colonial exploitation. They also require careful balancing of the participating nations' rights to control the land. Issues of enforcement have arisen with cases where the dept has been paid off but the land not protected. Nevertheless, despite these administrative issues, there is a growing appreciation of the need to conserve biodiversity, and creativity in the ways in which this might be achieved both effectively and equitably.

## Key points

- Mountains are characterized by a high degree of endemism and high levels of biodiversity as well as cultural diversity.
- Mountain protected areas range from small-scale national set-aside schemes to internationally designated World Heritage Sites and Biosphere Reserves. These different types of protected areas have different statuses with regard to legal protection.
- The impact of climate change and of population pressure and the need to continue to use mountain resources means that a system of core and buffer zones is the most effective model for protection. Links between areas by networks and corridors enhance ecosystems' capacity to respond to climate change.
- Trends in conservation comprise a change from single species conservation to a holistic ecosystem approach and the incorporation of human activity as a working part of the conservation project.
- Conflicts between conservation and development arise as environment and economics are strong competing forces in land-use strategies. Ethical issues are raised where the protection of nature conflicts with the economics of livelihoods.
- There is wide range of approaches to participatory management of forests and mountain regions. These aim to respect local needs as well as external needs, and to enhance food security and forest productivity in a sustainable manner. Revenue generation can enhance a conservationist attitude in local people.

- Biodiversity has been affected by the 'piracy' of pharmaceutical companies exploiting the knowledge and resources of poorer countries. There is a greater need for profit sharing and recognition of ownership and protection of rights to these resources.

# Chapter 12

# Mountain tourism

Tourism is the world's fastest-growing industry, with an annual turnover of US$444 billion, exceeding the GNP of the world's 55 poorest countries. There has been an average annual increase of 4.7 per cent between 1989 and 1998 and the growth rate is estimated to be 4.1 per cent per year to 2020 (Godde, 1999). John Muir predicted the growing importance of the mountains for lowland recreation in 1898 as an antidote to the stresses of urban life (Muir, 1975). His foresight has been fully vindicated as some 15–20 per cent of the world's tourism industry takes place in mountain regions. This is disproportionate to the contributions of mountain regions to national economies and represents the most significant flow of potential resources to mountains (most other aspects of mountains involve a draining of people, resource and even sediment out of the mountain system!).

Tourism has expanded rapidly since the Second World War with the increase in economic prosperity, disposable income and formalized leisure time, at least in the Western world. In the USA tourism has grown by over 500 per cent between 1945 and 1975, and in Nepal there were around 20 visitors in 1965, 3200 in 1972 and over 15 000 visitors per year to the Sagarmatha Park by the end of the 1980s (Brower, 1991). The recreational use of mountains is therefore a critical and dynamic aspect of the mountain environment and is linked to issues such as biodiversity conservation and economic development. It is marked by both opportunities and severe negative impacts, and conflicts with traditional livelihoods. It is an unpredictable economic resource for mountain people, and whilst it represents the greatest potential source for economic growth, it is also very uneven in the distribution of costs and benefits.

This chapter will first examine the nature of mountain tourism, then its various impacts and, finally, the sustainable development options which lie within its orbit.

## The nature and scale of mountain tourism

Tourism in mountains really began to develop with the explorations of mountaineers in the mid-twentieth century, first in the Alps and then further afield in the Himalayas. However, before then, there were several forms of tourism already occurring in mountains. First, the importance of pilgrimages to various sacred mountain peaks – for example, in the Indian Himalayas the state of

Uttar Pradesh has always been a destination of pilgrims to the Gangotri glacier, the source of the Ganges river. Today some 250 000 pilgrims as well as 25 000 trekkers and 75 mountain expeditions visit the area each year (Price *et al.*, 1997). This accounts for half of the GDP of the state. Other sites of pilgrimage include Mount Athos (Greece), Fujiyama (Japan), Mount Sinai (Egypt), Kilauea (Hawaii) and T'ai Shan (China).

In the case of the European Alps, before the nineteenth century any journey through the area was an unwelcome and arduous adventure, given the various environmental hazards and hostile, conservative populations. However, now 150 million people drive over the Alps each year, causing air pollution and requiring roads and other communications. There has been a history of visits in some form since the Roman times, with the establishment of spa towns (Kariel and Draper, 1992). As in many mountain regions, the traditional hospitality has been formalized and exploited on the sides of both guest and host, altering the sense of the 'gift culture' associated with marginal populations. There is a theory that the European Alps, particularly very popular areas such as Mont Blanc, are a constructed myth, developed in the nineteenth century as a result of the visits of artists, poets and literary figures, who all created an image of the mountains which is what many visitors nowadays seek out (Debarbieux, 1993). The sense of conquering and taming the landscape arises in both mountaineering achievements and the fixing of the image on canvas and page.

In the present day there is a general sense that this taming of the wild spaces has been achieved, leading to carelessness in development and in individuals. Thus the media continually brings stories of individuals lost on mountains through poor expedition preparation, and also accounts of natural disasters such as the 1999 avalanches in the European Alps which caused the loss of life and, in some quarters, bewilderment that such events were not under human control. Despite this, mountains continue to attract a wide variety of visitors, who seek mountains for a number of reasons. In many countries the legacy of the colonial era, such as that of British India, was to instil the idea of the attractiveness of the mountain climate as a refuge from the heat of the plains in summer (Singh, 1983). Each year whole households would make the trek from the plains to the mountains to towns such as Simla, Darjeeling, Mussoorie and Dalhousie, which grew up in a European form as summer service centres for the colonists. These towns were the first Himalayan stations with important administrative functions and the infrastructure remains little changed today when places such as Mussoorie receive 700 000 visitors each year (Singh, 1990).

For many Western tourists who spend their lives in a humdrum cycle of offices and housing estates, the appeal of mountains lies primarily in the fact that they are isolated, relatively wild and represent a sense of purity and tranquillity of another world. Despite the fact that many areas are largely human-altered landscapes, the attraction lies in the aesthetics of a lofty view of the world, surrounded by nature and by spectacular scenery. Contrasting cultural aspects offer an added ingredient of the unknown and of adventure. Accessibility is an interesting point, as many adventure tourists prefer areas 'off the beaten track', seeking to find undiscovered corners; the difficult road to the site and the basic, if not primitive accommodation, are part of the experience they have paid for. There are few other instances when relatively wealthy middle-class Western people happily pay considerable amounts for the pleasure of traipsing through mud and rain, sleeping with fleas on bare boards and washing only the extremities in icy glacier water! Finally, there is the idea of image: not just that of the adventurous trekker, but more obviously the up-market image associated with ski resorts such as Aspen, Colorado and Klosters in the Swiss Alps. These areas become targets for amenity migration (see Chapter 6), which raises land values high above the norm: in Aspen a medium-sized family home costs over US$1.5 million. The strong economic gradient marginalizes the original population by pricing them out of the market.

Tourism is a highly complex industry. Visitors can be divided into long-stay and weekenders or day-trippers. They may be luxury, standard or shoestring classes. Visits may be seasonally concentrated in either summer or winter, or, in some resorts, both. Visitors may be passive – sightseers and onlookers – or active – participating in anything from demanding 'professional' mountaineering and winter Olympic sports to skiing, snowboarding, rafting, trekking or just wandering. They may be foreign or local. They may be on themed holidays concentrating on architecture, culture or nature with knowledgeable guides, or seeking their own spiritual fulfilment through pilgrimage. In the current burgeoning tourist industry, more and more companies seek to offer their clients different types of experience and adventure, tailoring holidays and expeditions to individual requirements as well as seeking new niches, such as selling themselves as 'environmentally friendly' or subscribing to some form of '"sustainable" or "ecotourism"'.

The tourism industry is notoriously fickle in economic terms. Skiing may be 'in' at one resort and 'out' at another. Likewise fashions dictate the types of holiday and destination – South America is attracting more and more people, whereas older resorts in the Alps have become overfilled and less attractive to the modern generation. In addition to this, national and international eco-nomics affects tourist numbers. The German recession in 1981–1982 reduced the numbers of visitors to the Bavarian Alps by 12 per cent (Grötzbach, 1985). Here the income was largely generated by Germans on short weekend breaks, which were reduced during the recession. Likewise, civil conflict and threats of violence affect visitor numbers: the problems in Central Africa and scares over the tourists taken hostage in Yemen (December, 1998), kidnappings in Kenya (Spring, 1999) and disappearances in the Himalayas will put off individual travellers and curtail organized tours to such places.

## Environmental impacts of tourism

The environmental impacts of tourism are many and diverse and have been widely studied (Bayfield and Barrow, 1985). They include the impacts of development and, associated with it, the effects of pollution and waste genera-tion. Direct impacts on soils, vegetation, water resources and forests are also

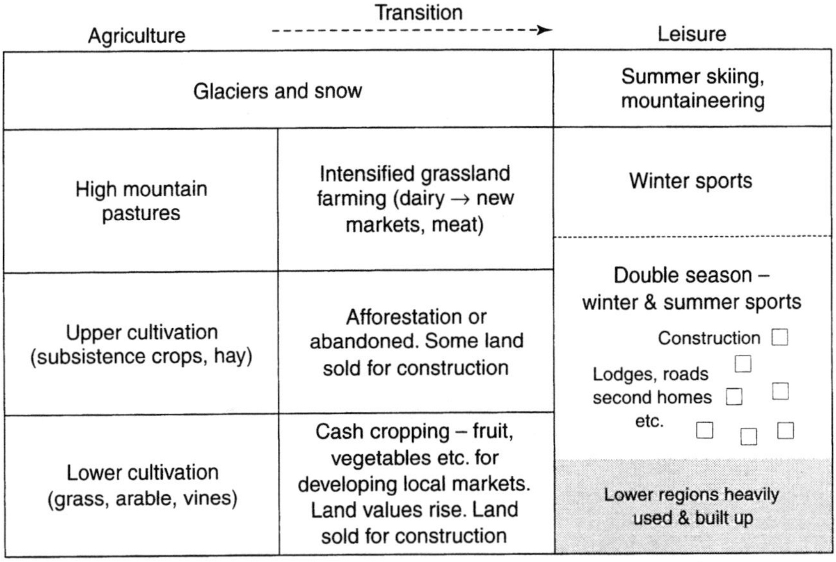

| Agriculture | Transition ----------------→ | Leisure |
|---|---|---|
| Glaciers and snow | | Summer skiing, mountaineering |
| High mountain pastures | Intensified grassland farming (dairy → new markets, meat) | Winter sports |
| Upper cultivation (subsistence crops, hay) | Afforestation or abandoned. Some land sold for construction | Double season – winter & summer sports  Construction ☐  Lodges, roads ☐  second homes ☐ ☐  etc. ☐ ☐ ☐ |
| Lower cultivation (grass, arable, vines) | Cash cropping – fruit, vegetables etc. for developing local markets. Land values rise. Land sold for construction | Lower regions heavily used & built up |

**Figure 12.1** The transition from an agricultural to a leisure-orientated society, showing some of the changes occurring in different altitudinal zones (adapted from Lichtenberger, 1988).

important and vary with the type of activity in a particular area. The inherent vulnerability of many mountain landscapes to highly localized and constant pressures means that the management of tourism becomes ever more important as more people travel to more remote regions.

Tourism developments in mountain regions often follow a similar pattern, with a vertical distribution of activities which may complement, rejuvenate or replace traditional agricultural activities (Lichtenberger, 1988; Figure 12.1). The development of valleys and slopes involves the construction of communication networks, hotels, cable cars and ski lifts, as well as an infrastructure of services. Originally, these were located in the lower parts of valleys and valley floors, which has created serious problems of flood damage. The valley floor is the natural energy absorption zone and this effect is greatly reduced if the area is covered with hard surfaces that increase runoff, and buildings which obstruct the natural expansion and contraction of mountain rivers in response to meltwater and precipitation. The floods in the European Alps in 1987 damaged almost exclusively buildings constructed in the last 40 years, including both industrial and tourist-related constructions (Schwarzl, 1990). With the advent of skiing, the pattern of construction has moved progressively upslope, with pistes, ski lifts, restaurants and lodges (Mosimann, 1991). The seeding of pistes with alien species as well as the natural change towards plants which are resistant to ski damage, such as some small herbs and grasses, means that the character of the upland meadows changes substantially.

Skiing, mountain biking and trekking on foot or on horseback create highly localized areas of severe erosion. Designated routes often have to be reinforced or even paved in order to take the pressure of so many feet, but this makes the pathways less aesthetically attractive and much careful thought is put into the design and execution of path surfaces which are both durable and 'invisible'. The feet of people and horses carry alien plant seeds and spores and these often take root along pathways, creating a different flora. If these introduced species are aggressive, they may cause substantial change in the natural flora of whole areas. Such effects have been studied – for example, in the Valley of the Flowers in the Indian Himalayas. Areas walked on about 20 times showed a displacement of stones and soil and fewer plants (Kaur, 1983). Trekkers can also spread diseases; in the spring of 2001 the British countryside was effectively closed in an effort to contain the spread of the foot and mouth virus. When paths become wet, erosion is increased and in an effort to find easier walking, people walk on the edges of the path, which becomes progressively wider (Singh, 1990). The construction process, especially where it is located on ever-steeper slopes, itself causes substantial erosion (Joshi and Pant, 1989). In Naintal, Uttar Pradesh, the number of hotels increased from 40 in 1980 to 94 in 1988.

It is often asserted that tourism increases rates of deforestation (Banskota and Sharma, 1995). Despite the fact that the influx of tourists needing food and warmth is just one factor, this idea persists and there is evidence to support it (Ives and Messerli, 1989; Shrestha, 1995). The Everest Trail is the best-known focus of such debate, although the increase in the local population is also cited as a cause of clearance (Andrews, 1983). Here trekkers move on foot with porters and animals to visit base camps and other areas, camping as they go. The evening spent around the fire is part of the experience and this requires wood. Camping grounds are often old *bari* terraces, which generate more income for the farmer than would a crop. It is now illegal for tourists to burn wood in Sagarmatha and they are asked to carry kerosene for cooking. Increasing use is also made of solar water heaters, various cookstoves and hydroelectricity. The construction of hotels on terraces reflects the growing economic importance of tourism and its gradual replacement of farming activity in these areas. The density of buildings has rapidly increased in the Hunza valley, particularly on the south-facing slopes of Karimabad, which offer warmth and spectacular views (Parish, 1999).

Water resources in mountains are often in short supply and increasing demands for tourist use conflicts with traditional irrigation and other local needs. The conflict in the Hunza valley between cash crop expansion and domestic, including tourist, demands, which take precedence over it, has already been mentioned (Chapter 8; Parish, 1999). The need for adequate sanitation for the large numbers of tourists, whether they are day trippers or long-stayers is very important. The development of dry toilets goes some way to conserving supplies, but Western tourists tend to be used to free access to unlimited quantities of hot and cold running water, and whilst cleanliness is admirable, there is often insufficient supply for this. Another aspect of water demands associated with tourism is the diversion of irrigation supplies to water golf courses and the seeding of clouds to create snow for ski resorts. In

1990 there were 100 golf courses in the European Alps, 250 in 1992 and over 500 by 1996 (Denniston, 1995). In Malaysia, the tropical montane cloud forests of Mount Kinabalu and the Cameron and Gentina Highlands are being progressively cleared for golf courses. The land is usually cleared forests with natural wetlands infilled and the application of insecticides, pesticides and fertilizers, not to mention the introduction of alien species to generate the smooth, rolling greens required. This not only pollutes the watercourses but also requires great quantities of water to keep the greens green – in Malaysia one medium-sized golf course requires the equivalent of 20 000 local people's needs in one year (Denniston, 1995). Ski resorts have high demands for water in the winter months when least is readily available. In Colorado some 2500 million litres of water is diverted to make snow, which is a fourfold increase on water use within a decade. It is not just the quantity of water that presents a problem, but the fact that it is taken from small streams which then risk drying up, with consequent damage to fauna and flora.

The generation of various forms of pollution and waste is an inevitable consequence of large influxes of Western tourists. In the West, most people are part of a 'disposable society', meaning that disposal rather than recycling is the order of the day. In many traditional societies it is only in recent years that manufactured articles, particularly plastics, have been introduced. The advent of plastic bags and plastic water bottles has left a trail of non-biodegradable debris in the wake of visitors; unfortunately, even putting the rubbish in bins rather than leaving it lying around does not solve the problem of waste – it still needs to be disposed of! Mountaineering expeditions have been much criticized for leaving litter, particularly around Mount Everest. This has received

## Box 12.1   Clearing up Mount Everest

In 1990 and several times since then, groups of people have volunteered for a clean-up operation on Mount Everest. Eleven stone holding areas for litter were constructed and nearly 1400 kg rubbish was removed to a lower-altitude landfill. This initiative demonstrated that environmental issues could be effectively addressed by such expeditions. Burning and decomposition at high altitude are slow processes and cause pollution to air and water. There are currently initiatives by the UIAA (International Association of Alpinist Associations) to establish a code of conduct, to promote planning and instil an obligation to leave the area with all the rubbish an expedition brought to it. The development of waste disposal and transport facilities helps this. Whilst mountaineering organizations might be cited as the cause of instigating interest in mountains and, indirectly, of littering the areas, they also play an important conservationist role, raising awareness of the nature of the environment, its dangers and vulnerability as well as its beauty and uniqueness.

*Sources*: McConnell, 1991; Morris, 1997.

considerable attention, not only because it is a national park, but also because of the increase in tourists visiting, whose reactions to piles of debris filters back. Tins, glass, oxygen cylinders, human waste and other debris have become a serious problem in the area, although it seems to offend Western visitors rather more than many locals (Fisher, 1990).

From the literature, it appears that most of the environmental impacts of tourism are a catalogue of disaster, but the influx of visitors to a region can give local people a new vision of the environment, particularly as a source of income. This can encourage a new awareness of the need to protect the environment which is translated into new sustainable management initiatives (see below).

## Social, cultural and economic impacts of tourism

Tourism is part of, and a catalyst for, wider economic development (Price et al., 1997) and, like all development, has both positive and negative effects on the local and regional economy and society. In traditional mountain communities economy and society are closely linked, so tourism development has ramifications in all areas of life. There is a process of acculturation to tourism as local people grow accustomed to the presence of visitors and, where necessary, the idea of providing services for them (Figure 12.2). The most immediate effect is the economic gain to locals, but whilst there are areas in which direct gains to local residents accrue, in all too many instances the revenues bypasses the locals, or is very concentrated in a limited geographical area.

Tourism as a strategy for economic development is affected by the attractions the area can offer and the feasibility of development. This requires external capital investment which can either bypass many of the locals by restricting participation, or stimulate local investment and involvement. If the initiative comes from central government or other external agencies, locals may be excluded from participation and may oppose such externally generated initiatives. The government of Pakistan, for example, earned US$390 000 in fees in 1995 but almost all of this benefited the urban elite. Many local entrepreneurs from the villages tend to become urban-based managers, although some revenue is ploughed back into the community in privately funded development projects (Parish, 1999). A direct economic opportunity arises where local food markets develop in association with hotels, as in several areas of Nepal (Bishop, 1998; Stevens, 1993), India and Ladakh (Singh and Kaur, 1985).

In Austria there was generally positive acceptance of external capital and tourist development as it is seen as the only viable alternative to farming. Alternatives within tourist developments have emerged, both to increase revenue and to spread the concentrations of visitors throughout the year. The development of glacier skiing in the last 30 years attracts visitors in spring and autumn, complementing the summer tourism season and winter snow skiers (Haimayer, 1989). Winter sports remain a preferred option as these visitors generally have more disposable income than summer visitors. However, even here there is a risk of marginalization of local initiatives, as international resort chains threaten to take over local concerns, or build their own. Planning

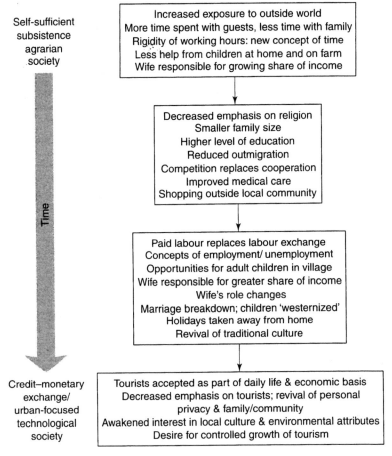

**Figure 12.2** Acculturation to tourism: changes in indigenous activities and attitudes as tourism develops in a mountain environment (adapted from Kariel, 1993).

controls can help to combat this, as do exploitation of niches such as health spas and resorts, après-ski entertainment and ski schools (Kariel, 1989). Many Tyrolean communities recognize that, despite their key share in the tourist market, this is not an ever-expanding prospect. Economic recession in the mid-1980s affected all kinds of visitors and, with the advent of much cheaper long-haul travel, the competing market increases each year.

Marketing is an important aspect of development, as is media coverage of areas. Government initiatives seeking to increase revenue often come into conflict with local attitudes and may overstretch available facilities. Nepal, which has a long-established reputation as an attractive destination, has been seeking to expand its market share in a bid to increase its revenue and to overcome negative images generated by the 1999 Indian Airlines hijack and threats of hotel workers' strikes. They have begun to open new areas, notably Mustang, and to open up to 50 new peaks to mountaineering expeditions, including Machapuchre. This latter proposal has met with an angry response by locals as the peak is a very sacred site, but the Ministry of Tourism knows that, to survive in the market, and to spread the costs and benefits of tourism developments throughout the country, it is important to open up new areas. In these days, anyone with sufficient funds can pay to scale peaks such as Everest (about £25 000) and treks to the base camps are almost commonplace. The Annapurna Conservation Area has been opened recently but only 7–10 per cent of the revenue stays in the local area (Shackley, 1994).

Attitudes to tourism development vary widely. In the European Alps is appears that the greater the distance from the area, the greater the negative response to developments and government investment. Many say that the Tyrol is overdeveloped as it is, but the attitudes of locals conflict with this as they rely on it as a viable source of livelihood (Haimayer, 1989). In Thailand, all tourist development of hill tribe and jungle tours are at the instigation of external forces – government and private enterprise (Cohen, 1996). The tribes are effectively a resource for national exploitation; they have had little say in their inclusion and direct benefits remain relatively small (see Box 12.2).

In 2000, Baltit Fort in Hunza was reported as the winner of the British Airways Tourism for Tomorrow award. This fort was restored in a project begun in 1990 using local skills and techniques and the media coverage is likely to ensure numbers of visitors to the area will increase (*The Sunday Times*, 31 December 2000). Nearby, in the village of Ganesh a Norwegian-sponsored project to restore the ancient mosque has received much less publicity, and this village remains, deliberately, outside the main focus of tourist activity around Karimabad and the Baltit Fort. This difference in attitude partly reflects the religious conservatism of the Shiites of Ganesh compared with the Ismailis of Karimabad (Parish, 1999) who view tourism as the latest of many changes in operation since the opening of the Karakoram Highway in 1978. Differences in attitudes to tourism by local people vary greatly according not only to cultural factors but also to whether they are marginalized from the activity or able to gain from it (Figure 12.3).

The government of Morocco has aggressively marketed the High Atlas region for trekkers and naturalists for some years (Berriane, 1993). This region was first opened up in the 1930s by the French, who established skiing resorts at Oukaimeden in the High Atlas and Mischliffen in the Middle Atlas. These areas remain the preserve of the Moroccan elite of Marrakech and Rabat respectively. Other activities, particularly trekking in the High Atlas, have grown. Villages such as Imlil, which lie on a motorable road, have been the focus of development, as have the roads leading over the two passes, Tiz 'n Test and the Tizi 'n Tichka, with their famous kasbahs. Here tourists are frequently assailed by insistent children after money or other gifts. Competition develops between various centres which can break down the traditional clan community. Tourism competes with cash cropping as the basis for the current economy in the area (Parish and Funnell, 1996; Parish and Funnell, 1999).

In Nepal, the opening of the Mustang region in March 1992 has been at the instigation of the government. Initially, the number of visitors was restricted

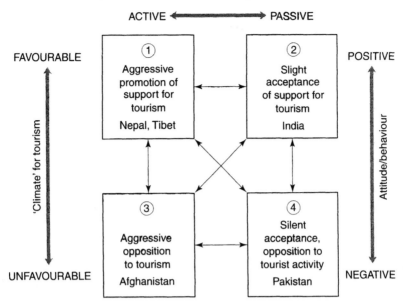

**Figure 12.3** The different attitudes to tourism, activity in response and 'climate' of favourability for tourism development in some Asian countries. The arrows indicate changes along the three scales. For example, Hunza in Pakistan might now be considered to be heading rapidly towards box 1 (after Allan, 1988).

to 200 per year but by November 1992 the number was 1000 and likely to rise (Shackley, 1994). With no infrastructure development, it could be argued that this region has already exceeded capacity. In addition, most of the US$700 + $70 per day per person charged is diverted to central coffers. Agencies arrange obligatory porterage and individuals must travel as part of a guided tour. As visitors cannot hire porters independently, little income accrues to local people.

The development of roads, usually for military purposes initially, has opened up many mountain regions to visitors. Some of these routes, such as the Karakoram Highway (opened in 1978 between Gilgit, Pakistan and Kashgar, China) become the attraction themselves, with visitors following them on organized and unorganized transport, stopping at scheduled places on the way. As with other forms of tourism, but most particularly trekking, locations on the route gain by their proximity to the traffic, whereas others often find it very difficult to gain a share in this trade. Such routes include the Thailand roads to the borders of Laos and Myanmar, the Srinagar–Leh road, which opened in 1974, and the Kathmandu–Lhasa road, which effectively opened to uses other than military in 1985 (Allan, 1988). This last route takes between two and four days on an organized tour, so the visits to individual places tend to be very brief. This means that local people have a short time in which to exploit the visitors, which may mean their attentions are more intense and persistent than otherwise. In the case of Afghanistan, the influx

## Box 12.2   Hill tribe tourism in Thailand

Trekking tours to the hill tribes of Thailand is an interesting example of the growing gap between reality and image which is prevalent in many cultural tourist attractions. The hill tribes did not have any say in their development as an attraction, but have been incorporated into a series of routes starting from bases such as Chiang Mai. They offer the tourist a glimpse of the 'real' Thailand, sleeping in huts and eating local meals. The emphasis is very much on the primitive and remote, on authenticity and unspoilt exoticism.

However, in actuality, the distances are not great and the routes often so well established now that the visitor sees a performance or an edited version of the real thing. Only selected villages are used, which represent the most aesthetic, or ones which are more welcoming than others. Whilst villages here have not yet adapted to developing a specific tourist space, they have adapted to the idea of payment for photographs and selling of craft work and adopt a 'commercial hospitality' approach to the groups. Some of this work is indeed local, but most of it is bought up and sold in the urban centres rather than to individual tourists. The latter is preferred as it can be sold at a higher price.

Interactions between villagers and tourists is very limited, in part by language. The guide is the guarantee that the visitors are bona fide, but the guide may be 'foreign' too – i.e. from a different ethnic group. Relationships between villagers and guides are complex and developed over time, but it is difficult for the guide to guarantee the behaviour of the visitors – however, a tolerance has grown up and trekkers are seen in a similar way to herds of animals, as a source of income. Payments are made to the trekking company so payments to villagers are few, and even the guides can be exploited.

Some of the problems include loss of dignity by the villagers in being commercial articles for inspection. Drugs are also an issue: opium may be available, but it is illegal and many young tourists are keen to experiment. The existence of a local market may encourage local addiction.

*Source*: Cohen, 1996.

of people in the 1960s and 1970s seeking what was, at that time, a legal trade in drugs led to a massive expansion of tourism. There was an increase in the cultivation of opium and hashish to feed the market and other drugs such as LSD were imported. However, this clashed with the conservative Islamic culture, with an inevitable backlash against Western cultural values. The drug trade distorted the local economy and the result was the Saur revolution and breakdown of society (Allan, 1988).

Thus, economic aspects are closely linked with cultural values and one of the most striking impacts of tourism is that of the clash between foreign values (e.g. urban, educated or Western) and traditional local values. The differences may be in behaviour, particularly between the sexes, dress (some forms of dress are offensive to certain religions), language, attitudes to resources and people (Moser, 1987; Stevens, 1993). Many tourists come to mountain areas seeking an authentic cultural experience, and thus expect traditional houses, dress, craftwork and festivals. These become commodified as local residents learn how to cash in on visitors' expectations. Initially, their own craft work, including heirlooms, is sold off, and then they generate items specifically to sell direct to tourists. In some places knick-knacks are imported from elsewhere, such as China, and bear little relevance to the local area.

Cultural festivals and traditional dances and songs may also become commodified. They may lose their cultural significance or be manufactured specifically for performance. In the Austrian Alps, one village recognized the importance of traditional dress and so reinvented their own in 1965 (Gamper, 1981). In Bhutan many monasteries are closed to visitors, and regulations applied to dress for visits to shrines, tombs, mosques and other sacred places. Some festivals may exclude foreigners in an effort to preserve their cultural integrity. Problems of theft, alcoholism and begging develop in some communities. The theft of valuable articles from temples in Mustang results from increased access and the fact that buildings are left unguarded due to winter depopulation (Shackley, 1994). However, the proposition to construct a museum to celebrate this area, which is rich in unaltered Tibetan art, architecture, religious and cultural traditions, was controversial as it involved removing these artefacts from their sacred places, dispossessing the people of their heritage.

Crime and begging are generally unknown and unnecessary in traditional cultures with an ethic of caring for all (Dearden, 1989), but one sign of tourist presence is the insistent demands of children for money, sweets and pens so prevalent in many mountain areas. Hosts' attitudes to visitors may become a form of tolerant performance as this is most likely to attract further custom from trekking companies and tips and disbursements for photographs, etc. from visitors. Special places may be constructed for tourists' use, thus separating family activities from tourists (Grötzbach, 1985; Stevens, 1993). This is particularly important in Muslim areas where the women would not normally interact with any male outside their immediate family. Young males may be attracted to employment as tour guides, porters, or in the service industry, which removes them from the demands of the agricultural cycle. The burden of work increasingly falls on women, who may need to take on roles previously barred by gender, but necessary because their menfolk are absent.

However, the existing social structures can operate to control the impact of visitors on local people. The construction of special tourist spaces has been mentioned, which limits interaction between individuals. Space controls may also be applied in terms of where favoured trekking routes go, or by the exclusion of tourists from some particularly sensitive areas. In other cases, internal social structures such as the local headman may act as a force for cultural conservation. The community may decide to exclude visitors from festivals or sites or buildings, or to charge for entry in order to generate revenue for repairs, as in Mustang (Shackley, 1994).

## Sustainable management of the 'capital of tourism' _____

The 'capital of tourism' refers to those features which attract visitors to mountains. They include the beauty of the landscape as well as its cultural aspects,
tranquillity and spiritual associations. In order to retain the attractions that
mountain regions have to offer it is necessary to manage the effects of tourism,
which may be accomplished in a number of ways, mainly by managing the
visitors, their impacts and development. The very rapid growth rate of the
tourist industry has occurred too fast for sustainable management to have
developed alongside it in many regions, and the pressure to expand has resulted and continues to result in overinvestment of capital and spoiling of the
landscape – as in the case of many Mediterranean and some Alpine resorts
where inappropriate architecture and congestion are rife. Even where there is a
policy of harmonious development, as in Switzerland, there is a substantial
gap between reality and policy (Krippendorf, 1984).

Many of the regions which attract the highest tourist figures are those
which are also protected areas, which makes their sustainable management all
the more important. Whilst the physical impacts need attention, it is also
imperative that the issue of exploitation of human resources such as the hill
tribes should be addressed, which requires an integrated approach involving
local participation in decision-making. The interdependence of people, landscape and tourism means that it is necessary to 'give the voice a choice' –
i.e. to include the voice of the mountain people and to facilitate alternative
options within tourism, not to invest heavily in one area (a modern 'monocrop').
This requires coordination of many aspects of planning as well as realistic
marketing and responsibility for environmental and social concerns in the part
of the tourist industry.

Tourism does not have to be synonymous with detrimental effects. Whilst
all have a right to enjoy mountain regions, all also have a responsibility to
tread lightly, and the growing trend for ecotourism and 'soft' options which
claim to be environmentally friendly and culturally respectful reflect this.
However, the economic power of the industry and highly competitive market
mean that these claims may be only faintly green-tinted. There is generally
more careful planning in advance of development, particularly where substantial environmental impacts may arise, and many countries have planning
regulations and requirements for Environmental Impact Assessments and appropriate Environmental Management Systems in place to assess potential damage
(Williams and Todd, 1997). This arises from the idea of self-regulating markets and fewer prescriptive controls in environmental management. The need
to protect the capital of tourism in order to maintain standing in the market
would operate to limit the degree of development and exploitation.

In the Bavarian Alps, most visitors are Germans coming for weekends.
Uncontrolled development until the recession in 1973 was followed by a
period of increasing regulation, culminating in the establishment of a zoning
system. Zone A permits new developments, Zone B allows only developments
meeting regulations and no developments are permitted in Zone C. However,

over the border in Austria, unregulated developments are permitted, and this has proved a particular problem in the management of the Berchtesgaden National Park which abuts the border (Grötzbach, 1985).

This 'Alpine Plan' has since been adopted by many other Alpine villages but with varying success. In Germany the strong legal framework throughout the country can enforce it, but in Austria the federal government has a mainly advisory role to strongly self-determined local communities, resulting in a piecemeal, regionally variable pattern of resource development (Barker, 1994). Conflicts arise in protected areas such as the Hohe Tauern National Park in Austria, which is one of the most popular sites in the Alps. The park was established in 1971 and in 1983–1991 three Länder (regions) agreed to maintain an unspoilt core but the surrounding buffer zone was subject to zoning in which some areas were protected and others given over to different uses. This was an effort to accommodate the need for protection, to exploit the recreation potential and to respect traditional agricultural uses of the area (Stadel *et al.*, 1996).

---

**Box 12.3    The Obergurgl model**

In the Austrian Tyrol the village of Obergurgl in the Otzal valley was made a focus of a study of the effects of tourism in the Man and Biosphere Project, beginning in 1971. The model which emerged from this study comprised four main elements: recreational demand, population and economic growth, farming and economic change, and land use and development. It identified a tourism 'trap' whereby in any one area, a potentially infinite recreational demand might be identified, but development would be limited to ever-decreasing areas of 'safe' land (i.e. safe from natural hazards) and would increasingly displace the agricultural base of the community, leading to a fall-off in demand due to overdevelopment, environmental degradation and communities with an oversupply of tourist facilities but diminishing returns.

In response to this, the local population sought to control the effects of developments and to protect and reinvigorate the local resident population. Areas were allocated different functions and zones where no developments of a certain kind were established. A reduction in quantity but improvement in quality of accommodation and facilities was made a focus of their strategy. New niches were developed, such as nature tourism, education of visitors and the wider world about the sensitivity of the mountain environment and its biodiversity. New sewage plants and sanitation systems were installed and traffic restrictions imposed to ease congestion and air pollution. Farmers received subsidies in the interests of maintaining something of the traditional agriculture which helps to sustain landscape stability as well as being part of the attraction of the place. Local music groups and youth education were established for the benefit of permanent residents.

*Source*: Moser and Peterson, 1988.

This need for planning to accommodate sustainable and diverse use of limited resources is common to all mountain areas where tourism is an integral part of the modern economy. In Svalbard in arctic Norway, local concerns over restrictions to snowmobiles in certain areas led to a relaxation of rules near settlements to enable residents to continue their everyday lives. The three Alpine Plan zones have been applied and tourism development has been welcomed as an alternative to the declining coal mining industry. The area attracts cruises, conferences and adventure tourists, but these are carefully controlled due to the particularly fragile nature of the arctic environment (Kaltenborn, 2000). Public opinion in the Canadian Rockies successfully influenced government decisions in Banff towards a more restricted development pattern, whilst in Alberta environmental groups have managed to maintain environmental precedence in decisions concerning tourist development. This still has to be balanced against local economic needs, particularly the replacement in the economy of declining mining and fur trading, as at Canmore (Kariel, 1988).

Recreation can be an important source of funding for conservation, and forms a focus of educating the wider world about mountain-based issues. In Ecuador, the salvation of the cloud forests may lie partly in appropriate ecotourism. This environmentally friendly type of recreation seeks to 'take only photographs and leave only footprints'. The Ecuador example seeks to give equal weighting to farmers' needs and the flora and fauna, and emphasizes the importance of the long-term environmental future rather than the shorter-term economic gains (Fair, 1996). Ecotourism is a viable, although often unreliable, alternative land use which is non-consumptive if well managed. It can, however, bring degradation, particularly in popular areas and around infrastructure concentrations, and it increases local consumption of resources. The inevitability and unstoppable nature of the industry are also an issue. Good management becomes not just a preference but a definite need (Good, 1995; Mountain Forum, 1998).

Ecotourism in the national parks of Central Africa is increasingly regarded as an important mechanism for protecting the endangered species. It brings foreign exchange into the region, increases surveillance and protection of the parks and boosts national tourism in general. These benefits need to be balanced with the risks of diseases brought into the region, behavioural disturbances to gorillas and an increasing dependence on an unreliable source of income. This has been brought strongly to the fore with the civil strife in Rwanda, Burundi and the Democratic Republic of Congo which has devastated the landscape and the people, threatened the ecosystems further and virtually halted the tourist economy. Whilst the gorillas were not directly affected by the war, and despite the intentions of all sides not to harm them, they were affected by the passage of tens of thousands of refugees moving through the region. In such circumstances the protection of nature becomes subsumed by more immediate needs of survival of refugees (Lanjouw, 1999).

The issue of sustainable, community-based tourism development was the subject of a Mountain Forum email conference in 1998. Table 12.1 provides

Table 12.1  Initiatives in community-based mountain tourism.

| Practice | Purpose | How it works | Examples |
|---|---|---|---|
| **Planning and assessment**<br>Local, regional and national strategic plans. Economic impact study. Educating local communities. Monitoring indicators and field studies. | Coordination for optimum sustainability. Increasing resource base, guiding development. Evaluate feasibility and impact, enabling informed choices of communities. Evaluate success and sustainability. Assess long-term impacts. | Plan long-term goals for people, culture and environment. State-guided cooperative planning. Conservation politics. Market research; public and financial analysis. Workshops with advisors inform of the advantages and disadvantages of options, document impacts. | Spirit Hawk, Canada; Budongo, Uganda; Alberta, Canada; Czech Inspiration; Vakavanua, Fiji; Tourism Norms, Mexico; HandMade in America, USA; Guandera, Ecuador; Re-thinking Tourism USA; Tourism Indicators, ICIMOD. |
| **Infrastructure and capacity building**<br>Roads/trails; restoration of original structures; alternative energy; waste management; skills training. Tourist information centres; women's education and training. | Access and marketing tool. 'New' attraction; revitalizes economy. Protect environment and aesthetics. Operating and management skills, vision, confidence. Enhance visitor experience, inform of cultural aspects. | Plan carrying capacity, erosion, looks, needs. Traditional design, techniques and materials. Kerosene subsidies. Clean-up campaigns and facilities. NGO workshops, tours, exchange of ideas. Panels, brochures, visitor and study centres. | Bouma Falls, Fiji; St-Martin, Switzerland; Baltit Fort, Hunza; Annapurna, Nepal; Mt Kenya, Kenya; Mt Everest, Nepal; Caucasus, Georgia; Langtang, Nepal; Oaxaca, Mexico; Uluru-Kata Tjuta, Australia. |
| **Institutional development**<br>Cooperatives; community unions. Tour operator association. Cross-cultural consortia; networks. | Participatory, community-orientated developments; local revenue and control; self-regulation; forum for sharing. | Discussions, planning and communal decisions; group sharing of tasks; promotion of shared resources. | Monteverde, Costa Rica; Dadia and Prespa Lakes, Greece; Czech Inspiration; Revelstoke, Canada; Indigenous Peoples' Biodiversity Network; Sa Pa, Vietnam. |

*(continued)*

Table 12.1 (cont'd)

| Practice | Purpose | How it works | Examples |
|---|---|---|---|
| **Zoning and regulation** Zoning in time and space for holistic management. Regulations on lodge size, numbers, position of facilities, pricing, quality control. Code of ethics. | Diversity of resource use and control of 'hot spots'; environmental recovery; sharing of profits, protection of landscape; limit impacts; retain revenue locally; protect sacred and culturally important sites, self-regulation of groups. | Regulate zones of use, visitor numbers, organize services in time and space. Rules enforced by community; legal or voluntary protection of sites. | Annapurna, Nepal; Ecotourism International, Nicaragua; Maori Rahui, New Zealand; Yiksam, India; Stevens Village, USA. |
| **Financial sustainability** Grants, loans, Cooperative subsidies; trust funds. Entrance and access fees; micro-enterprise; revenue distribution mechanisms. | Funds start-up, specific projects; long-term investments and sharing of profits and investments; generate revenue; investment in community. | Projects meet donor criteria; low interest rates; donor endowments with shared management; small enterprises; fee collection. | Yuendumu, Australia; Upper Mustang, Nepal; Gobi Gurvansaikhan, Mongolia; Sa Pa, Vietnam; Douiret, Tunisia. |
| **Promotion** Niche, targeted marketing; responsible promotion; WWW use. | Attract numbers/types of tourists; protect local cultures against misrepresentation; global dissemination of products. | Particular aspects targeted at different groups; true situation rather than 'rose-tinted'. | Tourism and Environment, Scotland; Ecotour International, Nicaragua; Pikes Park, USA. |

*Source:* From Godde, P. (ed.) (1999) *Community-Based Mountain Tourism: Practices for Linking Conservation with Enterprise. Synthesis of an Electronic Conference, April 13–May 18, 1998.* Mountain Forum and The Mountain Institute. Franklin, West Virginia. 56 pp. (Table 2).

a summary of the types of initiatives which are already in progress. There is a strong emphasis on egalitarian participation, environmentally friendly options, small-scale, enduring initiatives and an integrative, interdisciplinary approach. These initiatives are entitled 'promising examples' and reflect yet again the resilience of the mountain communities in adapting to change as well as the need for sound management practices.

## Key points

- Tourism is the fastest-growing industry in the world and, in many regions, by far the largest sector of the economy. It is global in its nature, but unreliable, having concentrated seasons of activity and being subject to economic fluctuations and civil conflict.
- Types of tourism vary widely, including winter and summer sports as well as cultural experiences. There may be organized groups or collections of independent travellers seeking adventure.
- The scenic beauty, cultural history and ecosystems of mountains, as well as their sacred aspects, have long made them attractive sites for visitors, and the recreational function of mountains is likely to be a major issue in the future.
- Tourism has environmental impacts: erosion and degradation, deforestation, waste as well as the infrastructure and communications developments needed to serve large influxes of people. This increasingly requires careful planning to avoid natural disasters and congestion problems. The collapse of traditional agricultural economies may also increase local degradation.
- Economic impacts range from diverse opportunities to develop local markets, accommodation services and employment to an overdependence on this fickle source of income, overinvestment in facilities causing blight of the landscape, and diversion of agricultural labour to other employment.
- Social and cultural issues occur, such as the clashes in values, behaviour, desecration of local monuments and disturbances of local customs. On the other hand, cultural traditions may be revived for display to tourists and rejuvenate a sense of local identity. External exploitation of indigenous people can be a problem.
- Attitudes of tourists and to tourists vary, and there are differences in the attitude to developments between governments seeking revenue, locals seeking a share in the profits and urban elites opposed to the desecration of pure mountain landscapes.
- Ecotourism is increasingly viewed as an environmentally and culturally sensitive form of tourism which can provide funds for local conservation, renew a conservationist view of landscapes by locals and protect cultural identities.
- Careful planning and management involving tourism companies, tourists, governments and local people are critical to sound, equitable developments.

# Epilogue

# Mountains: blessed or blighted futures?

## A piece of old rope?

Throughout this book, there has been a weaving together of different themes and aspects of mountain environments (Figure E.1). Like the individual strands of a rope, each aspect alone is relatively weak but the combination of social, cultural, economic, political, geological, geomorphic, climatic and ecological factors discussed in Parts 1 and 2 gives the whole a strength greater than the sum of its parts. The binding together of these factors gives mountain regions their unique and complex characteristics – resilience and fragility, continuity and change, diversity and commonality, as shown in the discussion of resources in Part 3. It is the particular combination of different factors which dictates the nature of the mountain environment at any one time and place (Plate E.1).

Taking the analogy further, the thickness and strength of the rope varies along its length. Some parts might be frayed where resources have flowed from

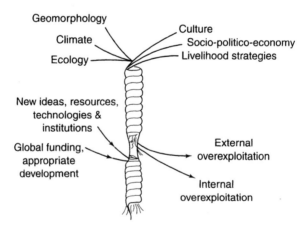

**Figure E.1** Representation of continual mountain development and change as rope, the different strands contributing to the structure of mountain environments. Partway down is a frayed area caused by overextraction (right-hand side) whilst on the left are potential mechanisms for mending the rope. The mended rope has a different character and composition from the old, but does the same job just as effectively – so too in mountain environments and communities, changes bring different patterns of life and activity, but ensure survival.

**Plate E.1**  All is not as it seems. In Hunza, there are small nucleated settlements distributed throughout terraced lands. On closer inspection, however, newer-style houses located on terraces reflect the fact that tourism and the growth of the hotel business are taking over precious agricultural land. The tall, slim poplar trees are valuable building timber, but privately owned and grown on individual plots to make up for decimated common supplies. Instead of the patchwork of different crops, the majority of these terraces now produce seed potatoes, whilst subsistence goods such as flour and grain are purchased and brought up from Gilgit by lorry. Even the musician in the foreground makes cassette tapes for sale to tourists. He sits on a piece of land on which lies the cemetery of one of the villages on the south-facing slopes, but this land is also claimed by the north-facing slope villages (to the left off the picture) and a dispute has been running for some years, with both sides lacking the resources to fund lawyers to settle it. These changes all reflect the adaptation of communities to changing conditions.

the mountains through overexploitation by internal or external sources; in these areas exploitation has exceeded the rate of regeneration of physical or human attributes, resulting in a weakening of the whole. Many regions have experienced this weakening to a greater or lesser extent. Often the community has recognized the weakening and been able to adapt to accommodate the recovery needed for long-term survival; for example, periodic increases in outmigration as a safety valve during times of resource scarcity, or an extension in exchange networks to accommodate deficits in subsistence needs.

Where the weakening or thinning of the rope has resulted from external exploitation, it is more difficult for communities to respond adequately to accommodate these major causes, stripping the rope of some of its layers. For example, the appropriation of local communal resources, especially forests and water, has been shown to place major constraints on the resilience of local livelihoods. As with old rope, a degree of fraying can be fixed, but if it goes too far, the rope snaps. The concept of the rope snapping was that envisaged by many workers in the 1960s and 1970s during the period of perceived crisis and collapse in the mountains discussed in Chapter 10.

This perception was paralleled by the global environmental movement which bordered on the hysterical with regard to the exhaustion of resources and over-population and the idea of a doomed planet. It may well be that the planet is doomed, but a closer examination of the rope reveals that it has a strong core – that of the struggle and tenacity of human endeavour for survival. This core provides the source of hope and expectation of recovery and regeneration of mountain environments, and its strength is demonstrated by the innovative mechanisms for sustainable development and economic change in the moun-tains which form the theme of Part 4. The nature of the core strengths is unique to each area at any one time and will determine whether or not the community or environment will survive.

When a rope frays it can be strengthened by the weaving in of new strands. Some of these may be of a different character – perhaps new synthetic rope materials as well as older natural fibres might be used. Some of the strands may be thinner or thicker and so are incorporated more or less invisibly into the rope's structure. Similarly, there is a continual incorporation of new ideas, technologies, social and cultural institutions and structures into mountain communities as they adapt to new opportunities and to different challenges. Thus changes in labour organization, landownership or in crop strategies might arise to take in new economic opportunities or to adapt to changing physical environmental conditions.

Just as there are external sources of pressure increasing exploitation, there are also external sources of new strands which can strengthen the community and its environment. National and international development initiatives, funds, legal and political measures to protect indigenous peoples, their cultures and environments may counteract some of the frayed areas. Some of these efforts will fit well into the old rope, whilst others will not. Thus the rope is continu-ally renewed, both from local resources and from the inputs of the wider world.

Two aspects of this final issue are important. First is the role that research and development has in understanding the structure of the rope and of gener-ating new strands to strengthen the frayed areas. The key to research and development therefore is finding a fit between the old and the new, creating new strands that blend seamlessly with the existing structure. The second issue is that of globalization and the incorporation of mountains into the wider world. This arises as part of the penetration of global capitalism into these regions, but also as a consequence of improved access to these regions. The development of tourism and the growth of concern over biodiversity and conservation are global issues, fixed firmly in the international arena. Moun-tains can no longer be considered as remote, inaccessible regions. They may still be marginalized but they are increasingly being given not just a voice, but also being recognized for their importance in the world. These two issues will be considered below before considering what value there is in developing a mountain-specific agenda.

## Trends in mountain research

There are many different approaches to the study of mountains according to the styles of different disciplines. Early studies of mountain environments tended to be regional, historical and descriptive. They sought an explanation and understanding of the complexity of mountain environments, resulting in the ideas of verticality and zonation of physical environments and of human activities. From this tradition sprang a variety of different approaches, all more or less specialized. The scientific approaches dominating the 1960s arose from the quantitative revolution, with the growth of the use of scientific data, modelling and quantification of parameters. This trend still exists in the global climate and economic modelling approaches to environmental change and to planning. These approaches had, and still have, a strong tendency towards fragmentation of disciplines through the specialization of workers in particular fields. Geomorphological studies are particularly true to this pattern, whereby studies of glacial processes, soils and even hazards tend to be examined in an isolated manner.

Cultural ecology derived from anthropological traditions based on human–environment relationships (Brush, 1976a). It sought to go beyond the older ideas of environmental determinism, where humans are controlled by their environments, into an approach based on the interaction of humans with their environments. The geoecological approaches of Troll (1968) and others were physically orientated and problem-driven, seeking a greater understanding of the complex interactions of climate, soils, fauna and flora. Geo-ecology formed part of the basis for the agro-ecology approach to subsistence systems, with its integration of the different aspects – cultivation, forests, livestock and other economic activities.

The trends in these 'softer' (as opposed to 'hard' science approaches) are towards integration and interdisciplinarity. The hardness of science here relates to its relative rigidity in terms of modelling and quantification of parameters, which requires a clear-cut approach, when compared to the relative flexibility of qualitative data. The nature of the data differs from hard scientific data in that much of it is qualitative and specific to individual cases, leading to comparative studies and dichotomies between generalization and simplification of patterns (such as the ubiquitous adaptations and characteristics of mountain regions listed in the Introduction), and the almost bewildering mass of information presenting the unique details of individual communities and environments and rejection of the appropriateness of simplification. For example, the Gujars of northern Pakistan and Afghanistan cannot strictly be considered to be only nomadic, as many have been forced to settle and adopt cultivation rather than livelihood strategies (Ehlers and Kreutzmann, 2000; see Box 5.2). Likewise, in Chapter 12, Figure 12.3, it was mentioned that Hunza communities were not all resistant to tourism developments, as some villages had bought into the tourist market in a major way.

The trends in research are currently firmly focused on integrative approaches. The need for integration emerged from various national and multinational research programmes which highlighted the need to consider the social and economic aspects of environment. Mapping, data gathering and modelling were the main focus, and still are in contemporary priority areas such as climate change and biodiversity research where compiling inventories of material continues to be a critical component of furthering understanding with a view to effective protection and utilization.

In the 1980s a sea-change occurred in approaches to mountain research. The assumptions embedded in the Theory of Himalayan Degradation were questioned, as was the idea of modernization being unidirectional and the need for highland–lowland linkages to be incorporated in mountain-focused research – i.e. mountains are not isolated entities but topographically connected with the wider world. The need for integration and the recognition of the interaction of many diverse elements in mountain environments is reflected in the contemporary ideas about complexity. Complexity is present in the nature of physical environmental change and the range of human responses. Apparently irrational decisions can be attributed to the complex interaction of the elements in the framework of operation in which the decisions are made and applied. In other words, a farmer's decisions are made within the framework of social, cultural, economic and physical and other factors which make up his or her life-world (Price and Thompson, 1997; Funnell and Parish, 1999). Complexity, and the application of cultural theory, have become a useful vehicle to carry the inexplicable and the incomprehensible (Price and Thompson, 1997; Parish, 1999).

The current themes in mountain research are primarily those associated with water, biodiversity and recreational functions of mountains, along with sustainable economic planning including agricultural development and management of forests. This combines elements of data accumulation, but is also orientated towards the creative solving of problems and enters even more clearly the realms of development studies. The focus is on household and community, on coordination and cooperation and on the exchange of ideas, solutions and issues within global fora such as the Mountain Forum website. The role of education of politicians, tourists, industry and the research and development community has grown in the current era of shared responsibility for globally recognized resources.

## Trends in mountain development

Development strategies in general have changed substantially over the last half-century, and this is reflected in the trends in development of mountains (Chapter 10). The transition from big, externally funded, imposed projects to the current emphasis on local self-determination, appropriate technology and choice of options has been driven by the almost cataclysmic failure of large-scale development planning and the need to find new ways of addressing the perceived problems.

Development failure had previously been assumed to be due to a lack of data. Certainly, until a body of data had been accumulated, it was difficult to assess the problems and their solutions. But with the overturning of the assumed linkages in the THD, and other failures, workers began to suggest that the lack of progress in development of mountains was not actually due to a lack of data, as more than enough existed, but of its translation into something useful for policy-makers (Maunch, 1983). Thus the blockage was one of communication and application. In addition, other causes of failures were propounded – a neglect of economic and social factors in environmental problems (Ehlers and Kreutzmann, 2000) and an overemphasis on 'how to' rather than 'where to' hit the development problem (Sanwal, 1989).

In order to discover 'how to' hit the problem, the emphasis has shifted further to 'who to' hit it. The role of various actors in development – agencies, individuals, government and non-government organizations and local people – has been reassessed. Whereas in the 1980s there was a wholesale adoption of participatory rural appraisal (Chambers, 1992), where local people were consulted as to their needs, aspirations and preferences, the current emphasis is on local peoples' self-determination of what development they want. The concept of 'knowing where to hit it' comes from Thompson and Warburton (1985) who used the analogy of a mechanic who, as a specialist, knows just what it wrong with a vehicle and what to do to fix it. In the case of sustainable development, it is now realized that the expert is not the technical or theoretical specialist but the indigenous farmer, who knows not just where to hit, but also what to hit.

This new acceptance on the part of the external development community that their role has changed, or perhaps should change, from a prescriptive approach to one of facilitation and response to indigenous propositions, has resulted in new trends and issues. The most vigorous of these is the management of indigenous knowledge. In the Rapid Rural Appraisal development approach (Chambers, 1992) the emphasis was on the accumulation of knowledge of the environment and society. In participatory approaches the emphasis shifted to interactive learning, where the development practitioner had to learn and understand the different ways and attitudes of indigenous populations in order to decide how and what might best fit the model. However, this required the management of indigenous knowledge, which, as we saw in Chapter 10, is difficult to fit into Western, more scientific, conceptions of how to manage development. Perhaps in the longer term, the emphasis on learning will shift still further as the holders of the knowledge have a stronger voice in the determination of their own futures.

There are two potential obstacles to this. First is the fact that it is very difficult for Western development practitioners to give up the power, and the control over others' livelihoods and futures. This may occur on a national scale with a conflict between local and national powers of governance and issues of regional autonomy, or internationally, between the developed world as facilitators of development which so often comes with strings attached (pre-cedence over other nations for access to genetic resources in return for funds, for example) and the developing world as targets for this tied development. The second issue is whether, on a global scale, the voice of the mountains can be heard above the din of states, NGOs and MNCs (multinational com-panies). At this point we touch on the place of mountains in the wider world agenda.

## Selling mountains to the wider world

Because mountains are a part of the whole world, and increasingly less isolated from it, and because issues such as biodiversity conservation are of interna-tional importance, it is ever more appropriate to place mountains as a specific issue to address on the international agenda. The potential for finding or funding solutions to the various problems lies in the global exchange of in-formation and ideas as well as funds. This exchange is facilitated by but should not be limited to mountain–mountain exchanges. Just as many of the processes operating in the physical and human environments are common to other areas of the world, so too are some of the problems, critical issues and solutions. Biodiversity and recreation are global in scope; so too is the issue of climate change. The effects of these changes vary between mountains and lowlands, as well as within mountain environments so there is a case for both global and regional treatment of these issues.

The growth in international awareness of mountains has been facilitated by the development of international projects and institutions. Perhaps the first significant mountain-focused project was in the Man and the Biosphere Programme, begun in 1973. Project 6, on the study of the impact of human activities on mountain ecosystems (which included the development of the Obergurgl model in Box 12.3), emphasized individual nations' efforts within a multinational framework of endeavour (Ives and Messerli, 1990). One spin-off from MAB-6 was the establishment of an international research centre for the Himalayan region – ICIMOD (the International Centre of Integrated Mountain Development) – in Kathmandu in 1983–1985. This was a new departure – a permanent institution with independent funding and a specific regional focus. Several institutions had interests in its establishment, includ-ing UNESCO; UNU (United Nations University); the IGU (International Geographical Union), which had established a Commission on Mountain Geoecology; and the IMS (International Mountain Institute) based in the USA.

The journal *Mountain Research and Development* was established by the UNU and IMS in 1980 and is a central publication forum for research and comment on mountain-related issues of all sorts. The UNU, IMS and IGU continue to be active in running workshops, international conferences and publishing on such issues. The Mountain Forum was set up by the Mountain Institute in the mid-1990s as a primarily electronic medium of exchange open to all interested in mountains. A very wide range of interested participants in a series of electronic conferences included mountaineers, local farmers, development practitioners, academics, members of NGOs, environmental groups and individuals from all over the world. The increasing access to the Internet and the relative cheapness of electronic means of information exchange have meant that it has been well used and the conferences provide a rich resource of papers, comments and other material archived on the net, with some available in summary form in hard copy. Regional groups, such as ICIMOD and the International Potato Centre (CIP), in Lima, Peru, as well as various European networks, are an important means of coordinating information at a supranational, subglobal scale.

The Rio Conference in 1992 saw the publication of Agenda 21 in which Chapter 13, one of the priority areas, was devoted to mountains. This inclusion was the result of strong lobbying by the mountain institutions, and of the Mountain Agenda based in Berne, formed specifically for this purpose. Publications such as Stone's *The State of the World's Mountains* (1992) were prepared specifically for Rio in order to support the case for the inclusion of mountains as a specific element, despite the fact that highland-related issues appear in many other Chapters of Agenda 21, including those for various sectors of development. Since then, there has been continued activity on the international mountain front; Messerli and Ives' book *Mountains of the World: A Global Priority* (1997) is a contribution to Agenda 21 Chapter 13, and a deliberate effort to maintain the momentum generated by Rio five years later, in 1997. The FAO has broadened its approach to watershed management to encompass conservation and development in a mountain context.

A critical component of these efforts has been the establishment of a spirit of cooperation and freedom of exchange across cultures, disciplines and professions. This remains the lifeblood of the various fora for exchange set up, but in many areas of research, cooperation can still be improved (Kreutzmann, 1998; Ehlers and Kreutzmann, 2000). This is particularly true of states and their machinery, especially in some areas of the world where conflict and national secrecy and dictatorship pose a threat to free movement of ideas and opinions. The role of institutions representing various groups of actors is an important part of successful development, and the interaction between these institutions at all levels is also critical. The conflicts between and within states continue to frustrate the efforts of NGOs of all sorts. This raises the question of the role of international law, to which states are subject, in protecting environments and peoples.

International issues such as conservation, management of the drugs problem and the conflicts between states are subject to different levels of international law. However, the enforcement of international law is difficult as there is no overarching body to which sovereign states are accountable. Sanctions and other indirect mechanisms of enforcement may take a long time to have any effect, and the international efforts at establishing peace in areas of conflict are notorious for their stalemates. The Kashmiri problem, for example, originated in the nineteenth century and continues today to be passed between India and Pakistan, fuelled by its location on the cease-fire line of the Indo-Pakistan wars following independence. The Dogra administration of the Kashmir state established by the Treaty of Amritsar in 1846 delayed accession to either side after partition. Since 1990 internal conflict has continued and complicated the issue, which continues unresolved (Kreutzmann, 1995). Such matters are subject to international law but it can do little to enforce a settlement.

Regional autonomy and regional policy are also arenas of conflict and cooperation. In Nepal, for instance, in the Annapurna Conservation Area, issues of jurisdictional authority are acute. The government transferred administrative responsibility for the region to the King Mahendra Trust, which has called for powers of jurisdiction over law enforcement, but this is resisted by the government because of concerns over a lack of political accountability. In Makalu-Baran this was resolved by joint agreements between NGO and government (Keiter, 1997). The issue of political accountability not only gives local groups a right to a voice, and a fair hearing, but also increases their level of responsibility for resources and makes them accountable locally and nationally. This cost is set against the benefits of retaining autonomy at a local level. Community responsibility does not require abandonment of central authority over legal issues, however, and it is important that a balance between national and local power remains, and that there should be transparency in legal proceedings. Many governments of mountainous countries lack the power or resources to enforce laws in remoter regions, and are often unable to harness the local powers.

In Switzerland, the cantons, which are effectively valley-regions, are fiercely protective of their autonomy, resulting in different policy objectives, law enforcement and development patterns across the country. In this case, a wider, Alpine-based regional focus can serve to unite these administrative units with others which cross national boundaries. The Alpine Convention is one of a series of regional initiatives in Europe which seeks to enhance the cooperation of alpine states and administrative units and has been widely cited as a success (Price, 1999). The campaign for the convention was begun in 1952 by CIPRA (International Commission for the Protection of the Alps) and relaunched in 1987. It was accepted by the European Parliament in 1988 and signatories have been accumulating since 1991.

The convention recognizes the importance and diversity of the whole alpine region across Europe, including the physical and human environments and the importance of the region as a 'working' one. It concentrates on environmental protection, transborder cooperation, monitoring and research. It acknowledges the fact that many alpine issues cannot be solved by national action alone but are best addressed by concerted efforts across the region. It thus serves as a mechanism for identifying and addressing mountain-specific issues. It also serves as a model for other regional-level mountain consortia.

Other European mountain-focused initiatives include recommendations for the implementation of Agenda 21 Chapter 13 presented to national governments and the European Union. The use of the EU is significant, as increasingly it has acquired jurisdiction over many issues which impinge on mountain development. Perhaps the most important is the status of many mountains as

Less Favoured Regions in the EU and thereby eligible for additional grants and funds for development. Whilst much of this development may in fact involve projects which are detrimental to the environment and of questionable benefit to the economy (Scott, 1998), the EU is a key actor in Europe with regard to managing the future of mountains – for example, through the Natura 2000 network of protected areas and the Common Agricultural policy as well as through regional development control.

Ives *et al.* (1997) propose six routes towards a twenty-first century mountain agenda. These are aimed at achieving desirable futures of mountain regions:

- political will and awareness of mountains and mountain issues – this goes beyond the scare crisis and celebrates the intrinsic, rather than instrumental value of mountains;
- guarantees of human rights and basic needs of mountain peoples – protecting their rights and traditions and establishing the mechanisms for hearing their voice – lie within institutional and legal spheres;
- appreciation of and support for indigenous knowledge and management systems of mountain communities;
- creation of 'montology' – a mountain-specific science;
- monetary compensation and ethical commitments as mechanisms of achieving appropriate sustainable development and avoiding exploitation;
- opening and maintaining a dialogue between stakeholder groups – making shared responsibility live up to its name.

## Mountain voices, mountain specificities

The Alpine Convention stresses the involvement of NGOs, states and 'social partnerships' of local groups. It enables regions, cantons and provinces to obtain funding for issues which fall under their own jurisdiction, thus giving the regions a voice without necessarily threatening the states, although issues of national sovereignty remain unresolved. This giving of a voice to regions overcomes their sense of isolation and powerlessness, and the cooperation with other alpine areas over similar issues imbues them with a new confidence in their own capacity for decision-making and in their cultures and identities. Methods of conflict resolution through an alpine tribunal or court or access to the International Court of Justice are two possible methods under discussion.

Euromontana represents the populations of European mountain regions and serves as a voice for these people to governments and the EU with regard to progress on various development and management objectives. It has contributed to a number of documents seeking to influence the pattern of mountain development, including a Green Paper on the Alps (Pils *et al.*, 1996) and studies for environment–agriculture integration for the European Commission (Euromontana, 1997).

Elsewhere, in India the establishment of the new state of Uttar Pradesh in 1998, which includes part of the Himalayas faced lengthy consideration by development planning, economic, financial and other departments since 1991 when the idea was first voiced (Mawdsley, 1999). However, little is heard of the voices of local people in determining their future – in the current age of local knowledge and its supremacy in development issues, this seems a not unexpected but lamentable omission, especially as their public protests were instrumental in persuading the government to go ahead with the proposal.

Thus, the strength of the mountain voice varies according to the security of the national government of different regions. Where it threatens national security, it is subdued. Nevertheless, other interesting developments represent small but significant footsteps along the way to audibility. The developmental shift towards community and shared responsibility for resources, which has stimulated the recognition of the value of local as well as national laws and powers, is increasingly visible (Lynch and Maggio, 2000). Specialized systems of conflict resolution, especially regarding environmental issues, are emerging (Carpenter and Kennedy, 1981). In Bolivia the election of the first peasant to the parliament placed a voice in support of local independence and self-determination right at the heart of power (Loayza and Rist, 2000). Finally, there are initiatives such as the mountain-to-mountain initiative which seeks to set up exchanges between 'ordinary' mountain dwellers to share ideas and exchange cultures. These operate both on a north–south and a south–south basis, allowing exchanges between, for example, Ecuador and Appalachia and Ecuador and the Philippines (Rhoades, 2000). This adds a new dimension to hearing local voices – taking them out of situ and allowing them to have the direct benefits of cultural interaction.

Mountains therefore not only face enormous challenges with regard to their environmental pressures and growing human needs, but they also begin to have the ability to tap into the resources of the whole world in order to address them.

Globalization can mean more than economic exploitation; it can include the global sharing of knowledge, responsibility, problems and solutions on a more democratic basis. The old rope will never be perfect and even, but it will not snap. It is continually renewing and reforming itself to meet the challenges of the future. It needs help, but of the right kind, and this is beginning to emerge in the theory and practice of mountain management and development.

## Some useful websites

FAO Mountain Division http://www.fao.org/waicent/faoinfo/forestry/Mountain/MNTPAG3.HTM
ICIMOD http://www.icimod.org
Mountain Forum http://www.mtnforum.org
The Mountain Institute http://www.mountain.org
World Conservation Monitoring Centre http://www.unep-wcmc.org/habitats/mountains/statistics.htm
World Heritage Sites http://www.unesco.org/whc/heritage.html

# Bibliography

Abelson, A.E. (1973) Altitude and fertility. *Human Biology*, **48**, 83–71.

Adarkwa, K. (1988) Basic needs – water delivery system in mountainous regions of Ghana: a case study of piped water supply on the Akwapim Ridge. *Mountain Research and Development*, **8**, 303–307.

Aegerter, S. and Messerli, P. (1983) The impact of hydroelectric power plants in a mountainous landscape. A technique for assessing environmental impacts. *Mountain Research and Development*, **3**, 157–175.

Agrawal, A. (1995) Dismantling the divide between indigenous and scientific knowledge. *Development and Change*, **26**, 413–439.

Aldrich, M., Billington, C., Edwards, M. and Laidlaw, R. (1997) Tropical montane cloud forests: an urgent priority for conservation. *WCMC Biodiversity Bulletin* No. **2**.

Alexandrian, D., Esnault, F. and Calabri, G. (1999) Forest fires in the Mediterranean area. *Unasylva*, **197**, **50**, 35–41.

al-Kasir, A. (1985) The impact of emigration on social structure in the Yemen Arab Republic. In: Pridham, B.R. (ed.) *Economy, Society and Culture in Contemporary Yemen*, Croom Helm, London, 122–131.

Allan, N.R.J. (1986) Accessibility and altitudinal zonation models of mountains. *Mountain Research and Development*, **6**, 185–194.

Allan, N.J.R. (1987) The impact of Afghan refugees on the vegetation resources of Pakistan's Hindukush-Himalaya. *Mountain Research and Development*, **7**, 200–204.

Allan, N.R.J. (1988) Highways to the sky: the impact of tourism on South Asian mountain culture. *Tourism Recreation Research*, **13**, 11–16.

Allan, N.R.J. (1991) From autarky to dependency: Society and habitat relations in the South Asian rimland. *Mountain Research and Development*, **11**, 65–74.

Allen, B.J. (1988) Adaptation to frost and recent political change in Papua New Guinea. In: Allan, N.R.J., Knapp, G.W. and Stadel, C. (eds) *Human Impact on Mountain Environments*, Rowman & Littlefield, Towota, NJ, 255–264.

Allison, R.J. and Thomas, D.S.G. (1993) The sensitivity of landscapes. In: Thomas, D.S.G. and Allison R.J. (eds) *Landscape Sensitivity*, Wiley, Chichester, 1–5.

Anderson, E.F. (1993) *Plants and People of the Golden Triangle: Ethnobotany of the Hill Tribes of Northern Thailand*, Discorides Press, Hong Kong.

André, M.-F. (1990) Geomorphic impact of spring avalanches in North-west Spitzbergen. *Permafrost and Periglacial Processes*, **1**, 97–110.

Andrews, C. (1983) Photographs and notes on tourism and deforestation in the Solu Khumbu, Nepal. *Mountain Research and Development*, **3**, 182–186.

Apffel-Marglin, F. (1997) Counter-development in the Andes. *The Ecologist*, **27**, 221–228.

Aris, M. (1990) Man and nature in the Buddhist Himalayas. In: Rustoniji, N.K. and Ramble, C. (eds) *Himalayan Environment and Culture*, Indian Institute of Advanced Study, Simla, 85–101.

Armand, A.D. (1992) Sharp and gradual mountain timberlines as a result of species interaction. In: Hansen, A.J. and di Castri, F. (eds) *Landscape Boundaries: Consequences for Biotic Diversity and Ecological Flows*, Ecological Studies, Vol. 92. Springer, New York, 360–378.

Arnold, J.E.M. and Campbell, J.G. (1986) Collective management of hill forests in Nepal: the Community Forestry Development Project. In: *Proceedings of a Conference on Common Property Resource Management, 21–26 April, 1985*, National Academy Press, Washington D.C.

Artz, N.E., Norton, B.E. and O'Rourke, J.T. (1986) Management of common grazing lands: Tamahdite, Morocco. In: *Proceedings of a Conference on Common Property Resource Management*, National Academy of Science, Board of Trade and Technology. National Academy Press, Washington D.C., 259–280.

Astrain, L.N., Stephens, M. and Thomas, N. (1997) *The Basques*, Ashley Drake Publishers, Deddington.

Aulitsky, H. (1994) Hazard mapping and zoning in Austria: methods and legal implications. *Mountain Research and Development*, **14**, 307–313.

Aulitsky, H., Heuberger, H. and Pratzelt, G. (1994) Mountain hazard geomorphology of Tyrol and Vorarlberg, Austria. *Mountain Research and Development*, **14**, 273–305.

Babikir, A.A.A. (1988) Vegetation, soil and land-use changes in Jebel Marra and other mountains in the Republic of the Sudan. *Mountain Research and Development*, **8**, 235–241.

Baer, A. (1959) L'extrémité occidentale du Massif de l'Aar. (Relation du socle avec la couverture). *Bulletin de la Société Neuchâteloise des Sciences Naturelles*, **82**, 1–160.

Bähr, J. (1985) Agriculture, copper mining, and migration in the Andean Cordillera of Northern Chile. *Mountain Research and Development*, **5**, 279–290.

Baied, C.A. and Wheeler, J.C. (1993) Evolution of High Andean *Puna* ecosystems: environment, climate and culture change over the last 12, 000 years in the Central Andes. *Mountain Research and Development*, **13**, 145–156.

Balikci, A. (1990) Tenure and transhumance: stratification and pastoralism amongst the Lakenkhel. In: Galaty, J.G. and Johnson D.L. (eds) *The World of Pastoralism: Herding Systems in Comparative Perspective*, Belhaven, London, 301–322.

Balland, D. (1988) Nomadic pastoralists and sedentary hosts in the Central and Western Hindukush Mountains, Afghanistan. In: Allan, N.R.J., Knapp, G.W. and Stadel, C. (eds) *Human Impact on Mountain Environments*, Rowman & Littlefield, Totowa, NJ, 265–276.

Banskota, K. and Sharma, B. (1995) *Mountain Tourism in Nepal: An Overview*, Discussion Paper 95/7, ICIMOD, Kathmandu.

Banskota, M. (1999a) *Water Issues in the Mountains: Social Aspects of the Prevailing Systems in the Hindu-Kush-Himalayan Communities*, ICIMOD, Kathmandu. http://www.icimod.org.sg/focus/water/

Banskota, M. (1999b) *Social Aspects and Local Water-Harvesting Systems: A Review of the Prevailing Systems in the Hindu-Kush-Himalayan Communities*, ICIMOD, Kathmandu. http://www.icimod.org.sg/focus/water/

Banskota, M. and Jodha, N.S. (1992) Mountain agricultural development strategies: comparative perspectives from the countries of the Hindu Kush-Himalayan region. In: Jodha, N.S., Banskota, M. and Partap, T. (eds) *Sustainable Mountain*

*Agriculture: Volume 1, Perspectives and Issues*, Intermediate Technology Publications, London, 83–114.

Barker, M.L. (1994) Strategic tourism planning and limits to growth in the Alps. *Tourism Recreation Research*, **19**, 43–49.

Barrow, C.J. (1999) *Alternative Irrigation: The Promise of Runoff Agriculture*, Earthscan, London.

Barry, R.G. (1992) Mountain climatology and past and potential future climatic changes in mountain regions: a review. *Mountain Research and Development*, **12**, 71–86.

Barsch, D. and Caine, N. (1984) The nature of mountain geomorphology. *Mountain Research and Development*, **4**, 287–298.

Barth, F. (1969) Ecologic relationships of ethnic groups in Swat, North Pakistan. In: Vayda, A.P. (ed.) *Environment and Cultural Behaviour: Ecological Studies in Cultural Anthropology*, The Natural History Press, New York, 362–375.

Bätzing, W., Perlik, M. and Dekleva, M. (1996) Urbanization and depopulation in the Alps. *Mountain Research and Development*, **16**, 335–350.

Baumann, M., Bell, J., Koechlin, F. and Pimbert, M. (eds) (1996) *The Life Industry: Biodiversity, People and Profits*, Intermediate Technology Publications, London.

Baumgartner, M.F. and Apfl, G. (1994) Monitoring snow cover variations in the Alps using alpine snow cover analysis system (ASCAS). In: Beniston, M. (ed.) *Mountain Environments in Changing Climates*, Routledge, London, 108–120.

Bayfield, N.G. and Barrow, G.C. (eds) (1985) *The Ecological Impacts of Outdoor Recreation on Mountain Areas in Europe and North America*, RERG Report No. 9. Wye College, UK.

Bebbington, A. (1996a) Debating 'indigenous' agricultural development: Indian organizations in the Central Andes of Ecuador. In: Collinson, H. (ed.) *Green Guerrillas: Environmental Conflict in Latin America and the Caribbean: A Reader*, Latin American Bureau, London, 51–60.

Bebbington, A. (1996b) Movements, modernizations and markets: Indigenous organizations and agrarian strategies in Ecuador. In: Peet, R. and Watts, M. (eds) *Liberation Ecologies*, Routledge, London, 86–109.

Bebbington, A. (1997) Social capital and rural intensification: local organizations and islands of sustainability in the rural Andes. *The Geographical Journal*, **163**, 189–197.

Bebbington, A. (1998) Sustaining the Andes? Social capital and policies for rural regeneration in Bolivia. *Mountain Research and Development*, **18**, 173–181.

Bedele, D.K. (1988) The changing use of mountains by a farming people – the Krobos of Ghana. *Mountain Research and Development*, **8**, 297–301.

Bencherifa, A. (1983) Land use and equilibrium of mountain ecosystems in the High Atlas of Western Morocco. *Mountain Research and Development*, **3**, 273–279.

Bencherifa, A. (1988) Demography and cultural ecology of the Atlas Mountains of Morocco: Some new hypotheses. *Mountain Research and Development*, **8**, 309–313.

Bencherifa, A. (1993) Migration extérieure et développement agricole au Maroc. *Revue Géographie du Maroc*, **15**, 51–90.

Bencherifa, A. and Johnson, D.L. (1990) Adaptation and intensification in the pastoral systems of Morocco. In: Galaty, J.G. and Johnson, D.L. (eds) *The World of Pastoralism: Herding Systems in Comparative Perspective*, Belhaven, London, 394–416.

Bencherifa, A. and Johnson, D.L. (1991) Changing resource management strategies and their environmental impacts in the Middle Atlas Mountains of Morocco. *Mountain Research and Development*, **11**, 183–194.

Beniston, M. (1994) Climate scenarios for mountain regions: an overview of possible approaches. In: Beniston, M. (ed.) *Mountain Environments in Changing Climates*, Routledge, London, 136–152.

Beniston, M. and Fox, D.G. (1996) Impacts of climate change on mountain regions. In: Watson, R.T., Zinyowera, M.C., Moss, R.H. and Dokken, D.J. (eds) *Climate Change 1995. Impacts, Adaptations and Mitigation of Climate Change: Scientific–Technical Analyses*, Cambridge University Press, Cambridge, 191–213.

Beniston, M., Ohmura, A., Rotach, M., Tschuck, P., Wild, M. and Marrinucci, M.R. (1995) *Simulation of Climate Trends over the Alpine Region: Development of the Physically Based Modelling System for Application to Regional Studies of Current and Future Climate*, Final Scientific Report No. 4031–33250 to the Swiss National Science Foundation, Bern, Switzerland.

Berkes, F., Davidson-Hunt, I. and Davidson-Hunt, K. (1998) Diversity of common property resource use and diversity of social interests in the Western Indian Himalaya. *Mountain Research and Development*, **18**, 19–33.

Berkhout, F. and Hertin, J. (2000) Socio-economic scenarios for climate impact assessment. *Global Environmental Change*, **10**, 165–168.

Berks, F. (ed.) (1989) *Common Property Resources: Ecology and Community-Based Sustainable Development*, Belhaven, London, 218–235.

Berriane, M. (1993) Le tourisme de montagne au Maroc. In: Bencherifa, A. (ed.) *African Mountains and Highlands: Resource Use and Conservation*, Colloques et Séminaires Series Vol. 29, Faculté des Lettres et des Sciences Humaines, Université Mohammed V, Rabat, 129–151.

Bharati, A. (1988) Mountain people and monastics in Kumaon Himalaya, India. In: Allan, N.J.R., Knapp, G.W. and Stadel, C. (eds) *Human Impact on Mountains*, Rowman & Littlefield, Towota, NJ, 83–95.

Bhuktan, J., Denning, G. and Fujisaka, S. (1999) Rice cropping practices in Nepal: indigenous adaptation to adverse and difficult environments. In: Prain, G., Fujisaka, S. and Warren, M.D. (eds) *Biological and Cultural Diversity: The Role of Indigenous Agricultural Experimentation in Development*, Intermediate Technology Publications, London, 6–31.

Biddulph, J. (1880) *Tribes of the Hindoo Koosh* (Reprinted, 1995), Ali Kamran Publishers, Lahore.

Birch-Thomsen, T. and Fog, B. (1996) Changes within small-scale agriculture. A case-study from Southwestern Tanzania. *Danish Journal of Geography*, **96**, 60–69.

Birnie, P.W. and Boyle, A.E. (1992) *International Law and the Environment*, Clarendon Press, Oxford.

Bishop, B. and Naumann, C. (1996) Mount Everest: Reclamation of the world's highest junk yard. *Mountain Research and Development*, **16**, 323–327.

Bishop, N.H. (1989) From zomo to yak: change in a Sherpa village. *Human Ecology*, **17**, 177–202.

Bishop, N.H. (1998) *Himalayan Herders*, Harcourt Brace, Fort Worth.

Bjønness, I.M. (1980) Animal husbandry and grazing, a conservation and management problem in Sagarmatha (Mt Everest) National Park, Nepal. *Norsk Geografisk Tidsskrift*, **34**, 59–76.

Blaikie, P. and Brookfield, H. (1987) *Land Degradation and Society*, Methuen, London.

Boehm, C. (1984) Mountain refuge area adaptations. In: Beaver, P.D. and Purrington, B.L. (eds) *Cultural Adaptations to Mountain Environments*, University of Georgia Press, Atlanta, 24–37.

Boggs, C.L. and Murphy, D.D. (1997) Community composition in mountain eco-systems: climatic determinants of montane butterfly distributions. *Global Ecology and Biogeography Letters*, **6**, 39–48.

Bohle, H.-G. and Adhikari, J. (1998) Rural livelihoods at risk: How Nepalese farmers cope with food insecurity. *Mountain Research and Development*, **18**, 321–332.

Bonnenfant, P. (1997) La maitrise de l'eau dans le Wadi Zabid, Yemen. *Bulletin de l'Association Géographique Française*, **1**, 12–24.

Bossio, D.A. and Cassman, K.G. (1991) Traditional rainfed barley production in the Andean highlands of Ecuador: soil nutrient limitations and other constraints. *Mountain Research and Development*, **11**, 115–126.

Boucher, K. (1990) Landscape and technology: the Gabcikovo–Nagymaros scheme. In: Cosgrove, D. and Petts, G. (eds) *Water, Engineering and Landscape: Water Control and Landscape Transformation in the Modern Period*, Belhaven Press, London, 174–187.

Bouderbala, N., Chiche, J., Herzennit, A. and Pascon, P. (1984) *La Question Hydraulique. 1 Petit et Moyen Hydraulique en Maroc*, Institut Agronomique et Vétérinaire, Université Hassan II, Rabat.

Boyazoglu, J. and Flamant, J.-C. (1990) Mediterranean systems of animal production. In: Galaty, J.G. and Johnson, D.L. (eds) *The World of Pastoralism: Herding Systems in Comparative Perspective*, Belhaven, London, 353–393.

Boyce, M.S. (1992) Intervention versus natural regulation philosophies for managing wildlife in national parks. *Oecologia Montana*, **1**, 49–50.

Bradburd, D.A. (1996) Toward an understanding of the economics of pastoralism: the balance of exchange between pastoralists and nonpastoralists in western Iran, 1815–1975. *Human Ecology*, **24**, 1–38.

Bragg, K. (1992) Akha ethnobotany. In: Walker, A.R. (ed.) *The Highland Heritage: Collected Essays on Upland North Thailand*, Suvarnabhumi Books, Singapore, 145–162.

Braun, D.D. (1989) Glacial and periglacial erosion of the Appalachians. *Geomorphology*, **2**, 233–256.

Brett, M. and Fentress, E. (1996) *The Berbers*, Blackwell, Oxford.

Bromley, D.W. (1991) *Environment and Economy: Property Rights and Public Policy*, Blackwell, Oxford.

Bromley, R.J. (1974) *Periodic Markets, Daily Markets and Fairs: A Bibliography*, Monash University.

Brookfield, H. (1999) Environmental damage: distinguishing human from geophysical causes. *Environmental Hazards: Human and Policy Dimensions*, **1**, 3–12.

Brower, B. (1990) Range conservation and Sherpa livestock management in Khumbu, Nepal. *Mountain Research and Development*, **10**, 34–42.

Brower, B. (1991) *Sherpa of Khumbu: People, Livestock and Landscape*, Oxford University Press, New Delhi.

Browman, D.L. (1983) Andean arid land pastoralism and development. *Mountain Research and Development*, **3**, 241–252.

Browman, D.L. (1990) High altitude camelid pastoralism of the Andes. In: Galaty, J.G. and Johnson, D.L. (eds) *The World of Pastoralism: Herding Systems in Comparative Perspective*, Belhaven, London, 323–352.

Brunsden, D. (1993) Barriers to geomorphological change. In: Thomas, D.S.G. and Allison, R.J. (eds) *Landscape Sensitivity*, Wiley, Chichester, 7–12.

Brunsden, D. and Allison, R.J. (1986) Mountains and highlands. In: Fookes, P.G. and Vaughan, P.R. (eds) *A Handbook of Engineering Geomorphology*, Surrey University Press, Guildford, 150–165.

Brunsden, D. and Thornes, J.B. (1979) Landscape sensitivity and change. *Trans. I.B.G.* **NS4**, 463–484.

Brunstein, F.C. and Yamaguchi, D.K. (1992) The oldest known Rocky Mountain bristlecone pines (*Pinus aristata Engelem*). *Arctic and Alpine Research*, **24**, 253–256.

Brush, S.B. (1976a) Introduction to cultural adaptations. *Human Ecology*, **4**, 125–133.

Brush, S.B. (1976b) Man's use of the Andean ecosystem. *Human Ecology*, **4**, 147–166.

Brush, S.B. (1984) The anthropology of highland peoples. In: Beaver, P.D. and Purrington, B.L. (eds) *Cultural Adaptations to Mountain Environments*, University of Georgia Press, Atlanta, 159–167.

Brush, S.B. (1988) Traditional agricultural strategies in the hill lands of tropical America. In: Allan, N.R.J., Knapp, G.W. and Stadel, C. (eds) *Human Impact on Mountain Environments*, Rowman & Littlefield, Towota, NJ, 116–126.

Brzeziecki, B., Kienast, F. and Wildi, O. (1994) Potential impacts of a changing climate on the vegetation cover of Switzerland: a simulation experiment using GIS technology. In: Price, M.F. and Heywood, D.I. (eds) *Mountain Environments and GIS*, Taylor & Francis, London, 263–279.

Bunch, R. (1999) Reasons for non-adoption of soil conservation technologies and how to overcome them. *Mountain Research and Development*, **19**, 213–220.

Buol, S.W., Hole, F.D. and McCracken, R.J. (1973) *Soil Genesis and Classification*, Iowa State University Press, Ames.

Butt, N. and Price, M.F. (eds) (2000) *Mountain People, Forests and Trees: Strategies for Balancing Local Management and Outside Interests*, The Mountain Institute, Harrisonburg, VA.

Butz, D. (1994) A note on crop distribution and micro-environmental conditions in Holshal and Ghoshushal villages, Pakistan. *Mountain Research and Development*, **14**, 89–97.

Byers, A. (1987a) Landscape change and man-accelerated soil loss: the case of the Sagarmatha (Mount Everest) National Parks, Khumbu, Nepal. *Mountain Research and Development*, 7, 209–216.

Byers, A. (1987b) An assessment of landscape change in the Khumbu region of Nepal using repeat photography. *Mountain Research and Development*, **7**, 77–81.

Byers, A. (1987c) *A Geomorphic Study of Man-Induced Soil Erosion in the Sagarmatha (Mount Everest) National Park, Khumbu, Nepal*, Unpublished Ph.D. Thesis, Department of Geography, University of Colorado, Boulder.

Byers, E. and Sainju, M. (1994) Mountain ecosystems and women: opportunities for sustainable development and conservation. *Mountain Research and Development*, **14**, 213–228.

Byrcyn, W.G. (1992) The history and present role of the Tatra National Park. *Mountain Research and Development*, **12**, 205–210.

Caine, N. (1983) *The Mountains of Northeastern Tasmania: A Study of Alpine Geomorphology*, Balkema, Rotterdam.

Campbell, J.G. (1997) Protected areas around Mount Everest. In: Messerli, B. and Ives, J.D. (eds) *Mountains of the World: A Global Priority*, Parthenon, Carnforth, 247.

Campbell, J.K. (1964) *Honour, Family and Patronage. A Study of Institutions and Moral Values in a Greek Mountain Community*, Oxford University Press, New York.

Camps, G. (1987) *Les Berbères: Mémoire et Identité*. 2nd edn. Editions Errance, Paris.

Caplan, L. (1991) From tribe to peasant? The Limbus and the Nepalese state. *Journal of Peasant Studies*, **18**, 305–321.

Carpenter, S.L. and Kennedy, W.J.D. (1981) Environmental conflict management: New ways to solve problems. *Mountain Research and Development*, 1, 65–70.

Carter, E.J. (1992) Tree cultivation on private land in the Middle Hills of Nepal: lessons from some villagers of Dolakha District. *Mountain Research and Development*, 12, 241–255.

Centeno, J.C. (1998) *Peruvian community fights back against mining company*, Email posted to Mountain Forum 22 October 1998. http://www.mtnforum.org

Cernea, M.M. (1987) Farmer organizations and institution building for sustainable development. *Regional Development Dialogue*, 8, 1–19.

Chakravarty-Kaul, M. (1998) Transhumance and customary pastoral rights in Himachal Pradesh: claiming the high pastures for Gaddis. *Mountain Research and Development*, 18, 5–17.

Chambers, R. (1992) *Rural Appraisal: Rapid, Relaxed and Participatory*, Institute of Development Studies, University of Sussex, Discussion Paper 311.

Chambers, R. (1993) *Challenging the Professions: Frontiers for Rural Development*, Intermediate Technology Publications, London.

Chambers, R. (1994) *The Poor and the Environment: Whose Reality Counts?* Institute of Development Studies, Working Paper 3. University of Sussex.

Chapin, F.S. and Körner, C. (eds) (1995) *Arctic and Alpine Biodiversity*, Ecological Studies 113, Springer Verlag, Berlin.

Charles, J.-P. (1984) Agricultural utilization of mountain grasslands and its ecological consequences. In: Brugger, E.A., Fürrer, G., Messerli, B. and Messerli, P. (eds) *The Transformation of Swiss Mountain Ranges*, Verlag Paul Haupt, Bern, 374–383.

Chattopadhyay, G.P. (1981) Landslide phenomena in the Darjeeling Himalaya: Some observations and analysis. In: Datye, V.S., Diddee, J., Jog, S.R. and Patil, C. *Explorations in the Tropics*, University of Pona, Pune, 198–205.

Chilton, R.R.H. (1981) *A Summary of Climatic Regimes of B.C.* Assessment and Planning Division, Ministry of Environment, British Columbia.

Chocarro, C., Fanlo, R., Filat, F. and Marin, P. (1990) Historical evolution of natural resource use in the Central Pyrenees of Spain. *Mountain Research and Development*, 10, 257–265.

Claassen, E.M. and Salin, P. (1991) *The Impact of Stabilization and Structural Adjustment Policies*, FAO Economic and Social Development Paper No. 90.

Clark, M.J., Gurnell, A.M., Milton, E.J., Seppala, M. and Kyostila, M. (1985) Remotely sensed vegetation classification as a snow depth indicator for hydrological analysis in sub-arctic Finland. *Fennia*, 163, 195–225.

Clark, R., Durón, G., Quispe, G. and Stocking, M. (1999) Boundary bunds or piles of stones? Using farmers' practices in Bolivia to aid soil conservation. *Mountain Research and Development*, 19, 235–240.

Clark, W.M. (1986) Irrigation practices: peasant-farming settlement schemes and traditional cultures. In: *Scientific Aspects of Irrigation Schemes*, The Royal Society, London, 229–243.

Cohen, E. (1996) *Thai Tourism: Hill Tribes, Islands and Open-Ended Prostitutes*, Studies in Contemporary Thailand #4. White Lotus Press, Bangkok.

Cole, J. (1986) Appalachia's moral life. In: Tobias, M. (ed.) *Mountain People*, University of Oklahoma Press, Norman.

Collins, J.L. (1983) Seasonal migration as a cultural response to energy scarcity at high altitude. *Current Anthropology*, 24, 103–104.

Connell, J., Dasgupta, B., Laishley, R. and Lipton, M. (1976) *Migration from Rural Areas: Evidence from Village Studies*, Oxford University Press, New Delhi.

Cooper, J.M. (1946) The Araucanians. In: Steward, J.H. (ed.) *Handbook of South American Indians, Volume 2, The Andean Civilizations*, Smithsonian Institution Bureau of American Ethnology, Bulletin No. 143, Washington D.C., 687–760.

Cooper, P.J.M. (1979) The association between altitude, environmental variables, maize growth and yields in Kenya. *Journal of Agricultural Science*, **93**, 635–649.

Council of Europe (1993) *The EUR-OPA Major Hazards Agreement of the Council of Europe*, Strasborg, Council of Europe AP/CAT (93) 30.

Crook, D.S. and Jones, A.M. (1999) Design principles from traditional mountain irrigation systems (*bisses*) in the Valais, Switzerland. *Mountain Research and Development*, **19**, 79–99.

Crook, J. and Osmaston, H. (eds) (1994) *Himalayan Buddhist Villages: Environment, Resources, Society and Religious Life in Zangskar, Ladakh*, University of Bristol Press, Bristol.

Dani, D.D. (1986) Population and society in Nepal: An overview. In: Joshi, S.C. (ed.) *Nepal Himalaya: Geo-Ecological Perspectives*, Himalayan Research Group, Delhi, 163–186.

Darbellay, C. (1984) Mountain agriculture in change. In: Brugger, E.A., Fürrer, G., Messerli, B. and Messerli, P. (eds) *The Transformation of Swiss Mountain Ranges*, Verlag Paul Haupt, Bern, 290–316.

Dearden, P. (1989) Tourism in developing countries: some observations on trekking in the Highlands of North Thailand. In: D'Amore, L.J. and Jafari, J. (eds) *Tourism – a Vital Force for Peace*. International Institute for Peace Through Tourism, Montreal, 207–216.

Debarbieux, B. (1993) Du haut en général et du Mont Blanc en particulier. *L'Espace Géographie*, **1**, 5–13.

de Haas, H.G. (1996) Socio-economic transformations and oasis agriculture in Southern Morocco. Paper presented at the *28th International Geographical Congress*, The Hague, 4–10 August 1996.

Denniston, D. (1995) *High Priorities: Conserving Mountain Ecosystems and Cultures.* Worldwatch Paper 123, Worldwatch Institute, Washington D.C.

De Scally, F.A. and Gardner, J.S. (1994) Characteristics and mitigation of the snow avalanche hazard in Kaghan valley, Pakistan Himalaya. *Natural Hazards*, **9**, 197–213.

Dessaint, W.Y. and Dessaint, A.Y. (1992) Economic systems and ethnic relations. In: Walker, A.R. *The Highland Heritage: Collected Essays on Upland North Thailand*, Suvarnabhumi Books, Singapore, 95–110.

Dewees, P.A. and Saxena, N.C. (1995) Tree planting and household land and labour allocation: case studies from Kenya and India. In: Arnold, J.M. and Dewees, R.A. (eds) *Tree Management in Farmer Strategies: Responses to Agricultural Intensification*, Oxford University Press, Oxford, 242–267.

Dhakal, D.N.S. (1990) Hydropower in Bhutan: a long-term development perspective. *Mountain Research and Development*, **10**, 291–300.

Dirksen, H. (1997) Solving problems of opium production in Thailand: lessons learned from the TG-HDP. In: McCaskill, D. and Kampe, K. (eds) *Development or Domestication? Indigenous Peoples of Southeast Asia*, Silkworm Books, Chiang Mai, 329–357.

Dixit, K.M. (1995) The porter's burden. *Himal*, **8**, 32–38.

Dollfus, O. (1982) Development of land-use patterns in the Central Andes. *Mountain Research and Development*, **2**, 39–48.

Dollfus, P. (1999) Mountain deities among the nomadic community of Kharnak (Eastern Ladakh). In: Van Beek, M., Bertelsen, K.B. and Pedersen, P. (eds) *Ladakh: Culture, History and Development between Himalaya and Karakoram*, Arhus University Press, Arhus, 92–118.

Dore, A., Sobik, M. and Migala, K. (1999) The role of orographic cap clouds in pollutant deposition in the Western Sudety Mountains. In: Beniston, M. (ed.) *Mountain Environments in Changing Climates*, Routledge, London, 89–91.

Dougherty, W. (1994) Linkages between energy, environment and society in the High Atlas Mountains of Morocco. *Mountain Research and Development*, **14**, 119–135.

Dove, M.R. (1988) The Kantu' system of land tenure: the evolution of tribal land rights in Borneo. In: Fortmann, L. and Bruce, J.W. (eds) *Whose Trees? Proprietary Dimensions of Forestry*, Westview Press, Boulder, CO, 86–95.

Dove, M.R. (1995) The shift of tree cover from forests to farms in Pakistan: a long and broad view. In: Arnold, J.E.M. and Dewees, P.A. (eds) *Tree Management in Farmer Strategies: Responses to Agricultural Intensification*, Oxford University Press, Oxford, 65–89.

Dower, N. (1988) *What is Development? A Philosopher's Answer*, Centre for Development Studies Occasional Paper 3. University of Aberdeen.

Downing, T.E. (1974) Irrigation and moisture-sensitive periods: a Zapotec case. In: Downing, T.E. and Gibson, McG. (eds) *Irrigation's Impact on Society*, University of Arizona Press, Tucson, 113–122.

Dresch, J. (1941) *Recherches sur l'évolution du relief dans le Massif Central du Grand Atlas, Haouz et le Sous*, Armand Colin, Paris.

Dresch, P. (1989) *Tribes, Government and History in Yemen*, Oxford University Press, Oxford.

Drewry, D. (1986) *Glacial Geologic Processes*, London, Arnold.

du Bois, F. (1994) Water rights and the limits of environmental law. *Journal of Environmental Law*, **6**, 71–84.

Durrenberger, E.P. (1983a) Changes in a Shan village. In: McKinnon, J. and Bhuruksasri, W. (eds) *Highlanders of Thailand*, Oxford University Press, Oxford, 113–122.

Durrenberger, E.P. (1983b) Lisu: Political form, ideology and economic action. In: McKinnon, J. and Bhuruksasri, W. (eds) *Highlanders of Thailand*, Oxford University Press, Oxford, 215–226.

Durrenberger, E.P. (1983c) The economy of sufficiency. In: McKinnon, J. and Bhuruksasri, W. (eds) *Highlanders of Thailand*, Oxford University Press, Oxford, 87–98.

Egziabher, T.B.G. (1988) Vegetation and environment of the mountains of Ethiopia: implications for utilization and conservation. *Mountain Research and Development*, **8**, 211–216.

Egziabher, T.B.G. (1991) Management of mountain environments and genetic erosion in tropical mountain systems: the Ethiopian example. *Mountain Research and Development*, **11**, 225–230.

Ehlers, E. (1995) Die Organisation von Raum und Zeit – Bevölkerungswachstum, Ressourcenmanagement und angepaßte Landnutzung im Bagrot/Karakoram. *Petermanns Geographische Mitteilungen*, **139**(2), 105–120.

Ehlers, E. and Kreutzmann, H. (2000) High mountain ecology and economy potential and constraints. In: Ehlers, E. and Kreutzmann, H. (eds) *High Mountain Pastoralism in Northern Pakistan*, Franz Steiner Verlag, Stuttgart, 9–36.

El-Daher, S. and Geissler, C. (1990) North Yemen: from farming to foreign funding. *Food Policy*, **16**, 531–535.

El Mdaghri, C.A. (1995) Women, environment and population: a Moroccan case study. *Institute of Development Studies Bulletin*, **26**, 61–65.

Elsasser, H., Seiler, C. and Scheurer, T. (1995) The regional economic impacts of the Swiss National Park. *Mountain Research and Development*, **15**, 77–80.

Elvidge, C.D. (1979) *Distribution and Formation of Desert Varnish in Arizona*, MSc. Thesis, Arizona State University.

England, P. (1997) UNCED and the implementation of forest policy in Thailand. In: Hirsch, P. (ed.) *Seeing the Forests for Trees: Environment and Environmentalism in Thailand*, Silkworm Books, Chiang Mai, 37–52.

English, P.W. (1968) The origin and spread of qanats in the Old World. *Proceedings of the American Philosophical Society*, **112**, 170–181.

Erickson, C.L. (1992) Prehistoric landscape management in the Andean Highlands: Raised field agriculture and its environmental impact. *Population and Environment: A Journal of Interdisciplinary Studies*, **13**, 285–300.

Euromontana (1997) *The Integration of Environmental Concerns in Mountain Agriculture*, Report for the European Commission DG XI, Environment, Safety and Civil Protection.

Fair, J. (1996) Can ecotourism save Ecuador's threatened cloud forests? In: Collinson, H. (ed.) *Green Guerillas: Environmental Conflict and Initiatives in Latin America and the Caribbean. A Reader*, Latin American Bureau, London, 115–120.

Felmy, S. (1997) *The Voice of the Nightingale: A Personal Account of the Wakhi Culture in Hunza*, Oxford University Press, Karachi.

Ferdmann, J. (1959) Davos und sein Wald. *Bünderwald*, **12**, 161–175.

Ferguson, R.I. (1984) The sediment load of the Hunza River. In: Miller, K.J. (ed.) *The International Karakoram Project*, Vol. 1, 581–598.

Fischlin, A. and Gyalistras, D. (1997) Assessing impacts of climatic change on forests in the Alps. *Global Ecology and Biogeography Letters*, **6**, 19–37.

Fisher, J.F. (1990) *Sherpas: Reflections on Change in Nepal*, University of California Press, Los Angeles.

Fisher, R.J. (1989) *Indigenous Systems of Common Property Forest Management in Nepal*, Environment and Policy Institute, East–West Centre, Working Paper No. 18, Honolulu.

Fisher, R.J. (1995) *Collaborative Management of Forests for Conservation and Development: Issues in Forest Conservation*, IUCN/WWF Report.

Flenley, J. (1979) *Equatorial Rain Forest: A Geological History*, Butterworth, London.

Fogg, W. (1935) Villages and *Suqs* in the High Atlas mountains of Morocco. *Scottish Geographical Magazine*, **51**, 144–151.

Föhn, P. (1991) Les hivers de demain seront-ils blancs comme neige ou vert comme les prés? WSL/FNP (ed.) *Argument de la Recherche*, **3**, 3–12.

Fontela, E. (2000) Bridging the gap between scenarios and models. *Foresight*, **2**, 11–14.

Fookes, P.G., Sweeney, H., Manby, C.N.D. and Martin, R.P. (1985) Geological and geotechnical engineering aspects of low-cost roads in mountainous terrain. *Engineering Geology*, **21**, 1–152.

Forsyth, T. (1998) Mountain myths revisited: integrating natural and social environmental science. *Mountain Research and Development*, **18**, 107–116.

Fortmann, L. and Bruce, J.W. (1988) The daily struggle for rights. In: Fortmann, L. and Bruce, J.W. (eds) *Whose Trees? Proprietary Dimensions of Forestry*, Westview Press, Boulder, CO, 337–341.

Fox, D.J. (1997) Mining in mountains. In: Messerli, B. and Ives, J.D. (eds) *Mountains of the World: A Global Priority*, Parthenon, Carnforth, 171–198.

Fox, J.L., Nurbu, C., Bhatt, S. and Chandola, A. (1994) Wildlife conservation and land-use change in the Transhimalayan region of Ladakh, India. *Mountain Research and Development*, **14**, 39–60.

Frank, R.C. and Lee, R. (1966) *Potential Solar Beam Irradiation on Slopes*, US Department of Agriculture, Forest Service Research Paper, RM 18.

Franz, H. (1979) *Ökologie der Hochgebirge*, Verlag Eugen Ulmer, Stuttgart.

Fricke, T. (1984) *Himalayan Households: Tamang Demography and Domestic Processes*, University of Colombia Press, New York.

Friedl, J. (1984) Education and social mobility: some parallels between Appalachia and the Alps. In: Beaver, P.D. and Purrington, B.L. (eds) *Cultural Adaptations to Mountain Environments*, University of Georgia Press, Atlanta, 38–49.

Frisancho, A.R. (1993) *Human Adaptation and Accommodation*, University of Michigan Press, Ann Arbor.

Fujisaka, S. (1995) Taking farmers' knowledge and technology seriously: upland rice cultivation in the Philippines. In: Warren, D.M., Slikkerveer, L.J. and Brokensha, D. (eds) *Cultural Dimensions of Development: Indigenous Knowledge Systems*, Intermediate Technology Publications, London, 354–370.

Fujisaka, S. (1999) Side-stepped by the Green Revolution: farmers' traditional rice cultivars in the uplands and rainfed lowlands. In: Prain, G., Fujisaka, S. and Warren, M.D. (eds) *Biological and Cultural Diversity: The Role of Indigenous Agricultural Experimentation in Development*, Intermediate Technology Publications, London, 50–63.

Fuller, S. and Gemin, M. (1995) *Proceedings of the Karakoram Workshop*, Skardu, IUCN, Pakistan.

Funnell, D.C. (1994) Intervention and indigenous management. *Land Use Policy*, **11**, 45–54.

Funnell, D.C. and Parish, R. (1995) Environment and economic growth in the Atlas Mountains, Morocco: a policy-orientated research agenda. *Mountain Research and Development*, **15**, 91–100.

Funnell, D.C. and Parish, R. (1999) Complexity, cultural theory and strategies for intervention in the High Atlas of Morocco. *Geografisker Annaler*, **81B**, 131–144.

Funnell, D.C. and Parish, R. (2000) Local knowledge, feedback and action. Problems in the analysis of mountain communities and development strategies. Paper presented at *Colloque International La Montagne et le Savoir*, 18–20 September, 2000, Le Pradel.

Fürer-Haimendorf, C. von (1964) *The Sherpas of Nepal: Buddhist Highlanders*, John Murray, London.

Fürer-Haimendorf, C. von (1975) *Himalayan Traders: Life in Highland Nepal*, Aris and Phillips, Warminster.

Fürer-Haimendorf, C. von (1984) *The Sherpas Transformed: Social Change in a Buddhist Society of Nepal*, Sterling Publications Ltd, New Delhi.

Gamachu, D. (1988) Some patterns of altitudinal variation of climatic elements in the mountainous regions of Ethiopia. *Mountain Research and Development*, **8**, 131–138.

Gamper, J.A. (1981) Tourism in Austria – a case study of the influence of tourism on ethnic relations. *Annals of Tourism Research*, **8**, 432–446.

Ganjanapan, A. (1996) The politics of environment in northern Thailand: ethnicity and highland development programmes. In: Hirsch, P. (ed.) *Seeing the Forests for Trees: Environment and Environmentalism in Thailand*, Silkworm Books, Chiang Mai, 202–222.

Ganjanapan, A. (1998) The politics of conservation and the complexity of local control of forests in the northern Thai highlands. *Mountain Research and Development*, **18**, 71–82.

Garcia-Gonzalez, R., Hidalgo, R. and Montserrat, C. (1990) Patterns of livestock use in time and space in the summer ranges of the Western Pyrenees: a case study in the Aragon valley. *Mountain Research and Development*, **10**, 241–255.

García-Ruiz, J.M. and Lasanta-Martínez, T. (1990) Land-use changes in the Spanish Pyrenees. *Mountain Research and Development*, **10**, 267–279.

García-Ruiz, J.M. and Lasanta-Martínez, T. (1993) Land-use conflicts as a result of land-use change in the Central Spanish Pyrenees: a review. *Mountain Research and Development*, **13**, 295–304.

Gardner, J.S. (1983) Accretion rates on some debris slopes in the Mt Rae area, Canadian Rocky Mountains. *Earth Surface Processes and Landforms*, **17**, 323–343.

Gardner, J.S. and Jones, N.K. (1993) Sediment transport and yield at the Raikot Glacier, Nanga Parbat, Punjab Himalaya. In: J.F. Schroder (ed.) *Himalaya to the Sea*, Routledge, London, 43–71.

Gaventa, J. (1984) Land ownership, power and powerlessness in the Appalachian Mountains. In: Beaver, P.D. and Purrington, B.L. (eds) *Cultural Adaptations to Mountain Environments*, University of Georgia Press, Atlanta, 142–155.

Geiger, R. (1965) *The Climate Near the Ground*, Harvard University Press, Cambridge, MA.

Gerrard, A.J. (1990) *Mountain Environments: an examination of the Physical Geography of Mountains*, Belhaven, London.

Gerrard, A.J. and Gardner, R. (2000) The nature and management implications of landsliding on irrigated terraces in the Middle Hills of Nepal. *International Journal of Sustainable Development and World Ecology*, **7**, 1–7.

Getahun, A. (1984) Stability and instability of mountain ecosystems in Ethiopia. *Mountain Research and Development*, **4**, 39–44.

Gigon, A. (1983) Typology and principles of ecological stability and instability. *Mountain Research and Development*, **3**, 95–102.

Gilles, J.L., Hammoudi, A. and Mahdi, M. (1986) Oukaimedene, Morocco: A high mountain *agdal*. In: *Proceedings of a Conference on Common Property Resource Management*. National Academy of Science, Board of Trade and Technology. National Academy Press, Washington D.C., 281–304.

Gilmour, D.A. (1988) Not seeing the trees for the forest: a re-appraisal of the deforestation crisis in two hill districts of Nepal. *Mountain Research and Development*, **8**, 343–350.

Gilmour, D.A. (1995) Rearranging the trees in the landscape in the Middle Hills of Nepal. In: Arnold, J.E.M. and Dewees, P.A. (eds) *Tree Management in Farmer Strategies: Responses to Agricultural Intensification*, Oxford University Press, Oxford, 21–42.

Gilmour, D.A. and Nurse, M.C. (1991) Farmer initiatives in increasing tree cover in central Nepal. *Mountain Research and Development*, **11**, 329–337.

Glick, T.F. (1996) The Berbers in Valencia: the case of irrigation. In: Chevedden, P.E., Kagay, D.J. and Padilla, P.G. (eds) *Iberia and the Mediterranean World of the Middle Ages*, E.J. Brill, Leiden, 91–206.

Godde, P. (ed.) (1999) *Community-based Mountain Tourism: Practices for Linking Conservation with Enterprise. Synthesis of an Electronic Conference of the Mountain Forum. April 13–May 18, 1998*, The Mountain Institute, Franklin, WV.

Godet, M. (1997) *Manuel de Prospective Stratégiques*, Dunod, Paris.

Godoy, J.C. (1998) The Meso-American biological corridor: a regional tool for transboundary cooperation and peace-keeping efforts. In: *Parks for Peace. Proceedings of an International Conference on Transboundary Protected Areas as a Vehicle for International Cooperation*, Capetown, 16–18 September, 1997. IUCN/WCMC, 248–253.

Goldstein, M.C. and Messerschmidt, D.A. (1980) The significance of latitudinality in Himalayan mountain ecosystems. *Human Ecology*, **8**, 117–134.

Goldstein, M.C., Tsarong, P. and Beall, C.M. (1983) High altitude hypoxia, culture and human fecundity/fertility: a comparative study. *American Anthropologist*, **85**, 28–49.

Good, R. (1995) Ecologically sustainable development in the Australian Alps. *Mountain Research and Development*, **15**, 251–258.

Goudie, A.J. (1995) *The Changing Earth: Rates of Geomorphological Processes*, Blackwell, Oxford.

Goudie, A.S., Brunsden, D., Collins, D.N., Derbyshire, E., Ferguson, R.I., Hashmet, Z., Jones, D.K.C., Perrott, F.A., Said, A., Waters, R. S. and Whalley, W.B. (1984) The geomorphology of the Hunza Valley, Karakoram mountains, Pakistan. In: Miller, K.J. (ed.) *The International Karakoram Project*, Vol. 1, Cambridge University Press, Cambridge, 359–410.

Grandstaff, T.B. (1988) Shifting cultivation in Northern Thailand: possibilities for development. In: Fortmann, L. and Bruce, J.W. (eds) *Whose Trees? Proprietary Dimensions of Forestry*, Westview Press, Boulder, CO, 318–327.

Graumlich, L.J. (1994) Long-term vegetation change in mountain environments: palaeological insights into modern vegetation dynamics. In: Beniston, M. (ed.) *Mountain Environments in Changing Climates*, Routledge, London, 167–179.

Gray, K. (1995) The ambivalence of property. In: Kirkby, J., O'Keefe, P. and Timberlake, L. (eds) *Sustainable Development*, Earthscan, London, 223–226.

Greenwood, D.J. (1975) *Unrewarding Wealth: The Commercialization and Collapse of Agriculture in a Spanish Basque Town*, Cambridge University Press, Cambridge.

Griffin, D.M. (1987) Implementation failure caused by institutional problems. *Mountain Research and Development*, **7**, 230–253.

Grosjean, M., Hofer, T., Liechti, R., Messerli, B., Weingartner, R. and Zumstein, S. (1995) Sediments and soils in the floodplain of Bangladesh: looking up to the Himalayas? In: Schreier, H., Shah, P.B. and Brown, S. (eds) *Challenges in Mountain Resource Management in Nepal: Processes, Trends and Dynamics in Middle Mountain Watersheds*, Proceedings of a workshop held in Kathmandu, 10–12 April, 1995. ICIMOD, Kathmandu, 25–32.

Grötzbach, E. (1984) Mobility of labour in high mountains and the socio-economic integration of peripheral areas. *Mountain Research and Development*, **4**, 229–235.

Grötzbach, E. (1985) The Bavarian Alps: Problems of tourism, agriculture and environment conservation. In: Singh, T.V. and Kaur, J. (eds) *Integrated Mountain Development*, Himalayan Books, New Delhi, 141–155.

Grötzbach, E. (1988) High mountains as human habitat. In: Allan, N.J.R., Knapp, G.W. and Stadel, C. (eds) *Human Impact on Mountain Environments*, Rowman and Littlefield, Towota, NJ, 24–35.

Grötzbach, E. and Stadel, C. (1997) Mounatin peoples and cultures. In: Messerli, B. and Ives, J.D. (eds) *Mountains of the World: A Global Priority*, Parthenon, Carnforth, 17–38.

Grove, J. (1988) *The Century Time-Scale*, Methuen, London.

Grove, J. (1996) The century time-scale. In: Driver, T.S. and Chapman, G.P. (eds) *Time-scales and Environmental Change*, Routledge, London, 39–87.

Gruber, G. (1983) Ecological endurance limits of mountain regions and current dangers through tourism. In: *Deutscher Alpenverenin, Himalaya Konferenz, 1983*, German Alpine Club, Munich, 57–67.

Guha, R. (1989) *The Unquiet Woods: Ecological Change and Peasant Resistance in the Himalaya*, Oxford University Press, Delhi.

Guhl, E. (1968) Los paramos circundantes de la sabaña de Bogata, 511 ecologia y su importancia para el regimen hidrologico de la misma. In: Troll, C. (ed.) *Geo-Ecology of the Mountainous Regions of the Tropical Americas*, Ferd. Dummlers Verlag, Bonn, 195–212.

Guillet, D. (1987) Terracing and irrigation in the Peruvian highlands. *Current Anthropology*, **28**, 409–430.

Guillet, D.W. (1984) Agro-pastoral land use and the tragedy of the commons in the Central Andes. In: Beaver, P.D. and Purrington, B.L. (eds) *Cultural Adaptations to Mountain Environments*, University of Georgia Press, Atlanta, 12–23.

Haagsma, B. (1995) Traditional water management and State intervention: the case of Santo Antao, Cape Verde. *Mountain Research and Development*, **15**, 39–56.

Haigh, M.J., Rawat, J.S., Rawat, M.S., Bartarya, S.K. and Rai, S.P. (1995) Interactions between forest and landslide activity along new highways in the Kumaun Himalaya. *Forest Ecology and Management*, **78**, 173–189.

Haimayer, P. (1989) Glacier-skiing areas in Austria: a socio-political perspective. *Mountain Research and Development*, **9**, 51–58.

Halpin, P.N. (1994) Latitudinal variation in the potential response of mountain ecosystems to climatic change. In: Beniston, M. (ed.) *Mountain Environments in Changing Climates*, Routledge, London, 180–219.

Hamilton, A.C. and Taylor, D. (1992) History of climate and forests in Tropical Africa during the last 8 million years. *Climatic Change*, **19**, 65–78.

Hamilton, L.S. (1995) Mountain cloud forest conservation and research: a synopsis. *Mountain Research and Development*, **15**, 259–266.

Hamilton, L.S. (1998) Guidelines for effective transboundary cooperation: philosophies and best practice. In: *Parks for Peace. Proceedings of an International Conference on Transboundary Protected Areas as a Vehicle for International Cooperation*, Capetown, 16–18 September, 1997. IUCN/WCMC. 27–36.

Hamilton, L.S. and Bruijnzeel, L.A.S. (1997) Mountain watersheds – integrating water, soils, gravity, vegetation and people. In: Messerli, B. and Ives, J.D. (eds) *Mountains of the World: A Global Priority*, Parthenon, Carnforth, 337–370.

Hamilton, L.S., Gilmour, D.A. and Cassells, D.S. (1997) Montane forests and forestry. In: Messerli, B. and Ives, J.D. (eds) *Mountains of the World: A Global Priority*, Parthenon, Carnforth, 281–311.

Hanks, L.M. (1983) The Yuan or Northern Thai. In: McKinnon, J. and Bhruksasri, W. (eds) *The Highlanders of Thailand*, Oxford University Press, Oxford, 101–111.

Hanna, S.R. and Strimantis, D.G. (1990) Rugged terrain effects on diffusion. In: Blumen, W. (ed.) *Atmospheric Processes over Complex Terrain, Meteorological Monograph*, **23** (45) 109–143, American Meteorological Society, Boston.

Hardin, G. (1968) The Tragedy of the Commons. *Science*, **162**, 1243–1248.

Harris, R.B. (1991) Conservation prospects for musk deer and other wildlife in southern Qinghai, China. *Mountain Research and Development*, **11**, 353–358.

Harris, R.B. and Shilai, M. (1997) Initiating a hunting ethic in Lisu villages, Western Yunnan, China. *Mountain Research and Development*, **17**, 171–176.

Haslett, J.R. (1997a) Mountain ecology: organism responses to environmental change, an introduction. *Global Ecology and Biogeography Letters*, **6**, 3–6.

Haslett, J.R. (1997b) Insect communities and the spatial complexity of mountain habitats. *Global Ecology and Biogeography Letters*, **6**, 49–56.

Hastenrath, S. (1981) The climate of Mount Kenya and Kilimanjaro. In: Allen, I. (ed.) *Guide to Mount Kenya and Kilimanjaro*, Mountain Club of Kenya, Nairobi, 36–38.

Heinen, J.T. and Yonzon, P.B. (1994) A review of conservation issues and programs in Nepal: from a single species focus toward biodiversity protection. *Mountain Research and Development*, **14**, 61–76.

Hess, C.G. (1990) 'Moving up – moving down': agro-pastoral land-use patterns in the Ecuadorian paramos. *Mountain Research and Development*, **10**, 333–342.

Hewitt, F. (1989) Woman's work, woman's place: the gendered life-world of a high mountain community in Northern Pakistan. *Mountain Research and Development*, **9**, 335–352.

Hewitt, F. (1998) All paths lead to the hot spring: conviviality, the code of honour, and capitalism in a Karakoram village, Pakistan. *Mountain Research and Development*, **18**, 265–272.

Hewitt, F. (1999) Women of the high pastures and the global economy: reflections on the impacts of modernization in the Hushe Valley of the Karakoram, Northern Pakistan. *Mountain Research and Development*, **19**, 141–151.

Hewitt, K. (1968) The freeze–thaw environment of the Karakoram Himalaya. *Canadian Geographer*, **12**, 85–98.

Hewitt, K. (1982) Natural dams and outburst floods of the Karakoram Himalaya. In: Glen, J.W. (ed.) *Hydrological Aspects of Alpine and High Mountain Areas*, International Association of Hydrological Sciences Publication 138, 21–30.

Hewitt, K. (1988) The study of mountain lands and peoples: an overview. In: Allan, N.R.J., Knapp, G.W. and Stadel, C. (eds) *Human Impact on Mountains*, Rowman and Littlefield, Towota, NJ, 6–23.

Hewitt, K. (1992) Mountain hazards. *GeoJournal*, **27**, 47–60.

Hewitt, K. (1997) Risk and disasters in mountain lands. In: Messerli, B. and Ives, J.D. (eds) *Mountains of the World: A Global Priority*, Parthenon, Carnforth, 371–408.

Hillman, J.C. (1988) The Bale Mountains National Park area, southeast Ethiopia, and its management. *Mountain Research and Development*, **8**, 253–258.

Hirsch, P. (1997) Environment and environmentalism in Thailand: material and ecological bases. In: Hirsch, P. (ed.) *Seeing the Forests for Trees: Environment and Environmentalism in Thailand*, Silkworm Books, Chiang Mai, 15–36.

Hofer, T. (1993) Himalayan deforestation, changing river discharge and increasing floods: myth or reality? *Mountain Research and Development*, **13**, 213–233.

Hofer, T. (1997) Meghalaya, not Himalaya. *Himal, South Asia*, **Sept/Oct**, 52–55.

Hofer, T. (1998) Do land use changes in the Himalayas affect downstream flooding? Traditional understanding and new evidences. *Memoir of the Geological Society of India*, **41**, 119–141.

Holmes, A. and Holmes, D.L. (1978) *Principles of Physical Geology*, 3rd edn, Van Nostrand Reinhold (UK) Co, Wokingham.

Holtmeier, F.-K. (1994) Ecological aspects of climatically caused timberline fluctuations: review and outlook. In: Beniston, M. (ed.) *Mountain Environments in Changing Climates*, Routledge, London, 220–233.

Honsari, M. (1989) *Qanats* and human eco-systems in Iran. In: Beaumont, P., Bonine, P. and McLachlan, K.S. (eds) *Qanat, Kariz and Khattara*, Menas Press, Wisbech.

Hooker, Sir J.D. (1969) *Himalayan Journals: Notes of a Naturalist* (reprint), Today and Tomorrow's Printers and Publishers, New Delhi.

Houghton, J.T., Meira Filho, L.G., Callander, B.A., Harris, N., Kattenberg, A. and Maskell, K. (1996) *Climate Change 1995: The Science of Climate Change*, Cambridge University Press, Cambridge.

Houston, C.S. (1987) Deforestation in Solu Khumbu. *Mountain Research and Development*, 7, 76.

Hughey, K. (1998) *Big Business and the Mountain Environment*, Email posted to Mountain Forum, 9 March 1998. http://www.mtnforum.org

Hulme, M., Conway, D., Kelly, P.M., Subak, S. and Downing, T.E. (1995) *The Impacts of Climate Change in Africa*, Stockholm Environmental Institute, Stockholm.

Humphreys, G.S. and Brookfield, H. (1991) The use of unstable steeplands in the mountains of Papua New Guinea. *Mountain Research and Development*, **11**, 295–318.

Hurni, H. (1989) Late Quaternary of Simen and other mountains in Ethiopia. In: Mahaney, W.C. (ed.) *Quaternary Environmental Research on East African Mountains*, Balkema, Rotterdam, 105–120.

Hynes, A.L., Brown, A.D., Grau, H.R. and Grau, A. (1997) Local knowledge and the use of plants in rural communities in the montane forests of northwestern Argentina. *Mountain Research and Development*, **17**, 263–271.

ICIMOD (1997) *Development of Mini- and Micro-Hydel Projects: Issues and Constraints*, Issues in Mountain Development Discussion Paper (1997/2). http://www.icimod.org.np

IPCC (2001) *Climate Change 2001*, http://www.ipcc.ch/

Isaacs, B.L. (1992) Long-term land surface processes: erosion, tectonics and climate history in mountain belts. In: Mather, P.M. (ed.) *TERRA-1: Understanding the Terrestiral Environment: The Role of Earth Observations from Space*, Taylor and Francis, New York.

IUCN (1990) *IUCN 1990 Red List of Threatened Animals*, IUCN Publications, Gland.

IUCN (1996) *Maintaining Biodiversity in Pakistan with Rural Community Development*, Annual Report, IUCN Pakistan.

IUCN (1997) *Maintaining Biodiversity in Pakistan with Rural Community Development*, Annual Report, IUCN Pakistan.

Ives, J.D. (1985) Mountain environments. *Progress in Physical Geography*, **9**, 425–433.

Ives, J.D. (1987a) The Mountain Lands. In: Clark, M.J., Gregory, K.J. and Gurrell, A.M. (eds) *Horizons in Physical Geography*, Macmillan, Basingstoke, 232–249.

Ives, J.D. (1987b) The theory of Himalayan degradation: its validity and application challenged by recent research. *Mountain Research and Development*, 7, 189–199.

Ives, J.D. (1988a) Development in the face of uncertainty. In: Ives, J.D. and Pitt, D.C. (eds) *Deforestation: Social Dynamics in Watersheds and Mountain Ecosystems*, Routledge, London, 54–74.

Ives, J.D. (1988b) Mapping of Mountain Hazards in Nepal. In: Allan, N.J.R., Knapp, G.W. and Stadel, C. (eds) *Human Impact on Mountains*, Rowman & Littlefield, Towota, NJ, 154–164.

Ives, J.D. and Messerli, B. (1989) *The Himalayan Dilemma: Reconciling Development and Conservation*, Routledge, London.

Ives, J.D. and Messerli, B. (1990) Progress in theoretical and applied mountain research 1973–1989, and major future needs. *Mountain Research and Development*, **10**, 101–127.

Ives, J.D., Messerli, B. and Rhoades, R.E. (1997) Agenda for sustainable mountain development. In: Messerli, B. and Ives, J.D. (eds) *Mountains of the World: A Global Priority*, Parthenon, Carnforth, 455–466.

Ives, J.D., Messerli, B. and Spiess, E. (1997) Mountains of the world – A global priority. In: Messerli, B. and Ives, J.D. (eds) *Mountains of the World: A Global Priority*, Parthenon, Carnforth, 1–15.

Jackson, R. (1996) *Using Megafauna as a Conservation Catalyst*, Mountain Forum Discission Archive Article #080896g. http://www2.mtnforum.org/mtnforum/archives/

Jacquement, M. (1996) From the Atlas to the Alps: Chronicle of a Moroccan migration. *Public Culture*, **8**, 377–388.

Jaroz, L. (1996) Defining deforestation in Madagascar. In: Peet, R. and Watts, M. (eds) *Liberation Ecologies*, Routledge, London, 148–164.

Jeník, J. (1997) The diversity of mountain life. In: Messerli, B. and Ives, J.D. (eds) *Mountains of the World: A Global Priority*, Parthenon, Carnforth, 199–236.

Jodha, N.S. (1990) Mountain agriculture: the search for sustainability. *Journal of Farming Systems and Research Extension*, **1**, 55–75.

Jodha, N.S. (1992a) *Common Property Resources: A Missing Dimension of Development Strategies*, World Bank Discussion Paper 169, World Bank, Washington, D.C.

Jodha, N.S. (1992b) Mountain perspective and sustainability. In: Jodha, N.S., Banskota, M. and Partap, T. (eds) *Sustainable Mountain Agriculture: Volume 1, Perspectives and Issues*, Intermediate Technology Publications, London, 41–80.

Joekes, S. (1995) Gender and livelihoods in Northern Pakistan. *Institute of Development Studies Bulletin*, **26**, 66–74.

Joffe, G. (1992) Irrigation and water supply systems in North Africa. *Moroccan Studies*, **2**, 47–55.

Johnstone, A. (1997) A flash flooding event in the High Atlas Mountains of Morocco. *Geography*, **82**, 85–90.

Jones, S. (1986) Nuristan: lost world of the Hindu Kush. In: Tobias, M. (ed.) *Mountain People*, University of Oklahoma Press, Norman, 77–81.

Joshi, M. and Pant, P.D. (1990) Causes and remedial measures for rockfalls and landslides on Naina Peak, Nainital, Kumaun Himalaya, U.P., India. *Mountain Research and Development*, **10**, 343–351.

Joshi, S.C. and Pant, P. (1989) Environmental implications of the recent growth of tourism in Nainital, Kumaun Himalaya, U.P., India. *Mountain Research and Development*, **9**, 347–351.

Jungerius, P.D., de Mas, P. and van der Wusten, H.H. (1986) Land evaluation: a part of the decision environment of the subsistence farmer in the Rif Mountains, Morocco. *International Institute for Land Reclamation and Improvement*, **40**, 298–309.

Juvik, J.O. and Juvik, S.P. (1984) Mauna Kea and the myth of multiple use: endangered species and mountain management in Hawaii. *Mountain Research and Development*, **4**, 191–202.

Kaltenborn, B.P. (2000) Arctic-Alpine environments and tourism: can sustainability be planned? Lessons learned on Svalbard. *Mountain Research and Development*, **20**, 28–31.

Kaltenborn, B.P., Riese, H. and Hundeide, M. (1999) National Park planning and local participation: some reflections from a mountain region in southern Norway. *Mountain Research and Development*, **19**, 51–61.

Kalter, J. (1991) *The Arts and Crafts of the Swat Valley: Living Traditions in the Hindu Kush*, Thames and Hudson, London.

Karan, P.P. (1987) Population characteristics of the Himalayan region. *Mountain Research and Development*, **7**, 271–274.

Karan, P.P. (1989) Environment and development in Sikkim Himalaya: a review. *Human Ecology*, **17**, 257–271.

Kariel, H.G. (1988) Tourism and recreation developments in the Rocky Mountains of Canada. In: Allan, N.R.J., Knapp, G.W. and Stadel, C. (eds) *Human Impact on Mountain Environments*, Rowman & Littlefield, Towota, NJ, 228–242.

Kariel, H.G. (1989) Socio-cultural impacts of tourism in the Austrian Alps. *Mountain Research and Development*, **9**, 59–70.

Kariel, H.G. (1993) Tourism and society in four Austrian Alpine Communities. *GeoJournal*, **31**, 446–456.

Kariel, H.G. and Draper, D.L. (1992) Outdoor recreation in mountains. *GeoJournal*, **27**, 97–104.

Kattelmann, R. (1987) Uncertainty in assessing Himalayan water resources. *Mountain Research and Development*, **7**, 279–286.

Kattenberg, A., Giogi, F., Grassl, H., Meehl, G.A., Mitchell, J.F.B., Stouffer, R.J., Tokioka, T., Weaver, A.J. and Wigley, T.M.L. (1996) Climate models – projections of future climate. In: Houghton, J.T., Meira Filho, L.G., Callander, B.A., Harris, N., Kattenberg, A. and Maskell, K. (eds) *Climate Change 1995: The Science of Climate Change*, Cambridge University Press, Cambridge, 285–357.

Kaur, J. (1983) The Valley of Flowers: Himalaya's youngest national park in the making. In: Singh, T.V. and Kaur, J. (eds) *Himalayas, Mountains and Men*, Oxford University Press, New Delhi, 333–347.

Keen, F.G.B. (1983) Land use. In: McKinnon, J. and Bhruksasri, W. (eds) *The Highlanders of Thailand*, Oxford University Press, Oxford, 293–306.

Keiter, R. (1997) *Jurisdictional Authority – National vs Local Power*, Email posted to Mountain Forum, 10 April 1997. http://www.mtnforum.org

Kessler, J.J. (1995) Mahjur areas: traditional rangeland reserves in the Dhamar montane plains (Yemen Arab Republic). *Journal of Arid Environments*, **29**, 395–401.

Kharel, F.R. (1997) Agricultural crop and livestock depredation by wildlife in Langtang National Park, Nepal. *Mountain Research and Development*, **17**, 127–134.

Kienholz, H. (1984) Natural hazards: a growing menace? In: Brugger, E.A., Fürrer, G., Messerli, B. and Messerli, P. (eds) *The Transformation of Swiss Mountain Ranges*, Verlag Paul Haupt, Bern, 386–404.

Kienholz, H., Hafner, H., Schneider, G. and Tamrakar, R. (1984) Mountain hazards mapping on Nepal's Middle Mountains. Maps of land use and geomorphic damages (Kathmandu-Kakani area). *Mountain Research and Development*, **3**, 195–220.

Kienholz, H., Schneider, G., Bichsel, M., Grunder, M. and Mool, P. (1984) Mapping of mountain hazards and slope stability. *Mountain Research and Development*, **4**, 247–266.

King, B.C. (1949) *The Napak Area of Southern Karamoja, Uganda: A Study of a Dissected Late Tertiary Volcano*, Geological Survey of Uganda, Kampala.

Kirch, P.V., Flenley, J.R., Steadman, D.W., Lamont, F. and Dawson, S. (1992) Ancient environmental degradation: prehistoric human impacts on an island ecosystem: Mangaia, Central Polynesia. *National Geographic Research & Exploration*, **8**, 166–179.

Klötzli, F. (1997) Biodiversity and vegetation belts in tropical and subtropical mountains. In: Messerli, B. and Ives, J.D. (eds) *Mountains of the World: A Global Priority*, Parthenon, Carnforth, 232–235.

Knight, J. (1997) *Tatari*: livelihood and danger in upland Japan. In: Seeland, K. (ed.) *Nature is Culture: Indigenous Knowledge and Socio-Cultural Aspects of Trees and Forests in Non-European Cultures*, Intermediate Technology Publications, London, 28–43.

Kondolf, G.M. (1994) Livestock grazing and habitat for a threatened species: land-use decisions under scientific uncertainty in the White Mountains, California, USA. *Environmental Management*, **18**, 501–509.

Körner, C. (1993) Scaling from species to vegetation: the usefulness of functional groups. In: Schulze, E.D. and Mooney, H.A. (eds) *Biodiversity and Ecosystem Function*, Springer, Berlin, 117–140.

Körner, C. (1994) Impact of atmospheric changes on high mountain vegetation. In: Beniston, M. (ed.) *Mountain Environments in Changing Climates*, Routledge, London, 155–166.

Körner, C. and Cochrane, P. (1983) Influence of plant physiognomy on leaf temperature on clear midsummer days in the Snowy Mountains, south-eastern Australia. *Acta Oecol., Oecol. Plant.*, **4**, 117–124.

Körner, C. and Renhardt, U. (1987) Dry matter partitioning and root length/leaf area ratios in herbaceous perennial plants with diverse altitudinal distribution. *Oecologia*, **74**, 411–418.

Korovkin, T. (1991) Peasants, grapes and corporations: the growth of contract farming in a Chilean community. *Journal of Peasant Studies*, **19**, 228–254.

Kreutzmann, H. (1988) Oases of the Karakoram: Evolution of irrigation and social organization in Hunza, Northern Pakistan. In: Allan, N.J.R., Knapp, G.W. and Stadel, C. (eds) *Human Impact on Mountains*, Rowman & Littlefield, Towota, NJ, 243–254.

Kreutzmann, H. (1991) The Karakoram Highway: the impact of road construction on mountain societies. *Modern Asian Studies*, **25**, 711–736.

Kreutzmann, H. (1993a) Challenge and response in the Karakoram: socio-economic transformation in Hunza, Northern Areas, Pakistan. *Mountain Research and Development*, **13**, 19–39.

Kreutzmann, H. (1993b) Development trends in the high mountain regions of the Indian subcontinent: a review. *Applied Geography and Development*, **42**, 39–59.

Kreutzmann, H. (1994) Habitat conditions and settlement processes in the Hindukush-Karakoram. *Petermanns Geographische Mitteilungen*, **138**, 337–356.

Kreutzmann, H. (1995) Globalization, spatial integration and sustainable development in northern Pakistan. *Mountain Research and Development*, **15**, 213–227.

Kreutzmann, H. (1998) From water towers of mankind to livelihood strategies of mountain dwellers: approaches and perspectives for high mountain research. *Erdkunde*, **52**, 185–200.

Kreutzmann, H. (2000) Livestock economy in Hunza: societal transformation and pastoral practices. In: Ehlers, E. and Kreutzmann, H. (eds) *High Mountain Pastoralism in Northern Pakistan*, Erdkundliches Wissen 132, Franz Steiner Verlag, Stuttgart, 89–120.

Krippendorf, J. (1984) The capital of tourism in danger: reciprocal effects between landscape and tourism. In: Brugger, E.A., Fürrer, G., Messerli, B. and Messerli, P. (eds) *The Transformation of Swiss Mountain Ranges*, Paul Haupt Verlag, Bern, 427–449.

Krippendorf, J. (1986) *Alpsegen Alptraum: Für eine Tourismus-Entwickling im Einklang mit Mensch und Natur*, Kümmerley & Frey, Bern.

Kubler, G. (1952) *The Indian Caste in Peru, 1795–1940. A Population Study Based upon Tax Records and Census Reports*, Smithsonian Institution Institute of Social Anthropology, Washington, D.C.

Kunstadter, P. (1983) Highland populations of Northern Thailand. In: McKinnon, J. and Bhruksasri, W. (eds) *Highlanders of Thailand*, Oxford University Press, Oxford, 15–45.

Kupfer, J.A. and Cairns, D.M. (1996) The suitability of montane ecotones as indicators of global climatic change. *Progress in Physical Geography*, **20**, 253–272.

Kushner, G. (1984) Reflections on mountain adaptations. In: Beaver, P.D. and Purrington, B.L. (eds) *Cultural Adaptations to Mountain Environments*, University of Georgia Press, Atlanta, 168–178.

Kutsch, H. (1982) *Principal Features of a Form of Water-Concentrating Culture on Small-Holdings with Special Reference to the Anti Atlas*, Trierer Geographische Studien, No 5, Trier.

Ladurie, E. Le Roy (1972) *Times of Feast, Times of Famine: A History of Climate Since the Year 1000*, Allen & Unwin, London.

Lama, S. (1997) *Where does Macro meet Micro?* Email posted to Mountain Forum, 13 October 1997. http://mtnmorum.org

Lamb, H.F., Damblon, F. and Maxted, R.W. (1991) Human impact on the vegetation of the Middle Atlas, Morocco, during the last 5000 years. *Journal of Biogeography*, **18**, 519–532.

Lamb, H.F., Eicher, U. and Switsur, V.R. (1989) An 18 000-year record of vegetation, lake-level and climatic change from Tigalmamine, Middle Atlas, Morocco. *Journal of Biogeography*, **16**, 65–74.

Lamb, H.H. (1988) *Weather, Climate and Human Affairs*, Routledge, London.

Lambton, A.K.S. (1991) *Landlord and Peasant in Persia: A Study of the Land Tenure and Land Revenue Administration*, I.B. Tauris, London.

Lampe, K.J. (1983) Rural development in mountainous areas: why is progress so difficult? *Mountain Research and Development*, **3**, 125–129.

Lanjouw, A. (1999) Mountain gorilla tourism in Central Africa. *Mountain Forum Bulletin*, **3**, 7–8.

Laouina, A. (1994) L'érosion en milieu mediterranéen: une crise environmentale? In: *La Dynamique de l'Environnement en Afrique*, Actes du 2ème Congrès des Géographes Africains, Rabat, Morocco. Colloques de Rabat et Agadir 19–24 April, 1993, 191–200.

Lauer, W. (1981) Ecoclimatological conditions of the *paramo* belt in the tropical high mountains. *Mountain Research and Development*, **1**, 209–221.

Lauer, W. (1993) Human development and environment in the Andes: a geoecological overview. *Mountain Research and Development*, **13**, 157–166.

Le Houérou, H.N. (1987) Vegetation wildfires in the Mediterranean Basin: evolution and trends. *Ecologia Mediterranea*, XIII, 12.

Leach, M., Joekes, S. and Green, C. (1995) Gender relations and environmental change. *Institute of Development Studies Bulletin*, **26**, 1–8.

Leach, M., Mearns, R. and Scoones, I. (1997) *Environmental Entitlements: A Framework for Understanding the Institutional Dynamics of Environmental Change*, Working Paper 3, Institute of Development Studies, Brighton.

Lees, S.H. (1974) The State's use of irrigation in changing peasant societies. In: Downing, T.E. and Gibson, McG. (eds) *Irrigation's Impact on Society*, University of Arizona Press, Tucson, 123–128.

Leibundgut, C. (1984) Hydrologic potential – changes and stresses. In: Brugger, E.A., Fürrer, G., Messerli, B. and Messerli, P. (eds) *The Transformation of Swiss Mountain Ranges*, Verlag Paul Haupt, Bern, 168–193.

Lewis, P. (1992) Basic themes in Akha culture. In: Walker, A.R. (ed.) *The Highland Heritage: Collected Essays on Upland North Thailand*, Suvarnabhumi Books, Singapore, 207–224.

Lewis, P. and Lewis, E. (1984) *Peoples of the Golden Triangle: Six Tribes in Thailand*, River Books, Bangkok.

Lichtenberger, E. (1988) The succession of an agricultural society to a leisure society. In: Allan, N.R.J., Knapp, G.W. and Stadel, C. (eds) *Human Impact on Mountain Environments*, Rowman & Littlefield, Towota, NJ, 218–227.

Liniger, H. (1988) Water conservation for rain-fed farming in the semi-arid foothills west and northwest of Mount Kenya: consequences for soil productivity. *Mountain Research and Development*, **8**, 201–209.

Loayza, R. and Rist, S. (2000) The struggle for local independence and self-determination; an interview with Don Roman Loayza, the first peasant elected to the Bolivian government. *Mountain Research and Development*, **20**, 16–19.

Long, A. (1995) The importance of tropical montane cloud forests for endemic and threatened birds. In: Hamilton, L.S., Juvík, J.O. and Scatena, F.N. (eds) *Tropical Montane Cloud Forests*, Springer Verlag, New York, 79–106.

Lorimer, E.O. (1938) *Language Hunting in the Karakoram* (reprinted 1989), Indus Publications, Karachi.

Luckman, B.H. (1978) Geomorphic work of snow avalanches in the Canadian Rocky Mountains. *Arctic and Alpine Research*, **10**, 261–276.

Luckman, B.H. (1988) Dating the moraines and recession of Athabasca and Dome Glaciers, Alberta, Canada. *Arctic and Alpine Research*, **20**, 40–54.

Luckman, B.H. (1994) Using multiple high-resolution proxy climate records to reconstruct natural climate variability: an example from the Canadian Rockies. In: Beniston, M. (ed.) *Mountain Environments in Changing Climates*, Routledge, London, 42–59.

Lynch, O.J. and Maggio, G.F. (eds) (2000) *Mountain Laws and Peoples: Moving Towards Sustainable Development and Recognition of Community-Based Property Rights. A General Overview of Mountain Laws and Policies with Insights from the Mountain Forum's Electronic Conference on Mountain Policy and Law*, The Mountain Institute, Harrisonburg, VA.

MacDonald, K.I. (1996) Population change in the Upper Braldu Valley, Baltistan, 1900–1990: All is not as it seems. *Mountain Research and Development*, **16**, 351–366.

Mahdi, M. (1986) Private rights and collective management of water in a High Atlas Berber tribe. In: *Proceedings of a Conference on Common Property Resource Management*, National Academy of Science, Board of Trade and Technology, National Academy Press, Washington D.C., 181–197.

Maro, P.S. (1988) Agricultural land management under population pressure: the Kilimanjaro experience, Tanzania. *Mountain Research and Development*, **8**, 273–282.

Maunch, S.P. (1983) Key processes for stability and instability of mountain ecosystems: is the bottleneck really a data problem? *Mountain Research and Development*, **3**, 113–119.

Maurer, G. (1992) Agriculture in the Rif and Tell mountains of North Africa. *Mountain Research and Development*, **12**, 337–347.

Maurer, G. (1996) L'homme et les montagnes atlasiques au Maghreb. *Ann. Geo.*, **587**, 47–72.

Mawdsley, E. (1999) A new Himalayan State in India: popular perceptions of regionalism, politics and development. *Mountain Research and Development*, **19**, 101–112.

Maxwell, G. (1966) *Lords of the Atlas: The Rise and Fall of the House of Glaoua 1893–1956*, Longman, London.

McConnell, R.M. (1991) Solving environmental problems caused by adventure travel in developing countries: the Everest environmental expedition. *Mountain Research and Development*, **11**, 359–366.

McNamee, K. (1994) A retrospective on the Tatshenshini campaign: World Class Wilderness versus World Class Ore in northern protected areas and wilderness. In: Peepre, J. and Jickling, B. (eds) *Canadian Parks and Wilderness Society*, 379.

McNeill, J.R. (1992) Kif in the Rif: a historical and ecological perspective of marijuana, markets and manure in northern Morocco. *Mountain Research and Development*, **12**, 389–392.

McNeill, J.R. (1992) *Mountains of the Mediterranean World*, Cambridge University Press, Cambridge.

Mehta, M. (1995) *Cultural Diversity in the Mountains: Issues of Integration and Marginality in Sustainable Development*, Paper prepared for the Consultation on the Mountain Agenda, Lima, Peru, February 22–27.

Mellor, M. (1979) Ice and snow at high altitudes. In: Webber, P.J. (ed.) *High Altitude Geoecology*, Westview Press, Boulder, CO, 75–95.

Menzies, N. (1988) A survey of customary law and control over trees and wildlands in China. In: Fortmann, L. and Bruce, J.W. (eds) *Whose Trees? Proprietary Dimensions of Forestry*, Westview Press, Boulder, CO, 51–62.

Mesoscale Alpine Programme. http://www.map.ethz.ch/form/w4h/map.html accessed 1 April 2000.

Messerli, B. and Hofer, T. (1995) Assessing the impact of anthropogenic land use change in the Himalayas. In: Chapman, G.P. and Thompson, M. (eds) *Water and the Quest for Sustainable Development in the Ganges Valley*, Mansell Publishing Ltd, New York, 64–89.

Messerli, B. and Ives, J.D. (1997) (eds) *Mountains of the World: A Global Priority*, Parthenon, Carnforth.

Messerli, B. and Winiger, M. (1980) The Saharan Uplands and East African Uplands during the Quaternary. In: Williams, M. and Faure, H. (eds) *The Sahara and the Nile*, Balkema, Rotterdam, 87–132.

Messerli, B. and Winiger, M. (1992) Climate, environmental change, and resources of the African mountains from the Mediterranean to the Equator. *Mountain Research and Development*, **12**, 315–336.

Messerli, B., Grosjean, M. and Vuille, M. (1997) Water availability, protected areas and natural resources in the Andean desert altiplano. *Mountain Research and Development*, **17**, 229–238.

Messerli, B., Grosjean, M., Bonani, G., Burgi, A., Geyh, M.B., Graf, K., Ramseyer, K., Romero, H., Schotterer, U., Schreier, H. and Vuille, M. (1993) Climate change and natural resource dynamics of the Atacama Altiplano during the last 18 000 years: a preliminary synthesis. *Mountain Research and Development*, **13**, 117–127.

Metz, J.J. (1990) Conservation practices at an upper-elevation village of West Nepal. *Mountain Research and Development*, **10**, 7–15.

Miehe, S., Cramer, T., Jacobsen, J.-P. and Winiger, M. (1996) Humidity conditions in the Western Karakoram as indicated by climatic data and corresponding distribution patterns of the montane and alpine vegetation. *Erdkunde*, **50**, 190–204.

Miller, J.A. (1984) *Imlil: A Moroccan Mountain Community in Change*, Westview Press, Boulder, CO.

Miller, K.R. (1997) *Approaches to Biodiversity Policy*, Email posted to Mountain Forum, 6 June 1997. http://www,mtnforum.org

Milton, K. (1996) *Environmentalism and Cultural Theory: Exploring the Role of Anthropology in Environmental Discourse*, Routledge, London.

Moldenhauer, W.C. and Hudson, N.W. (1988) (eds) *Conservation Farming on Steep Lands*, Soil and Water Conservation Society, Ankeny.

Moore, D. and Sehgal, D. (1999) Forests for the future: an indigenous, integrated approach to managing temperate watershed resources in Oregon. In: Wolvekamp, P. (ed.) *Forests for the Future: Local Strategies for Forest Protection, Economic Welfare and Social Justice*, Zed Books, London, 167–187.

Moore, D.S. (1996) Marxism, culture, and political ecology: environmental struggles in Zimbabwe's eastern highlands. In: Peet, R. and Watts, M. (eds) *Liberation Ecologies*, Routledge, London, 125–147.

Moore, H.M., Fox, H.R., Harrouni, M.C. and El Alami, A. (1998) Environmental challenges in the Rif mountains, northern Morocco. *Environmental Conservation*, 25, 354–365.

Morris, D. (1997) The mountaineer's role. In: Messerli, B. and Ives, J.D. (eds) *Mountains of the World: A Global Priority*, Parthenon, Carnforth, 257–258.

Moser, W. (1987) Chronik von MAB-6 Obergurgl. In: Patzelt, G. (ed.) *MAB-Projekt Obergurgl*, Veröffentlichungen des Oesterreichischen MAB-Programms 10. Unisversitätsverlag Wagner, Innsbruck, 7–24.

Moser, W. and Peterson, J. (1988) Limits to Obergurgl's growth: an alpine experience in environmental management. In: Allan, N.R.J., Knapp, G.W. and Stadel, C. (eds) *Human Impact on Mountain Environments*, Rowman & Littlefield, Towota, NJ, 201–212.

Mosimann, T. (1991) *Beschneiungsanlagen in der Schwiez*, Geographisches Institut, Hannover.

Moss, L.A.G. (1987) *Santa Fe, New Mexico, Post-industrial Culture Based Town: Myth or Model?* Department of Economic Development and Trade, Government of Alberta, Edmonton and International Cultural Resources Institute, Santa Fe.

Moss, L.A.G. (1994) Beyond tourism: the amenity migrants. In: Mannermaa, M., Inayatullah, S. and Slaugther, R. (eds) *Chaos in Our Uncommon Futures*, University of Economics, Turku. 121–128.

Mountain Agenda (1997) *Mountains of the World: Challenges for the 21ˢᵗ Century*, Contribution to Chapter 13, Agenda 21, Paul Haupt Verlag, Bern.

Mountain Forum (1998) Snow leopard conservation, livestock depredation and compensation schemes. *Mountain Forum Bulletin*, 1, 8–13.

Mountain Forum (1999a) Environment and tourism in Ladakh. *Mountain Forum Bulletin*, 3, 8–10.

Mountain Forum (1999b) Mountains of the world: tourism and sustainable mountain development. *Mountain Forum Bulletin*, 3, 1–5.

Mriouah, D. (1992) Planification de resources en eau au Maroc. *Revue Canadienne d'études du développement, Numéro Spécial*.

Muir, J. (1975) *John of the Mountains*, University of Wisconsin Press, Madison.

Müller-Böker, U. (1988) Spatial organization of a caste society: the example of the Newar in the Kathmandu Valley, Nepal. *Mountain Research and Development*, 8, 23–31.

Mundy, M. (1989) Irrigation and society in a Yemeni valley: on the life and death of a bountiful resource. *Peuples mediterranéens*, 46, 97–128.

Munro, L.T. (1989) Technology choice in Bhutan: labour shortage, aid dependence and a mountain environment. *Mountain Research and Development*, 9, 15–23.

Murthy, Y.K. (1981) Water resource potentials of the Himalaya. In: *The Himalaya: Aspects of Change*, Oxford University Press, Delhi, 152–171.

Narayana, V.V.D. (1987) Downstream impacts of soil conservation on the Himalayan region. *Mountain Research and Development*, 7, 287–298.

Naval Intelligence Division (1942) *Morocco, Volume II*, Geographical Handbook Series. B.R. 506A. HMSO, London.

Nayab, D.-E. and Ibrahim, S. (1994) The appropriateness of the community-based programme: a case-study of the AKRSP in two villages in Gilgit District. *The Pakistan Development Review*, 33, 251–254.

Negi, A.K., Bhatt, B.P., Todaria, N.P. and Saklani, A. (1997) The effects of colonialism on forests and the local people in the Garhwal Himalaya, India. *Mountain Research and Development*, 17, 159–168.

Negi, G.C.S. (1994) High yielding vs. traditional crop varieties: a socio-agronomic study in a Himalayan village in India. *Mountain Research and Development*, **14**, 251–254.

Netting, R.M. (1972) Of men and meadows: strategies of alpine land use. *Anthropological Quarterly*, **45**, 132–144.

Netting, R.M. (1974) The system nobody knows: village irrigation in the Swiss Alps. In: Downing, T.E. and Gibson, McG. (eds) *Irrigation's Impact on Society*, University of Arizona Press, Tucson, 67–75.

Netting, R.M. (1981) *Balancing on an Alp: Ecological Change and Continuity in a Swiss Mountain Community*, Cambridge University Press, Cambridge.

Norberg-Hodge, H. and Page, J. (1983) Unscientific observations. In: Kantowsky, D. and Sander, R. (eds) *Recent Research on Ladakh*, Weltforum, Munich, 263–268.

Nusser, M and Clemens, J. (1996) Impacts on mixed mountain agriculture in the Rupal Valley, Nanga Parbat, Northern Pakistan. *Mountain Research and Development*, **16**, 117–133.

Oberlander, T.M. (1994) Rock varnish in deserts. In: Abrahams, A.D. and Parsons, A.J. (eds) *Geomorphology of Desert Environments*, Chapman and Hall, London, 106–119.

OFDA (US Office of Foreign Disaster Assistance) (1988) *Disaster History: Significant Data on Major Disasters Worldwide, 1900-Present*, Agency for International Development, Washington D.C.

Olsen, C.S. and Helles, F. (1997) Medicinal plants, markets and margins in the Nepal Himalaya: trouble in Paradise. *Mountain Research and Development*, **17**, 363–374.

Oniang'o, R. (1995) The impact of out-migration on household livelihoods and on the management of natural resources: A Kenyan case study. *Institute of Development Studies Bulletin*, **26**, 54–60.

Osborn, G.D. and Luckman, B.H. (1988) Holocene glacier fluctuations in the Canadian Cordillera (Alberta and British Columbia). *Quaternary Science Reviews*, 7, 115–128.

Osmaston, H. (1994) The geology, geomorphology and Quaternary history of Zangskar. In: Crook, J. and Osmaston, H. (eds) *Himalayan Buddhist Village: Environment, Resources, Society and Religious Life in Zangskar, Ladakh*, University of Bristol Press, Bristol, 1–36.

Osmaston, H., Frazer, J. and Crook, S. (1994) Human adaptation to environment in Zangskar. In: Crook, J. and Osmaston, H. (eds) *Himalayan Buddhist Village: Environment, Resources, Society and Religious Life in Zangskar, Ladakh*, University of Bristol Press, Bristol, 37–110.

Osmaston, H.A. (1989a) Glaciers, glaciations and equilibrium line altitudes on Kilimanjaro. In: Mahaney, W.C. (ed.) *Quaternary Environmental Research on East African Mountains*, Balkema, Rotterdam, 17–30.

Osmaston, H.A. (1989b) Glaciers, glaciations and equilibrium line altitudes on the Ruwenzori. In: Mahaney, W.C. (ed.) *Quaternary Environmental Research on East African Mountains*, Balkema, Rotterdam, 31–105.

Ostrom, E. (1985) The rudiments of a revised theory of the origins, survival and performance of institutions for collective action. Working Paper 32, *Workshops on Political Theory and Policy Analysis*, Indiana University, Bloomington.

Ostrom, E. (1991) *Governing the Commons: The Evolution of Institutions for Collective Action*, Cambridge University Press, Cambridge.

Ott, S. (1992) 'Indarra': some reflections on a Basque concept. In: Peristiany, J.G. and Pitt-Rivers, J. (eds) *Honour and Grace in Anthropology*, Cambridge University Press, Cambridge.

Pacey, A. and Cullis, A. (1986) *Rainwater Harvesting: The Collection of Rainfall and Runoff in Rural Areas*, Intermediate Technology Publications, London.

Parish, R. (1999) The unseen, unknown and misunderstood: complexities of development in Hunza, Pakistan. *International Journal of Sustainable Development and World Ecology*, **6**, 1–16.

Parish, R. and Funnell, D.C. (1996) Land, water and development in the High Atlas and Anti Atlas Mountains of Morocco. *Geography*, **81**, 142–154.

Parish, R. and Funnell, D.C. (1999) Climate change in mountain regions: some possible consequences in the Moroccan High Atlas. *Global Environmental Change*, **9**, 45–58.

Parker, T.A. and Carr, J.L. (1992) (eds) *Status of Forest Remnants in the Cordillera de la Costa and Adjacent Areas of Southwestern Ecuador (Rapid Assessment Programme)*, Conservation International, Washington D.C.

Partap, T. (1999) Sustainable land management in marginal mountain areas of the Himalayan region. *Mountain Research and Development*, **19**, 251–260.

Patzelt, G. (1974) Holocene variations of glaciers in the Alps. In: *Colloques internationaux du Centre National de la Recherche Scientifique, No 219: Les méthodes quantitatives d'étude des variations du climat au cours du pléistocène*. CNRS, Paris.

Pawson, I.G. and Jest, C. (1978) The high-altitude areas of the world and their cultures. In: Baker, P.T. (ed.) *The Biology of High Altitude Peoples*, Cambridge University Press, Cambridge, 17–45.

Pearce, F. (1991) Building a disaster: the monumental folly of India's Tehri Dam. *The Ecologist*, **21**, 123–128.

Pendle, G. (1967) *A History of Latin America*, Penguin, Middlesex.

Penz, H. (1988) The importance, status and structure of *Almwirtschaft* in the Alps. In: Allan, N.R.J., Knapp, G.W. and Stadel, C. (eds) *Human Impact on Mountain Environments*, Rowman & Littlefield, Towota, NJ, 109–115.

Perez, M.R. and Saez, A.V. (1990) Transhumance with cows as a rational land use option in the Gredos Mountains (Central Spain). *Human Ecology*, **18**, 187–202.

Perla, R.I. and Martinelli, M. (Jr) (1976) *Avalanche Handbook*, Agricultural Handbook 489. US Department of Agriculture (Forest Service), Washington D.C.

Peters, R.L. and Darling, J.D.S. (1985) The greenhouse effect and nature reserves: global warming would diminish biological diversity by causing extinctions among reserve species. *Bioscience*, **35**, 707–717.

Petley, D.N. (1998) Geomorphological mapping for hazard assessment in a neotectonic terrain. *The Geographical Journal*, **164**, 183–201.

Petts, G. (1990) Water, engineering and landscape: development, protection and restoration. In: Cosgrove, D. and Petts, G. (eds) *Water, Engineering and Landscape: Water Control and Landscape Transformation in the Modern Period*, Belhaven Press, London, 188–208.

Pfister, C. (1983) Changes in stability and carrying capacity of lowland and highland agro-systems in Switzerland in the historical past. *Mountain Research and Development*, **3**, 291–297.

Pfister, C. (1986) Bevölkerung, Wirtschaft und Ernährung in den Berg- und Talgebieten des Kantons Bern 1760–1860. In: Mattmüller, M. (ed.) *Wirtschaft und Gesellschaft in Berggebieten*, Itinera Fac 5/6 Basle, 361–391.

Pfister, C. (1994) Climate in Europe during the late maunder minimum period (1675–1715). In: Beniston, M. (ed.) *Mountain Environments in Changing Climates*, Routledge, London, 60–90.

Phillips, J. (1997) Resource access, environmental struggles and human rights in Honduras. In: Johnston, B. (ed.) *Life and Death Matters: Human Rights and the Environment at the End of the Millennium*, Altamira Press, London, 173–184.

Pils, M., Glauser, P. and Siegrist, D. (1996) *The Alps, a Touchstone for Europe*, Green Paper on the Alps, Nature Friends International, Vienna.

Pinter, N. and Brandon, M.T. (1997) How erosion builds mountains. *Scientific American*, April, 60–65.

Planck, U. (1987) Issues of water in agrarian reform legislation of the Near East. *Land Reform, Land Settlement and Co-operation*, 1–2, 58–82.

Polunin, O. and Walkers, M. (1985) *A Guide to the Vegetation of Britain and Europe*, Oxford University Press Oxford.

Poudel, D.D., Nissen, T.M. and Midmore, D.J. (1999) Sustainability of commercial vegetable production under fallow systems in the uplands of Mindanao, the Philippines. *Mountain Research and Development*, **19**, 41–50.

Powell, J.W. (1895) Physiographic Features. *National Geographic Society Monograph*, **1**, 34–40.

Preston, D. (1969) The revolutionary landscape of Highland Bolivia. *The Geographical Journal*, **135**, 1–16.

Preston, D. (1998) Post-peasant capitalist graziers: the 21$^{st}$ century in southern Bolivia. *Mountain Research and Development*, **18**, 151–158.

Preston, D., Macklin, M. and Warburton, M. (1997) Fewer people, less erosion: the twentieth century in southern Bolivia. *The Geographical Journal*, **163**, 198–205.

Preston, L. (1997) *Innovative Mechanisms and Promising Examples for Financing Conservation and Sustainable Development*, Synthesis of a Mountain Forum Electronic Conference in Support of the Mountain Agenda, The Mountain Institute, Franklin, WV.

Price, L.W. (1981) *Mountains and Man: A Study of Process and Environment*, University of California Press, Berkeley.

Price, M.F. (1985) Impacts of recreational activities on alpine vegetation in western North America. *Mountain Research and Development*, **5**, 263–277.

Price, M.F. (1991) An assessment of patterns of use and management of mountain forests in Colorado, USA: implications for future policies. *Mountain Research and Development*, **11**, 57–64.

Price, M.F. (1992) Patterns of the development of tourism in mountain environments. *GeoJournal*, **27**, 87–96.

Price, M.F. (1996) People in Biosphere Reserves: an evolving concept. *Society and Natural Resources*, **9**, 645–654.

Price, M.F. (1999) *Cooperation in the European Mountains. 1. The Alps*, Environmental Research Series, IUCN, Cambridge.

Price, M.F. and Barry, R.G. (1997) Climate change. In Messerli, B. and Ives, J.D. (eds) *Mountains of the World: A Global Priority*, Parthenon, Carnforth, 409–445.

Price, M.F. and Thompson, M. (1997) The complex life: human land uses in mountain ecosystems. *Global Ecology and Biogeographical Letters*, **6**, 77–90.

Price, M.F., Moss, L.A.G. and Williams, P.W. (1997) Tourism and amenity migration. In Messerli, B. and Ives, J.D. (eds) *Mountains of the World: A Global Priority*, Parthenon, Carnforth, 249–280.

Quiroz, C. (1999) Farmer experimentation in a Venezuelan Andean group. In: Prain, G., Fujisaka, S. and Warren, M.D. (eds) *Biological and Cultural Diversity: The Role of Indigenous Agricultural Experimentation in Development*, Intermediate Technology Publications, London, 113–124.

Rald, J. and Rald, K. (1988) Rural organization in Bukoba District, Tanzania. In: Fortmann, C. and Bruce, J.W. (eds) *Whose Trees? Proprietary Dimensions of Forestry*, Westview Press, Boulder, CO, 96–105.

Rana, J.M. (1993) *Review of the Floods of Bangladesh: A Case Study*, CERG, University of Geneva.

Rapp, A (1960) Recent development of mountain slopes in Karkevegge and surroundings, northern Scandinavia. *Geografisker Annaler*, **42**, 65–200.

Ratzel, F. (1882) *Anthropo-Geographie, oder Grunzuge der Andwendung der Erdkunde auf die Gesichte*, Engehorn Stuttgart.

Rawat, D.S. and Sharma, S. (1997) The development of a road network and its impact on the growth of infrastructure: a study of Almora District in the central Himalaya. *Mountain Research and Development*, **17**, 117–126.

Refass, M.A. (1992) Historical migration patterns in the Eastern Rif Mountains. *Mountain Research and Development*, **12**, 383–388.

Renard, R. (1997) The making of a problem: narcotics in mainland southeast Asia. In: McCaskill, D. and Kampe, K. (eds) *Development or Domestication? Indigenous Peoples of Southeast Asia*, Silkworm Books, Chiang Mai, 307–328.

Renaud, F. (1997) Financial cost-benefit analysis of soil conservation practices in northern Thailand. *Mountain Research and Development*, **17**, 11–18.

Renaud, F., Bechstedt, H.-D. and Nakorn, U.N. (1998) Farming systems and soil-conservation practices in a study area of northern Thailand. *Mountain Research and Development*, **18**, 345–356.

Retzer, J.L. (1974) Alpine soils. In: Ives, J.D. and Barry, R.G. (eds) *Arctic and Alpine Environments*, Methuen, London, 771–802.

Rhoades, R. (1985) *Traditional Potato Production in Farmers' Selection of Varieties in Eastern Nepal*, Potatoes in Food Systems Research Series, Report No 2. International Potato Center, Lima, Peru.

Rhoades, R. and Bebbington, A. (1995) Farmers who experiment: an untapped resource for agricultural research and development. In: Warren, D.M., Slikkerveer, L.J. and Brokensha, D. (eds) *Cultural Dimensions of Development: Indigenous Knowledge Systems*, Intermediate Technology Publications, London, 296–307.

Rhoades, R.E. (1992) Thinking globally, acting locally: technology for sustainable mountain agriculture. In: Jodha, N.S., Banskota, M. and Partap, T. (eds) *Sustainable Mountain Agriculture. Volume 1: Perspectives and Issues*, Intermediate Technology Publications, London.

Rhoades, R.E. (2000) Integrating local voices and visions into the global mountain agenda. *Mountain Research and Development*, **20**, 4–9.

Rhoades, R.E. and Thompson, S.I. (1975) Adaptive strategies in alpine environments: beyond ecological particularism. *American Ethnologist*, **3**, 535–552.

Richter, M., Pfeifer, H. and Fickert, T. (1999) Differences in exposure and altitudinal limits as climatic indicators in a profile from Western Himalaya to Tian Shan. *Erdkunde*, **53**, 89–107.

Rieder, P. (1984) A revaluation of highland agriculture. In: Brugger, E.A., Fürrer, G., Messerli, B. and Messerli, P. (eds) *The Transformation of Swiss Mountain Ranges*, Verlag Paul Haupt, Bern, 452–642.

Rinschede, G. (1988) Transhumance in European and American mountains. In: Allan, N.R.J., Knapp, G.W. and Stadel, C. (eds) *Human Impact on Mountain Environments*, Rowman & Littlefield, Towota, NJ, 96–108.

Robinson, D.A. and Williams, R.B.G. (1992) Sandstone weathering in the High Atlas, Morocco. *Zeitschrift für Geomorphologie N.F.*, **36**, 413–429.

Robinson, H. (1967) *Latin America. A Geographical Survey*, Praeger, New York.

Robinson, N.A. (1987) Marshalling environmental law to resolve the Himalaya-Ganges problem. *Mountain Research and Development*, **7**, 305–315.

Roche, P. (1965) L'irrigation et le statut juridique des eaux au Maroc. *Revue Juridique et Politique de l'Outre Mer*, **19**, 55–120.

Rocheleau, D.E. (1995) Gender and biodiversity: a feminist political ecology perspective. *Institute of Development Studies Bulletin*, **26**, 9–16.

Roder, W. (1997) Slash-and-burn rice systems in transition: challenges for agricultural development in the hills of northern Laos. *Mountain Research and Development*, **17**, 1–10.

Rognon, P. (1987) Late Quaternary climatic reconstruction for the Maghreb (North Africa). *Palaeogeography, Palaeoclimatology, Palaeoecology*, **58**, 11–34.

Roling, N. and Brouwers, J. (1999) Living local knowledge and sustainable development. In: Prain, G., Fujisaka, S. and Warren, M.D. (eds) *Biological and Cultural Diversity: The Role of Indigenous Agricultural Experimentation in Development*, Intermediate Technology Publications, London, 50–63.

Ron, Z.Y.D. (1985) Development and management of irrigation systems in the mountain regions of the Holy Land. *Transactions of the Institute of British Geographers*, N.S., **10**, 149–169.

Rosman, A. and Rubel, P.G. (1995) *The Tapestry of Culture: an Introduction to Cultural Anthropology*, 10th edn. McGraw Hill, New York.

Rosqvist, G. (1990) Quaternary glaciations in Africa. *Quaternary Science Reviews*, **9**, 281–297.

Rostom, R.F. and Hastenrath, S. (1994) Variations in Mount Kenya's glaciers 1987–1993. *Erdkunde* **48**, 174–180.

Rougerie, G (1990) *Les montagnes dans la biosphère*. Armand Colin, Paris.

Roy, P. (1990) Changing patterns of settlement: a case study of the Trishuli Valley. In: Rustoniji, N.K. and Ramble, C. (eds) *Himalayan Environment and Culture*, Indian Institute of Advanced Study, Simla, 197–206.

Rundel, P.W., Smith, A.P. and Menzier, F.C. (eds) (1994) *Tropical Alpine Environments: Plant Form and Function*, Cambridge University Press, New York.

Rusten, E.P. and Gold, M.A. (1995) Indigenous knowledge systems and agro-forestry projects in the central hills of Nepal. In: Warren, D.M., Slikkerveer, L.J. and Brokensha, D. (eds) *Cultural Dimensions of Development: Indigenous Knowledge Systems*, Intermediate Technology Publications, London, 87–111.

Sahlins, M. (1997) The original affluent society. In: Rahnema, M. and Bawtree, V. (eds) *The Post-Development Reader*, Zed Books, London, 3–21.

Saleem, M., Tetlay, K.A. and Iqbal, J. (1994) *Socio-Economic Analysis of Community-managed Micro-Hydel Projects in Northern Pakistan*, AKRSP, Gilgit, Pakistan.

Salzano, F.M. (1968) Survey of the unacculturated Indians of Central and South America. In: *Biomedical Challenges Presented by the American Indian*, Pan American Health Organization, Scientific Publication No 165, PAHO/WHO Washington D.C., 59–66.

Sandberg, A. (1998) Against the wind: on reintroducing commons law in northern Norway. *Mountain Research and Development*, **18**, 95–106.

Sands, P. (1995) *Principles of International Environmental Law. Volume 1: Frameworks, Standards and Implementation*, Manchester University Press, Manchester.

Sandwith, T. (1998) The Drakensberg-Maloti transfrontier conservation area: experiences and lessons learned. In: *Parks for Peace. Proceedings of an Inernational Conference on Transboundary Protected Areas as a Vehicle for International Cooperation*, Capetown, 16–18 September, 1997. IUCN/WCMC, 121–132.

Sanwal, M. (1989) What we know about mountain development: common property, investment priorities, and institutional arrangements. *Mountain Research and Development*, **9**, 3–14.

Sarmiento, L., Monasterio, M. and Montilla, M. (1993) Ecological bases, sustainability, and current trends in traditional agriculture in the Venezuelan High Andes. *Mountain Research and Development*, **13**, 167–176.

Schaaf, T. (1997). *Legal Framework for Biosphere Reserves*, Email posted to Mountain Forum, 21 March 1997. http://www.mtnforum.org

Schelling, D. (1988) Flooding and road destruction in Eastern Nepal. *Mountain Research and Development*, **8**, 78–79.

Schemenauer, R.S. and Cereceda, P. (1992) Water from fog-covered mountains. *Waterlines*, **10**, 10–13.

Schickhoff, U. (1995) Himalayan forest-cover changes in historical perspective: a case-study in the Kaghan Valley, Northern Pakistan. *Mountain Research and Development*, **15**, 3–18.

Schneider, J. (1999) Varietal diversity and farmers' knowledge: the case of the sweet potato in Irian Jaya. In: Prain, G., Fujisaka, S. and Warren, M.D. (eds) *Biological and Cultural Diversity: The Role of Indigenous Agricultural Experimentation in Development*, Intermediate Technology Publications, London, 158–162.

Schuler, M. (1984) Migration patterns in the Swiss mountain areas. In: Brugger, E.A., Fürrer, G., Messerli, B. and Messerli, P. (eds) *The Transformation of Swiss Mountain Ranges*, Verlag Paul Haupt, Bern, 243–260.

Schumm, S.A. (1979) Geomorphic thresholds: the concept and its applications. *Trans.I.B.G.*, **NS4**, 485–515.

Schwarzl, S. (1990) Causes and effects of flood catastrophes in the Alps – examples from summer 1987. *Energy and Buildings*, **15–16**, 1085–1103.

Schweizer, G. (1984) Traditional distribution systems under the influence of recent development processes: periodic markets in the Yemen Arab Republic as an example. *Applied Geography and Development*, **24**, 24–37.

Schweizer, G. (1985) Social and economic change in the rural distribution system: weekly markets in the Yemen Arab Republic. In: Pridham, B.R. (ed.) *Economy, Society and Culture in Contemporary Yemen*, Croom Helm, London, 107–121.

Scott, J. (1998) *EC Environmental Law*, Longman, London.

Seale, R. (1998) Parks at the edge: the case of Uganda. In: *Parks for Peace. Proceedings of an International Conference on Transboundary Protected Areas as a Vehicle for International Cooperation*, Capetown, 16–18 September, 1997. IUCN/WCMC, 83–87.

Selby, M.J. (1985) *Earth's Changing Surface*, Oxford University Press, Oxford.

Semple, E.C. (1923) *Influences of Geographic Environment in the Basis of Ratzel's System of Anthropo-Geography*, Holt, London.

Shackley, M. (1994) The Land of Lo, Nepal/Tibet. The first eight months of tourism. *Tourism Management*, **15**, 17–26.

Shackley, S., Young, P., Parkinson, S. and Wynne, B. (1998) Uncertainty, complexity and concepts of good science in climate change modelling: are GCMs the best tool? *Climate Change*, **38**, 159–205.

Shibusawa, A.H. (1987) Co-operation in water resources development in the Ganges–Brahmaputra basins. *Mountain Research and Development*, **7**, 319–322.

Shine, C. (1998) Legal mechanisms to strengthen and safeguard transboundary protected areas. In: *Parks for Peace. Proceedings of an International Conference on Transboundary Protected Areas as a Vehicle for International Cooperation*, Capetown, 16–18 September, 1997. IUCN/WCMC, 37–48.

Shoup, J. (1990) Middle Eastern sheep Pastoralism and the hima system. In: Galaty, J.G. and Johnson, D.L. (eds) *The World of Pastoralism: Herding Systems in Comparative Perspective*, Belhaven, London, 195–215.

Shrestha, T. (1995) *Mountain Tourism and Environment in Nepal*, Discussion Paper MEI95/4. ICIMOD, Kathmandu.

Sidky, M.H. (1993) Subsistence, ecology, and social organization among the Hunzakut: a high-mountain people in the Karakorams. *Eastern Anthropologist*, **46**, 145–170.

Sillitoe, P. (1998a) It's all in the mound: fertility management under stationary shifting cultivation in the Papua New Guinea Highlands. *Mountain Research and Development*, **18**, 123–134.

Sillitoe, P. (1998b) What, know natives? Local knowledge in development. *Social Anthropology*, **6**, 203–220.

Singh, T.V. (1983) Tourism in the Himalaya, benefit or burden? In: *Deutscher Alpenverenin, Himalaya Konferenz, 1983*, German Alpine Club, Munich.

Singh, T.V. (1990) Tourism in the Garhwal Himalayas: Problems of resource use and conservation. In: Rustomji, N.K. and Ramble, C. (eds) *Himalayan Environment and Culture*, Indian Institute of Advanced Study. Simla, No. 781.

Singh, T.V. and Kaur, J. (1985) In search of holistic tourism for the Himalayas. In: Singh, T.V. and Kaur, J. (eds) *Integrated Mountain Development*, Himalayan Books, New Delhi, 365–401.

Sinha, A.C. (1990) The Indian north-east frontier and the Nepalese immigrants. In: Singh, T.V. and Kaur, J. (eds) *Integrated Mountain Development*, Himalayan Books, New Delhi, 217–236.

Skeldon, R. (1985) Population pressure, mobility, and socio-economic change in mountainous environments: regions of refuge in comparative perspective. *Mountain Research and Development*, **5**, 233–250.

Slaymaker, O. (1990) Climate change and erosion processes in mountain regions of Western Canada. *Mountain Research and Development*, **10**, 171–182.

Slocombe, D.S. (1992) The Kluane/Wrangell-St Elias National Parks, Yukon and Alaska: seeking sustainability through Biosphere Reserves. *Mountain Research and Development*, **12**, 87–96.

Smith, K. (1992) *Environmental Hazards: Assessing Risk and Disaster*, Routledge, London.

Sobik, M. and Migala, K. (1993) The role of cloudwater and fog deposits on the water budget in the Karkonosze mountains. *ALPEX Regional Bulletin*, **21**, 13–15.

Spear, T. (1997) *Mountain Farmers: Moral Economies and Land and Agricultural Development in Arusha and Meru*, James Currey Ltd, Oxford.

Spoor, G. and Berry, R.H. (1990) Dryland farming tillage and water-harvesting guidelines for the Yemen Arab Republic. *Soil and Tillage Research*, **16**, 233–244.

Stadel, C. (1992) Altitudinal belts in the Tropical Andes: their ecology and human utilization. *Benchmark 2000: Conference of Latin Americanist Geographers*, **17/18**, 45–60.

Stadel, C. (1997) The mobilization of human resources by non-governmental organizations in the Bolivian Andes. *Mountain Research and Development*, **17**, 213–228.

Stadel, C., Slupetsky, H. and Kremser, H. (1996) Nature conservation, traditional living space or tourist attraction? The Hohe Tauern National Park, Austria. *Mountain Research and Development*, **16**, 1–16.

Steinmann, S.H. (1993) Effects of international migration on women's work in agriculture: the case of the Todghra Oasis, Southern Morocco. *Revue Géographie du Maroc*, **15**, 105–124.

Stevens, S.F. (1993) *Claiming the High Ground: Sherpas, Subsistence and Environmental Change in the Highest Himalaya*, University of California Press, Berkeley.

Steward, J.H. and Faron, L.C. (1959) *Native Peoples of South America*, McGraw-Hill, New York.

Stone, P.B. (1992) (ed.) *The State of the World's Mountains: A Global Report*, Zed Books, London.

Strahler, A.H. and Strahler, A.N. (1992) *Modern Physical Geography*, 4th edn. Wiley, New York.

Summerfield, M.A. (1991) *Global Geomorphology*, Longman, Harlow.

Swagman, C.F. (1988) *Development and Change in Highland Yemen*, University of Utah Press, Salt Lake City.

Swanson, J.C. (1979) *Emigration and Economic Development: The Case of the Yemen Arab Republic*, Westview Press, Boulder, CO.

Swanson, J.C. (1985) Emigrant remittances and local development: co-operatives in the Yemen Arab Republic. In: Pridham, B.R (ed.) *Economy, Society and Culture in Contemporary Yemen*, Croom Helm, London, 132–146.

Swanston, D.N. and Swanson, F.J. (1976) Timber harvesting, mass erosion and steepland forest geomorphology in the Pacific north-west. In: Coates, D.R. (ed.) *Geomorphology and Engineering*, Dowden, Hutchinson & Ross, Stroudberg, PA, 199–221.

Swearingen, W.D. (1988) *Moroccan Mirages: Agrarian Dreams and Deceptions, 1912–1986*, I.B. Tauris & Co, London.

Szabo, A. and Barfield, T.J. (1991) *Afghanistan: An Atlas of Indigenous Domestic Architecture*, University of Texas Press, Austin.

Tan-Kim-Yong, U. (1997) The Karen culture: a co-existence of two forest conservation systems. In: McCaskill, D. and Kampe, K. (eds) *Development or Domestication? Indigenous Peoples of South East Asia*, Silkworm Books, Chiang Mai, 219–236.

Taylor-Ide, D., Byers, A.C. and Campbell, J.G. (1992) Mountains, nations, parks and conservation. *GeoJournal*, **27**, 105–112.

Thapa, G.B. (1996) Land use, land management and environment in a subsistence mountain economy in Nepal. *Agriculture, Ecosystem and Environment*, **57**, 57–71.

Thompson, M. and Warburton, M. (1985) Knowing where to hit it: a conceptual framework for the sustainable development of the Himalaya. *Mountain Research and Development*, **5**, 203–220.

Thompson, M. and Warburton, M. (1988) Uncertainty on a Himalayan scale. In: Ives, J.D. and Pitt, D.C. (eds) *Deforestation: Social Dynamics in Watersheds and Mountain Ecosystems*, Routledge, London, 1–53.

Thorsell, J. (1990) *Parks on the Borderline: Experience in Transfrontier Conservation*, IUCN, Gland, 98.

Thorsell, J. (1997) Protection of nature in mountain regions. In: Messerli, B. and Ives, J.D. (eds) *Mountains of the World: A Global Priority*, Parthenon, Carnforth, 237–248.

Thorsell, J.W. and Harrison, J. (1992) National parks and nature reserves in mountain environments and development. *GeoJournal*, **27**, 113–126.

Tiffin, M., Mortimore. M. and Gichiki, F. (1994) *More People, Less Erosion: Environmental Recovery in Kenya*, Wiley, Chichester.

Tobin, G.A. (1999) Sustainability and community resilience: the holy grail of hazards planning? *Environmental Hazards: Human and Policy Dimensions*, **1**, 13–26.

Trakarnsuphakorn, P. (1997) The wisdom of the Karen in natural resource conservation. In: McCaskill, D. and Kampe, K. (eds) *Development or Domestication? Indigenous Peoples of South East Asia*, Silkworm Books, Bangkok, 219–236.

Treacey, J.M. (1989) Agricultural terraces in Peru's Colca Valley: promises and problems of an ancient technology. In: Browder, J.O. (ed.) *Fragile Lands of Latin America*,
Westview Press, Boulder, CO, 209–229.

Troll, C. (1968) (ed.) *Geo-Ecology of the Mountainous Regions of the Tropical Americas*,
Ferdinand Dummlers Verlag, Mexico City.

Troll, C. (1971) Geoecology and biogeoecology. *Geoforum*, **8**, 43–46.

Troll, C. (1972) Geoecology and world-wide differentiation of high-mountain ecosystems. In: Troll, C. (ed.) *Geoecology of the High Mountain Systems of Eurasia*, Franz
Steiner Verlag, Wiesbaden, 1–16.

Troll, C. (1988) Comparative geography of the high mountains of the world in the
view of landscape ecology. A development of three and a half decades of research
and organization. In: Allan, N.J.R., Knapp, G.W. and Stadel, C. (eds) *Human
Impact on Mountains*, Rowman & Littlefield, Towota, NJ, 36–55.

Tromp, H. (1980) Hundert Jahre forstliche Panung in der Schweiz. *Mitteilungen
EAFV, Birmensdorf*, **56**, 253–267.

Tucker, G.B. (1954) Mountain cumulus. *Weather*, **9**, 198–200.

Tucker, R.P. (1988) The depletion of India's forests under British imperialism:
planters, foresters and peasants in Assam and Kerala. In: Worster, D. (ed.) *The Ends
of the Earth: Perspectives on Modern Environmental History*, Cambridge University
Press, Cambridge, 118–140.

Uhlig, H. (1969) Hill tribes and rice farmers in the Himalayas and South-East Asia:
problems of the social and ecological differentiation of agricultural landscape types.
*Agricultural Landscapes, Himalayas and South East Asia*, 1–23

Uhlig, H. (1978) Geoecological controls on high-altitude rice cultivation in the
Himalayas and mountain regions of Southeast Asia. *Arctic and Alpine Research*, **10**,
519–529.

Uhlig, H. (1988) Problems of land use and recent settlement. In Allan, N.J.R.,
Knapp, G.W. and Stadel, C. (eds) *Human Impact on Mountains*, Rowman &
Littlefield, Towota, NJ, 185–200.

Uhlig, H. (1995) Persistence and change in high mountain agricultural systems. (Post-
humously edited by H. Kreutzmann.) *Mountain Research and Development*, **15**,
199–212.

UN (1994) United Nations Commission on Human Rights, Sub-Commission on
Prevention of Discrimination and Protection of Minorities: Draft United Nations
Declaration on the Rights of Indigenous Peoples. Adopted August 26, 1994. *Inter-
national Legal Materials*, **34**, 541–545.

UNESCO (1986) Report of the Scientific Advisory Panel on Biosphere Reserves. In:
*Final Report, 9$^{th}$ Session, International Co-ordinating Council of the Programme on
Man and the Biosphere*, MAB Report Series No. 60, UNESCO, Paris, 6–79.

UNESCO (1995) *Statutory Framework of the World Network of Biosphere Reserves*,
UNESCO, Paris.

USDS (1996) International narcotics control strategy report, March 1996. US Department of State, US Information Service, American Embassy, Stockholm.

Van der Hammen, T. (1974) The Pleistocene change of vegetation and climate in
Tropical South America. *Journal of Biogeography*, **1**, 3–26.

van der Ploeg, J.D. (1993) Potatoes and knowledge. In: Hobart, M. (ed.) *An Anthropological Critique of Development: The Growth of Ignorance*, Routledge, London,
209–227.

Vander Velde, E.J. (1989) *Irrigation Management in Pakistan Mountain Environments*,
Country Paper No 3. International Irrigation Management Institute, London.

Varisco, D.M. (1983) *Sayl* and *Ghayl*: the ecology of water allocation in Yemen. *Human Ecology*, **11**, 365–383.

Viazzo, P.P. (1989) *Upland Communities: Environment, Population and Social Structure in the Alps since the Sixteenth Century*, Cambridge University Press, Cambridge.

Viazzo, P.P. and Albera, D. (1986) Population, resources and homeostatic regulation in the Alps: the role of nuptuality. In: Mattmuller, M. (ed.) *Wirtschaft und Gesellschaft in Berggebieten*, Itinera Fac. 516, Basel, 182–231.

Vincent, L. (1995) *Hill Irrigation: Water and Development in Mountain Agriculture*, Intermediate Technology Publications, London.

Vita-Finzi, C. (1969) *The Mediterranean Valleys: Geological Changes in Historical Times*, Cambridge University Press, Cambridge.

Vogel, H. (1987) Terrace farming in Yemen. *Journal of Soil and Water Conservation*, **Jan–Feb,** 18–21.

Vogel, H. (1988a) Deterioration of a mountainous eco-system in the third world due to emigration of rural labour. *Mountain Research and Development*, **8**, 321–329.

Vogel, H. (1988b) Impoundment-type bench terracing with underground conduits in Jibal Haraz, Yemen Arab Republic. *Transactions of the Institute of British Geographers*, N.S., **13**, 29–38.

Vogt, E.Z. (1990) *The Zinacantecos of Mexico: A Modern Maya Way of Life*, Harcourt Brace Jovanovitch College Publishers, Fort Worth.

Walker, A.R. (1983) The Lahu people: an introduction. In: McKinnon, J. and Bhruksasri, W. (eds) *The Highlanders of Thailand*, Oxford University Press, Oxford, 227–237.

Walker, A.R. (1992) North Thailand as geo-ethnic mosaic – an introductory essay. In: Walker, A.R. (ed.) *The Highland Heritage: Collected Essays on Upland North Thailand*, Suvarnabhum Books, Singapore, 1–93.

Wallace, M.B. (1983) Managing resources that are common property: from Kathmandu to Capitol Hill. *Journal of Policy Analysis and Management*, **2**, 220–237.

Walter, H. (1973) *Vegetation of the Earth and Ecological Systems of the Geo-Biosphere*, 3rd edn. Translated by J. Weiser. English Universities Press, London.

Wanner, H. and Fürger, M. (1990) The Bise – climatology of a regional wind north of the Alps. *Met. Atmos. Phys.* **43**, 105–115.

Warburton, J. and Beecroft, I. (1993) Use of meltwater stream material loads in the estimation of glacial erosion rates. *Zeitschrift für Geomorphologie*, **37**(1), 19–28.

Warner-Merl, K. (1999) The Russian Altai: potential environmental impacts on mountain areas from anthropogenic air pollution. In: Price, M. (ed.) *Global Change in the Mountains*, Parthenon, Carnforth, 155–157.

Warren, A. (1987) Geography and conservation: The application of ideas about people and environment. In: Clark, M.J., Gregory, K.J. and Gurnell, A.M. (eds) *Horizons in Physical Geography*, Macmillan, Basingstoke, 322–336.

Warsinsky, S. (1997) Cheese production in the Beaufort Valley. In: Byers, A. (ed.) *Investing in Mountains: Innovative Mechanisms and Promising Examples for Financing Conservation and Sustainable Development*, Mountain Forum e-conference, 1996, The Mountain Institute, Franklin, 25.

WCED (1987) *Our Common Future*, World Commission on Environment and Development: the Brundtland Report, Oxford University Press, Oxford.

WCMC (2000) World Conservation Monitoring Centre: http://www.unep-wcmc.org/habitats/mountains/statistics.htm

Weathers, K.C., Lovett, G.M. and Likens, G.E. (1995) Cloud deposition to a spruce forest edge. *Atmos. Envir.*, **29**, 665–672.

Whalley, W.B., McGreevy, J.P. and Ferguson, R.I. (1984) High altitude rock weathering processes. In: Miller, K.J. (ed.) *The International Karakoram Project*, Volume 1, 365–382.

Whetton, P.H., Haylock, M.R. and Galloway, R. (1996) Climate change and snow cover duration in the Australian Alps. *Climate Change*, **32**, 447–479.

White, I.D., Mottershead, D.N. and Harrison, S.J. (1992) *Environmental Systems; An Introduction*, 2nd edn, Chapman & Hall, London.

White, S. and Maldonado, F. (1991) The use and conservation of natural resources in the Andes of Ecuador. *Mountain Research and Development*, **11**, 37–55.

Whiteman, P.T.S. (1985) The mountain environment: an agronomist's perspective with a case study from Jumla, Nepal. *Mountain Research and Development*, **5**, 151–162.

Whiteman, P.T.S. (1988) Mountain agronomy in Ethiopia, Nepal and Pakistan. In: Allan, N.R.J., Knapp, G.W. and Stadel, C. (eds) *Human Impact on Mountain Environments*, Rowman & Littlefield, Towota, NJ, 57–82.

Wiegandt, E. (1977) Inheritance and demography in the Swiss Alps. *Ethnohistory*, **24**, 133–148.

Williams, P.W. and Todd, S.E. (1997) Towards an environmental management system for ski areas. *Mountain Research and Development*, **17**, 75–90.

Wilson, R.T. (1997) Livestock, pastures, and the environment in the Kyrgyz Republic, Central Asia. *Mountain Research and Development*, **17**, 57–68.

Winiger, M. (1981) Zur thermisch-hygrischen Gliederung des Mt. Kenya. *Erdkunde*, **35**, 248–263.

Winiger, M. (1983) Stability and instability of mountain systems: Definitions for evaluation of human systems. *Mountain Research and Development*, **3**, 103–111.

Witmer, U., Filliger, P., Kunz, S. and Kung, P. (1986) *Erfassung, Bearbeitung und Kartiering von Schneedaten in der Schweiz*, Geographica Bernensia G25. University of Bern, Bern, Switzerland.

Wolde-Mariam, M. (1991) *Suffering Under God's Environment: A Vertical Study of the Predicament of Peasants in North-Central Ethiopia*, Geographica Bernensia, University of Bern, Bern, Switzerland.

Wolf, E.R. (1970) The inheritance of land amongst Bavarian and Tyrolese peasants. *Anthropologia NS*, **12**, 99–114.

Wongsprasert, S. (1983) Lahu agriculture and trade in North Thailand. In: McKinnon, J. and Bhuruksasri, W. (eds) *Highlanders of Thailand*, Oxford University Press, Oxford, 238–241.

Young, A. and Saunders, I. (1986) Rates of surface processes and denudation. In Abrahams, A.D. (ed.) *Hillslope Processes*, Allen & Unwin, Boston, 1–27.

Zimmerer, K.S. (1992) Biological diversity and local development: 'popping beans' in the Central Andes. *Mountain Research and Development*, **12**, 47–61.

Zimmerer, K.S. (1992) The loss and maintenance of native crops in mountain agriculture. *GeoJournal*, **27**, 61–72.

Zimmerer, K.S. (1994) Human geography and the 'New Ecology': the prospect and promise of integration. *Annals of the Association of American Geographers*, **84**, 108–125.

Zuckerman, L. (1998) *The Potato*, Basingstoke, Macmillan.

Zurick, D. (1986) The household as adaptive unit in the space economy of Western Nepal. In: Joshi, S.C. (ed.) *Nepal Himalaya: Geoecological Perspectives*, Himalayan Research Group, Delhi, 249–254.

Zurick, D.N. (1989) Historical links between settlement, ecology and politics in the mountains of West Nepal. *Human Ecology*, **17**, 229–255.

# Index

Note: page numbers in **bold** indicate chapters; *passim* is used for references that are not continuous